TRUE GENIUS

Also by Lillian Hoddeson:

*Crystal Fire: The Invention of the Transistor and
the Birth of the Information Age*
(with Michael Riordan)

TRUE GENIUS

THE LIFE AND SCIENCE OF JOHN BARDEEN

The Only Winner of Two Nobel Prizes in Physics

Lillian Hoddeson

Vicki Daitch

Joseph Henry Press
Washington, D.C.

Joseph Henry Press • **500 Fifth Street, N.W.** • **Washington, D.C. 20001**

The Joseph Henry Press, an imprint of the National Academies Press, was created with the goal of making books on science, technology, and health more widely available to professionals and the public. Joseph Henry was one of the founders of the National Academy of Sciences and a leader in early American science.

Any opinions, findings, conclusions, or recommendations expressed in this volume are those of the authors and do not necessarily reflect the views of the National Academy of Sciences or its affiliated institutions.

Library of Congress Cataloging-in-Publication Data

Hoddeson, Lillian.
 True genius : the life and science of John Bardeen / Lillian Hoddeson and Vicki Daitch.
  p. cm.
 Includes bibliographical references and index.
 ISBN 0-309-08408-3 (hardcover)
 1. Bardeen, John. 2. Physicists—United States—Biography. 3. Superconductivity. I. Daitch, Vicki. II. Title.
 QC16.B27 H63 2002
 530'.092—dc21

2002007967

Printed in the United States of America

In memory of Jane Bardeen and Betsy Gretak Bardeen

Contents

Preface

The idea of writing a biography of John Bardeen emerged at the University of Illinois in the weeks after Bardeen's sudden death in January 1991. Physicists there were seeking ways to preserve the legacy of their most famous and well-loved colleague. One of us, Lillian, an historian of physics, was encouraged to conduct oral history interviews and write a biography of Bardeen. It was an honor that turned into a labor of love.

Within a few months, Vicki joined Lillian in the effort. She was the first of about a dozen history graduate students who would help with the research. She however stayed with the project throughout its duration, even after completing her doctorate in 2000. As the chapters slowly came into existence and passed repeatedly between us, the individual marks of authorship faded and the work became a true collaboration. Our coauthorship developed in ways we could not have anticipated in 1991.

The original plan called for a scholarly biography focused on scientific contributions. Given the intricacy of Bardeen's physics, we often questioned whether we were the right historians for this formidable task. Seven years into the writing we received a surprising confirmation. Vicki came upon a letter in the Bardeen papers from a Mr. Jefferson Bushman. Writing in May 1989, Bushman asked Bardeen to cooperate with him in writing Bardeen's biography. Bardeen declined, politely: "While there are no present plans for writing a biography, I expect that when I have time to do some work on it, it will be done by Lillian Hoddeson." We burst into laughter. Characteristically, Bardeen told no one of his plan.

We were gratified but also saddened by the loss of a great opportunity. Our work would have benefitted immensely had we had more of Bardeen's own reflections on his life and science. He would probably have guided us through his massive body of contributions to physics. In the end we could treat only a small part of his physics, leaving the rest for future historians to address. We are painfully aware that this book merely scratches the surface of its subject.

We would like to have known what role Bardeen wanted his biography to play. We are quite sure he would have wished it would contribute to the history of physics, an area in which he had a strong interest in his later years. We believe that he would have enjoyed reaching a popular audience, including nonscientists, for he cared about society and education, especially education of the young.

He is likely to have winced, at least initially, at this biography's secondary goal, to shed light on the meaning of true scientific genius. This particular objective emerged late in our work. In drafting the introduction, we found ourselves grappling with a question raised some years earlier in the popular press by the writer Iris Chang: Why is the first person ever to win two Nobel prizes in the same field a "John Who?" to the general public? We concluded that he is unknown largely because he did not fit the popular image of a great scientist. We present our views on this issue in the first and last chapters.

Our efforts to understand the myth of genius and to characterize Bardeen's very different profile opened a Pandora's box of literature on genius and creativity dating back hundreds of years. It was unnerving to survey the as yet unwritten history of genius and discover the roots of the popular myth of genius, perhaps most effectively expressed by the nineteenth-century Romantic writer Mary Shelley in her classic novel *Frankenstein*. The charismatic Dr. Frankenstein, the unbalanced scientist whose talents are innate and who works in isolation on otherworldly problems, is the prototype of a figure who remains vital even today, and against whom Bardeen appears rather bland. It was no surprise to learn that Chang's earlier suggestion of writing a biography of Bardeen had encountered the reaction that he was too ordinary to interest a wide body of readers. We disagree. Bardeen's life offers insights for everyone.

We also believe it is important to inform the widest possible audience about the distinction between true genius and its popular myth. Confusing the two notions can damage the motivation of young people whose creativity holds the promise of great future achievement. Those young people, and all those who mentor and support them, need to understand what really fuels creative work. Bardeen's story can help make this clear. Because Bardeen related easily and genuinely to children on their own terms, we suspect that he would have come to endorse our secondary goal.

Lillian Hoddeson
Urbana, Illinois

Vicki Daitch
Canterbury, New Hampshire

1

The Question of Genius

J ohn Bardeen walked slowly down the corridor of the physics
building, his arms swinging oddly, as though he were paddling
the air. He appeared lost in thought. It was the first of November
1956. He had been a professor of physics at the University of
Illinois for five years now.

Everything about him projected modesty. He was of moderate
height and solid build. His dark hair was thinning slightly. He wore
thick glasses with plain, beige-colored rims. His bland, kind
features matched his inexpensive blue suit, plain white shirt, and
conservative tie, which he wore neatly tucked into his belt.

He was still struggling to absorb the morning's news: that he
and two colleagues, William Shockley and Walter Brattain, had won
the Nobel Prize for Physics. When he heard the announcement he
dropped the frying pan with which he was cooking eggs for breakfast, scattering its contents across the kitchen floor.

The prize was for the invention of the transistor, the tiny semiconductor device that would lead the way to what is now called the
Information Age. The invention, in December of 1947, was the
result of teamwork at Bell Telephone Laboratories, the research
and development arm of the American Telephone and Telegraph
Corporation. The company had wanted to replace vacuum tube
amplifiers and relays in telephone circuits with a cheaper and more
reliable technology.

Bardeen was deeply pleased by the recognition, but he harbored reservations about it. Close friends and colleagues gradually learned about them over the years.

One reservation concerned Shockley, the leader of the semiconductor group who had been absent from the invention itself. In late 1945, not long after Bardeen arrived at Bell Laboratories, Shockley had asked Bardeen to investigate why a particular design for a silicon amplifier did not work. Shockley had sketched the design some months earlier in his laboratory notebook. He had applied the best-available quantum mechanical theories; according to them his device should have amplified signals, but it didn't. The explanation that Bardeen developed—that electrons on the surface can be trapped in surface states—led the team into a productive two-year period of intensive research into surface states that culminated in Bardeen and Brattain's invention, subsequently named the transistor.

After the invention, Shockley had pushed Bardeen and Brattain rudely aside, so that he could design the second-generation transistor without them. Shockley had also revised the story of the invention to highlight his own contributions and downplay those of Bardeen and Brattain. It upset Bardeen that Bell Labs had, at least initially, supported Shockley. And it was Shockley, rather than Bardeen and Brattain, who received wide recognition for the discovery. Even today, popular magazines sometimes credit Shockley alone with the invention.

One reason for this error was the glamour that Shockley projected. A product of Hollywood High School in the era of the great silent movies, he became an articulate speaker and in many ways an eccentric character who enjoyed representing himself as a charismatic genius. Later in his career, Shockley would revel in the publicity attracted by his theories correlating race with intelligence. Bardeen rarely expressed anger openly, but those who knew him could read his contempt for the behavior of Shockley, whose brilliance as a physicist Bardeen respected.

More than that, Bardeen never felt sure the transistor deserved the world's highest physics award. While he considered the physics behind transistors interesting and recognized its technological importance, he thought of the device mainly as a useful gadget, not as a major scientific leap. In 1956 the transistor had not yet revolutionized communication and commerce, nor had it yet ushered in the Information Age. That it would do so was hard to imagine then. Bardeen also felt a little embarrassed to be awarded a Nobel Prize before his teachers, Eugene Wigner and John Van Vleck, had received theirs. He was keenly aware of how much they had done to shape his professional course.

From a research point of view, the timing of the award was poor. Bardeen was pretty sure that he and two younger collaborators—Leon Cooper, a postdoctoral fellow, and J. Robert Schrieffer, a graduate student—were at the brink of developing a theory for superconductivity, the exotic low-temperature state of matter in which all traces of electrical resistance vanish. That, Bardeen thought, would be worthy of a Nobel, for superconductivity was the most significant solid-state physics problem of the decade, perhaps since the 1920s. Bardeen had been working on the problem since the late 1930s. He was just then frantically preoccupied with leading his small team to the solution. He fretted in the knowledge that other extremely capable physicists, including Richard Feynman, were also on the trail. Although the Nobel celebrations of 1956 cut into their work, Bardeen, Cooper, and Schrieffer solved the problem several months after Bardeen's return from Stockholm.

Today we can hardly imagine life without the transistor. The device transformed society in the last quarter of the twentieth century. As the fundamental building block of the microchip, it has become the "nerve cell" (Shockley's phrase in 1949) of modern electronic technologies. We take for granted countless devices that were the stuff of science fiction half a century ago—from personal computers, cellular telephones, automatic teller machines, and microwave ovens to facsimile machines and satellites. Every day, billions of transistors are at work in the lives of almost everyone in the industrialized world.

The Bardeen-Cooper-Schrieffer (BCS) theory of superconductivity was momentous in other ways. The theory was a pioneering step in the creation of the present quantum mechanical picture of many-body systems, liquids and solids whose behavior is determined by the interactions between their electrons and other microscopic components. The theory has also deepened our understanding of nuclear physics, elementary-particle physics, and astrophysics. For example, its notion of spontaneous symmetry breaking underlies the explanation of the fundamental and elusive question of why particles behave as if they have mass. Although superconductivity has not yet had as wide an impact as the transistor on the day-to-day lives of most people, superconducting materials have begun to enter a few important technologies; for example, the cables of power grids and cellular telephone towers. Other superconducting technologies, such as commercial high-speed trains levitated by superconducting magnets, are still in the future.

In the competitive world of theoretical physics, the BCS theory was the triumphant solution of a long-standing riddle. Between 1911 and 1957, all the best theorists in the world, among them Feynman, Albert Einstein, Niels Bohr, Werner Heisenberg, Wolfgang Pauli, and Lev Landau, had tried and failed to explain superconductivity. Felix Bloch's tongue-in-cheek theorem—"Every theory of superconductivity can be disproved"—suggests the frustration of the many researchers who struggled unsuccessfully with superconductivity. When the Nobel Committee awarded Bardeen, Cooper, and Schrieffer the 1972 physics prize, it was the first time an individual was awarded a second Nobel Prize in the same field.

One could argue that Bardeen's physics changed the modern world as much as Einstein's, if only through the transistor's wide range of applications. Yet while every schoolchild knows of Einstein, few people (other than solid-state physicists) have heard of Bardeen. The *Chicago Tribune* aptly titled Iris Chang's profile of Bardeen: "To scientists he's an Einstein. To the public he's . . . John Who?" Why is a "father of the Information Age" so unknown?

In the popular mind, the word "genius" evokes the image of a man (rarely a woman) born with superhuman talents. He needs no training. His insights come magically from a place beyond normal experience. He is unbalanced, a bit mad perhaps, a recluse who works in isolation and whose personal relationships are troubled. Exemplified by such figures as Dr. Frankenstein or "the nutty professor," the scientific genius is a well-known myth.

Consider, for example, the brilliant young mathematician portrayed in the popular 1997 film *Good Will Hunting*. An untrained and underprivileged janitor at the Massachusetts Institute of Technology, Will Hunting takes a break from his nightshift maintenance duties to surreptitiously complete a mathematical proof posed by a famous math professor. Will is depicted as nonconforming, unstable, and outrageous. He works in solitude. He is alienated from ordinary society. And it is no wonder he is emotionally unstable.

This notion of genius, from which Bardeen differed in almost every way, dates back to antiquity when the Muses were thought to breathe creativity into men. Persisting during the Middle Ages and Renaissance, the superhuman and sublime characterization of genius achieved full realization during the Romantic period. Writers such as Johann Wolfgang von Goethe, Samuel Taylor Coleridge, William Wordsworth, Edgar Allan Poe, Percy Shelley, and Mary Shelley, the author of *Frankenstein*, depicted the genius as solitary, unbalanced, untrained, and typically possessed by (rather than endowed with) creative power. More like spirits than people, as the historian Simon Schaffer described them, and often evil, these "Illuminati" merged with the awesome powers over which they brooded. Long after scholars became convinced that genius is an empty notion (a process that Francis Galton may have begun when he tried to connect genius with heredity), the romantic image of the genius sustained itself in popular iconography.

This myth is dangerous, for it can extinguish the confidence and enthusiasm of those who aspire to excellence. In particular, as historical studies show, the stereotype does not fit true geniuses in science. Among the great physicists of the twentieth century who have changed the domains of physics, or created new ones, is Richard Feynman. He deserves to be called a genius. Yet in his outstanding biography, *Genius: The Life and Science of Richard Feynman*, James Gleick stressed the contingencies that contributed to Feynman's success as a physicist, noting that "the history of science is a history not of individual discovery but of multiple, overlapping, coincidental discovery." Although some people may be smarter than others in particular contexts, Feynman himself pointed out that "we are not *that* much smarter than each other" (emphasis added).

The public is confused about genius partly because a few of the greatest scientists, including both Einstein and Feynman, have enjoyed playing to its popular image, sometimes in the process obscuring the significance of their own work. The public is less aware of Feynman's quantum electrodynamics than of his bongo drumming, or his entertaining tales of accidentally cracking scientific riddles as easily as he cracked safes filled with national security secrets. Similarly, the wild-haired Albert Einstein, who mugged for the camera with his tongue sticking out, engaged reporters less with his revolutionary physics than with his eccentricities and controversial politics.

The particular genius of Bardeen did not dramatically shatter or create new domains of physics, as did some of the work of Feynman and Einstein. Bardeen's work did create new ways of solving problems and powerful ways to conceptualize real materials. As a creative scientist, he fit a less familiar profile, that of a genius grounded in the world.

Bardeen's public (and private) persona had nothing outrageous, otherworldly, or magical about it. The mumbling Midwesterner was a calm, balanced family man and friend. He was more likely to be found playing on the golf course, or rooting at a University of Illinois home game, than expounding for reporters. He preferred picnicking with his family or working quietly in his office. Utterly unassuming in his personal life, Bardeen was uninterested in appearing other than ordinary. That is perhaps why journalists have found him difficult to write about, or even to comprehend. Most ignored him.

It did not help that Bardeen was not adept at verbal communication. He "doesn't say much, but when he says something you really have to listen," a close friend cautioned. He spoke sparingly, usually in a low warble that some found difficult to understand or even to hear. Some of his students called him "Whispering John" because he would lecture in a low murmur. And like a ventriloquist, he barely opened his mouth when he spoke.

Bardeen's frugality with words extended to his emotional expression. Uncomfortable with flourishes of any kind, he appeared "flat" to some of his coworkers. Only close friends and colleagues could read his irritation from the subtle shake of his head or slight hiss that would occasionally enter his voice.

Even when Bardeen's words were audible, their meaning often remained puzzling. And when asked a question, he often lapsed into a sort of trance while pondering his response. His companions would be uncertain about what to do during the awkward silence that could last many minutes.

When colleagues or students asked Bardeen to clarify an explanation, he would typically repeat an earlier one. Some eventually learned that this was actually a backhanded compliment; for Bardeen knew how to tailor his explanations to his notion of what the other could grasp. At the Naval Ordnance Laboratory, where Bardeen worked during the Second World War, he is said to have tirelessly rephrased his explanations until they could be understood by each member of his group. Some of Bardeen's students eventually found speaking with him to be "very easy." James Bray came to experience the long silences in conversations with Bardeen as "very pleasant and very productive, and relaxing."

Bardeen differed from the mythical scientific genius in many other ways. For instance, he was by no means self-trained. He spent much of his life engaged in strenuous professional studies through which he developed his legendary encyclopedic knowledge of physics. He learned a great deal from important leaders of modern physics. At least five of them also became Nobel laureates—Van Vleck, Wigner, Paul A. M. Dirac, Percy W. Bridgman, and Peter Debye. From such mentors Bardeen learned much about how to solve problems, how to recognize problems that are ripe, and how to find colleagues who were working at the cutting edge of physics. Nor did Bardeen usually work in solitude, like the mythical genius. He preferred to work in collaboration, usually as a member of a small team that included both theorists and experimentalists. And solutions did not come to him in a sudden flash. He often spent years working on problems, stubbornly holding on to them like a bulldog gnawing a bone. He differed from the stereotype as well in his sometimes plodding approach to complex problems. Rather than solving them as a whole, he would repeatedly break them down into manageable pieces.

The profile that emerges from this biography of John Bardeen differs greatly from the popular image of a creative genius, but its features are common to the profiles of many, perhaps most, real geniuses. Noted throughout the book and discussed more fully in the last chapter are the features of this profile. They include perseverance, motivation, passion, talent, confidence, focus, and effective problem solving. Educators, psychologists, and other scholars have often discussed them in the vast and contentious literature on creativity. All these features can be cultivated—as Bardeen's life story illustrates.

2

Roots

Ten-year-old John Bardeen teetered breathlessly before crashing down from the pantry shelf. Just before falling, his attention had been focused on having another piece of the delicious cake his mother had stowed away after lunch on the top shelf. He heard a crack as he landed and knew he was in trouble. His right arm was broken.

His parents soothed him and took care of his throbbing arm, but they also used the accident as an example for all four of their children. "He did it right after luncheon on a full stomach of cake, which was inexcusable," wrote John's mother Althea. "It has had a splendid moral effect on the entire family."

The normally diffident lad had not intended to be disobedient. It was just that he focused with single-minded intensity on whatever captured his interest. Althea remarked on John's determination in her frequent letters to his grandfather. "John just hangs on and won't let go." That unwavering attachment to ideas and problems would be a hallmark of Bardeen's later scientific work and one of the reasons for his success.

The Bardeen heritage had been built on hard work and high expectations. The American roots of the family trace back to Plymouth Colony, where a young Englishman named William Barden immigrated in 1638, probably from Yorkshire. Barden began his life in the New World as an indentured servant bound for seven years to Thomas Boardman, from whom he was to learn the carpenter trade. Seven months later, however, before he had done so, Barden was transferred to John Barker, a proprietor in the town of Marshfield, some 10 miles north of Plymouth. From Barker, Barden learned how to lay bricks and run a ferry service across the Jones River. Eventually Barden married Barker's daughter, Deborah, who was much younger than he. They built a house in Middleboro, Massachusetts, which still stands, and had 13 children. Their descendants scattered across the continent.

John Bardeen was a tenth-generation son of the Barden-Bardeen clan. His early years were molded by three features of the landscape of his youth: the American progressive movement, which swept the nation at the turn of the twentieth century; the rise of the American industrial laboratory between 1900 and 1920; and the maturation of American physics in the 1920s and 1930s. Bardeen's involvement with industrial research and physics is discussed later, but the progressive ideals of his family shaped his values. They need to be briefly examined here. Had the ideals of Bardeen's family not impressed John, it would still have been almost impossible for him not to absorb progressive ideas simply by growing up in Madison, as the city was then the Midwestern center of the progressive movement.

Exactly what constituted the progressive movement, in the years from roughly 1890 to 1920, or indeed whether progressivism was coherent enough to be called a movement at all, is a subject on which historians disagree. But there is a consensus that the closing years of the nineteenth century and the opening years of the twentieth fostered the growth of certain American attitudes about what ought to be the relationship of the individual to society and of the citizen to government.

Reformers of the era felt that the turmoil and distress of the new industrial economy should be addressed, somehow, by government or private institutions. There were almost as many programs and agendas for doing so as there were people who considered themselves progressives. Activists formed countless national, state, and local organizations devoted to bringing about change of one kind or another: the National Association of Manufacturers, the American Federation of Labor, the National Child Labor Committee, the American Society of Civil Engineers, the National Conservation

Association, the Women's Christian Temperance Union, and hundreds more. These organizations looked to experts for help in solving social problems. One result was a wave of professionalization in fields such as engineering, medicine, and law. The groups organized themselves to define standards for training, performance, and ethics. A new scientific and bureaucratic elite stood ready to direct change aimed at the betterment of American society.

The University of Wisconsin, founded in 1848, had been built in Madison, the state's capital, at one end of State Street, the central avenue running along the half-mile-long isthmus between two prominent lakes of the region, Lake Mendota and Lake Monona. The campus expanded to hug the southern shore of the larger Lake Mendota. By the 1880s the rapidly growing university had become a center for the expression of a philosophy sometimes called the "Wisconsin Idea." Never clearly defined or formalized, this philosophy called for the application of professional research to the political, economic, and social problems of the day.

John Bascom, who became president of the University of Wisconsin in 1874, drew up one of the first versions of the Wisconsin Idea. He proposed applying university talents and resources toward the social good. For example, he promoted the application of research to solving everyday problems in state government and civil service.

A student of Bascom's, Robert M. La Follette popularized the ideas of his mentor. After 1900, when La Follette became governor and later a United States senator, "Fighting Bob" La Follette gained national fame for the flamboyant oratory with which he and his followers in the progressive wing of the local Republican Party battled powerful monopolies and government corruption in an attempt to overhaul the political system and improve social welfare programs in Wisconsin and throughout the nation.

Another student influenced by Bascom's ideas about education and government was Charles Van Hise. A geologist by training, Van Hise focused his research on mining engineering. In 1903 with the support of La Follette, Van Hise became president of the University of Wisconsin. It was Van Hise who brought John Bardeen's father, Charles Russell Bardeen, to the University of Wisconsin.

Van Hise worked to make the academic offerings at the university reflect the interests and needs of citizens who believed in political democracy and the sharing of public wealth. His motto was "the boundaries of the University are the boundaries of the state." Research conducted at the university brought improved methods of dairy farming and agriculture, as well as a program to conserve the natural resources of the state. Van Hise may have exaggerated the power of the university to influence government, but the relationships he advocated were real enough to become clichéd in phrases such as "the expert on tap, not on top," "the service university," "applying the scientific method to legislation," and "the democratization of knowledge."

Among Van Hise's plans for the University of Wisconsin was to establish a medical school. In 1903, almost immediately upon becoming president, he traveled to the East Coast to identify the proper man to lead this task. He found Charles Russell Bardeen at Johns Hopkins. A thirty-two-year-old associate professor of anatomy, Bardeen already had a solid reputation based on his research in anatomy and his teaching at Hopkins. The promise of spreading his influence beyond his scientific community appealed to the young doctor.

Bardeen's father, Charles William Bardeen—whom later generations of Bardeens would refer to as C. W.—was a progressive educator who considered social service the highest ideal to which a man could aspire. Born in Massachusetts in 1847 to an abolitionist family, C. W. had left school at the age of fourteen to enlist in the Civil War, where he signed up as a drummer boy. He was a poor drummer, however, and therefore spent the war as a fifer. He then returned to school and eventually graduated from Yale University. Afterwards he devoted his career to improving the American education system. In 1874 C. W. established his own publishing company, School Bulletin Publications. The *School Bulletin* magazine, which he edited and published for fifty years, became a forum for expressing his strong views on the importance of high-quality education. He also took positions of national leadership in the National Education Association and the Educational Press Association of America. His reputation for scholarship in education brought invitations for membership in the American Association for the Advancement of Science, the American Geographical Society, and the American Social Science Association.

In later years, the Charleses, father and son, frequently exchanged letters in which they discussed values and philosophical issues about education, work, or life in general. In a letter written on

Charles Russell's forty-fifth birthday, C. W. reflected, "It is the useless life that grows old." The "highest happiness," he wrote, was having "just a little more than one could do of the kind of work one wants to do. So long as that holds out, I don't see why a man should feel old." He described love and work as the building blocks of a good life, and "you and I are both unusually fortunate in both."

The values of education, hard work, and a purposeful life were major themes during the childhood of Charles Russell, who was born in 1871 in Kalamazoo, Michigan. Charles spent most of his youth in Syracuse, New York, where C. W. had moved the family in 1874. All C. W.'s children, including Charles, were sent to Europe to study for a year at the Teichmann School in Leipzig, Germany, after completing their course work at Syracuse High School. By the time Charles entered the Teichmann School, he had passed Harvard's examinations in Latin, advanced Greek, and geometry, and he expected, while in Leipzig, to finish studies for additional requirements. As a teenager, he described himself as "a fair scholar, being able to reason better than memorize, and I have a slight mechanical bend [sic]." He also prided himself on his athletic ability. While at Leipzig, Charles wrote to his father: "It seems to me that the only real happiness of existence is derived from working for those one loves and with those one loves toward a common purpose."

In 1893, the year after he graduated from Harvard, Charles entered the first class of the new medical school at Johns Hopkins University. The facilities were inadequate and the instruction so casual that half the members of that first class transferred to another school. Charles remained and in 1897, because his surname began with the letters "Ba," became the first person to graduate from the Hopkins medical program. He decided to stay on at Hopkins, where he rose quickly to the rank of associate professor while teaching and carrying out postdoctoral research. He wrote to his father C. W., that his goal was to "live an effective life" and put "to useful account the education which you so generously gave me."

It must therefore have been easy for him to accept Van Hise's offer to create a new medical school in Madison. On a wintry day early in 1904, Charles boarded a train to Madison to take up his position as professor of anatomy. A fellow alumnus of the Johns Hopkins Medical School met him at the Northwestern Railroad Station. The two doctors rode the trolley to the end of State Street and trudged up Bascom Hill to meet with President Van Hise. During the next several years Charles taught anatomy while working with Van Hise to convince the board of regents and the state legislature to fund Wisconsin's new two-year medical program.

Before the end of Charles's first year in Madison, a new interest came into his life. He began to court an independent career woman who lived in Chicago. Just under thirty, Althea Harmer had classic features and long, dark hair that she sometimes pulled back into an elegant French twist. Photographs reveal a slender body complemented by a melancholy face atop a long, graceful neck. Her delicate appearance concealed an adventurous spirit, high intelligence, and an iron will. Charles was utterly taken with her.

Althea had broken relations with her conservative, eastern Pennsylvania family to strike out on her own and study art, first at the Pratt Institute in New York and then in Chicago. Her mother had died young of tuberculosis. Her father opposed her pursuit of a career. A relative in California, Alexander Francis Harmer, may have served as an inspiration for Althea; he too had left home at an early age to pursue an art career.

Althea had been supporting herself by teaching home economics at the progressive experimental school established by John Dewey, who the historian Scott Montgomery described as "among the greatest reformers of sensibility in American history." Two teachers, Katherine Mayhew and Anna Edwards, later published a history of this Laboratory School of the University of Chicago. The school's purpose, they wrote, was "to work out with children an educational experience more creative than that provided by even the best of the current systems."

Dewey's Laboratory School recruited faculty from the University of Chicago, including Dewey himself, who at that time served as head of the departments of philosophy, psychology, and pedagogy. In his introduction to Mayhew and Edwards's book, Dewey remarked:

> The school whose work is reported in this volume was animated by a desire to discover in administration, selection of subject-matter, methods of learning, teaching, and discipline, how a school could become a cooperative community while developing in individuals their own capacities and satisfying their own needs.

Dewey's ideas on the teaching of science probably influenced Althea Harmer. Opposed to the rote learning of facts, ideas, and structures, he believed that creative scientific thinking was inherently joined with inspiration. It followed that an important goal of science education at the Dewey School was to enhance motivation.

The actual building of scientific knowledge in the child, Dewey argued, should be based on "explorations where the child learns to solve problems he or she develops." He also stressed learning by the "cooperate" approach. He felt that working in groups to develop scientific thinking was not only effective but excellent training for a life of productive and responsible work. By educating children to work cooperatively on solving problems by the scientific method, he hoped to create a generation of adults ready to collaborate in applying such principles to social problems. According to the school's literature, "the essence of its philosophy of social welfare was its development of social individuals who could carry on intelligent social action."

Althea encouraged her students to define their own problems and to seek creative answers. She emphasized problem solving in real, concrete situations. In a course segment on the development of the textile industry, she designed a hands-on, practical experiment in which the children analyzed various fibers and learned to use simple cards, spindles, and other implements of early textile manufacture. Based on their own practical experience with these materials, they were expected to come to their own conclusions about the qualities and uses of various materials and tools. In her words:

> Again this method involves the exercise of judgment. The ability to think and the method of thinking are part and parcel of all the reinventing and the rediscovering. Thinking does not occur for its own sake; it is not an end in itself. It arises from the need of meeting some difficulty, in reflecting upon the best way of overcoming it, and thus leads to planning, to projecting mentally the end to be reached, and deciding upon the steps necessary. The tool and method of going to work are always seen to be dependent upon the material on the one side and the result to be attained on the other. These being given, to find the third term is the problem, surely as logical an exercise as any in geometry. It has the added advantage of being concrete and of calling the constructive imagination into play.

Decades after the end of Althea's tenure at the Dewey School, her son John would become internationally known for solving problems using a cooperative experiment-based approach built on overcoming specific difficulties, setting concrete goals, visualizing their achievement, and "calling the constructive imagination into play."

In 1904, two years after the Dewey School merged with the Chicago Institute (under the direction of the University of Chicago's School of Education), Dewey had a falling out with the president of the university. In the heat of the administrative conflicts, "Dewey-ites" came to feel unwelcome at the school. When Dewey left that year for Columbia University, Althea was out of a job.

She then embarked on an unusual venture for a single woman of this period. She began a small interior decorating business. Struggling in a world dominated by men, especially in her efforts to collect payment from her clients, Althea nevertheless finished her first year as a businesswoman "just about even," as Charles proudly reported to his father. "She had no capital to start with and it was slow getting started because her work is so highly original that it had to create its own market. Another year and she would have made a great financial success."

According to one Bardeen family story, Althea and Charles met in the context of her decorating business. She found herself working with him on the decor of the University of Wisconsin faculty club. During the spring of 1905 he came to Chicago to see her as often as possible. "The more I see of Althea the more I believe I have won a prize," Charles wrote to C. W. after she had given in to his courting. They were married in a civil ceremony in Chicago on August 6, 1905. Afterwards, they traveled by train to Madison for a quiet time of settling in together.

There is no evidence that Althea made any attempt to continue her business in Madison. With Charles's full support, she bent her skills toward creating an elegant home for the two of them. She decorated their first small apartment with Oriental rugs and other rich furnishings. Charles told his father that their aim was "to have an exceptionally beautiful home, not furnished with many things, but with each and every thing beautiful." Althea did, however, resume art lessons. She became active in Madison's art world, helping to organize exhibits and occasionally giving lectures. Having specialized in Japanese art in her earlier studies, she particularly enjoyed the opportunity to organize an exhibit and deliver a lecture on Japanese block printing at Madison's State Historical Library.

Charles continued his activities at the university, following the

family tradition of working in the public interest. With characteristic self-discipline he applied himself to his research and to establishing the medical school. The University of Wisconsin School of
Medicine opened in 1907. He was named dean.

He then created a small local university hospital, where medical students could complete their clinical requirements without
having to transfer to another school. Not until 1925 did Charles
succeed in establishing a full teaching hospital. The politics
required constantly pushing the sluggish Wisconsin legislature for
support. In early 1906 Charles wrote to C. W. that politics had
"made it impossible to get the city to do what it should," namely,
to fund a new building for the city hospital. He was forced to solicit
funds "at a time when the rich fear the poor are going to tax them
too far and the poor have nothing to give."

Charles thought constantly about how to enhance American
health care and education while trying to improve his own school
of medicine. He lectured on the primacy of the patient in hospitals
and stressed the importance of using hospitals as a forum for the
education of doctors, nurses, and other health care professionals.
Emphasizing the value of clinical care and public health initiatives
in preventing and curing disease, he decried the crass commercialization of medicine as being against the interests of patients.

As a professor, Charles's students and colleagues found him
stubborn, reserved, and unusually even tempered. Paul Clark later
wrote in his history of the Wisconsin Medical School that Charles
"may have been boiling inside, but outwardly he remained calm,
marshalling his facts and gathering advocates for his cause." He
was not generally regarded as a good teacher. Most of his students
relied on the teaching assistants and had little contact with Charles
until examination time. But one close associate in the Anatomy
Department insisted that "in his informal, often blundering way,
in lectures and laboratory, he succeeded in directing students to
the understanding of essentials and I am sure stimulated them by
his obvious interest in the subject." He added that Bardeen's habit
of "treating students as adults and allowing them to manage their
own affairs" encouraged them to become "better students and I am
sure better men in later life than they would have become under a
more regimented system."

Althea and Charles appeared reserved with one another, at least
in public. In one letter to C. W., Althea mentioned that her sister-

in-law Betty "didn't consider us a very loving pair, but thought that Charles meant well and that perhaps it agreed with my temperament." They do appear to have had a happy marriage, with a strong bond of shared values and goals. A few months after they were wed, Charles wrote his father, "Althea is unusually affectionate, devoted and unselfish. She is strong and intelligent and I believe few women are so well fitted to be real helpmates as well as wives."

Women in the Bardeen clan were expected to contribute to the family both through service and by nurturing children. Charles wrote to his father with regard to one of his unmarried sisters that, "while it is unnatural for a healthy woman not to have children to look after and protect, present social conditions seem to make their number constantly greater. I suppose so long as these conditions last many women will have to make an unselfish philosophy do for love. There is always the opportunity to adopt children."

One job that Althea gradually took over from her husband was writing to C. W. Having been an only child and separated from her own family, she was pleased to now be part of a large and close-knit family. In one of her early letters to C. W. she wrote that Charles had told her that "you were his best friend." She told him that she, too, hoped "to become a friend of yours and I cannot consider that I have fully entered into this new life until I really become a part of Charles' family."

In April 1906, Althea gave birth prematurely to twins—a tiny boy they named William, after Charles's grandfather, and a still-born boy they called Prince. They grieved for the lost infant and turned their love and energy to William, who was sickly and under-weight. Doctors advised Althea to feed him every hour for the first few weeks. She was exhausted for months after his birth. They tried without success to find temporary nurses and housemaids. Charles gave up "what spare time I could to staying at home while Althea goes out for fresh air and exercise." He wrote his father, "A girl is at a disadvantage if she has no old friends or relatives in town when a baby comes. She has to stick close to the baby until he gets old enough to trust to a maid and she hasn't advice in times of colic and the little ills the experienced mother knows so well how to handle."

With the new addition, the Bardeens' apartment seemed too small. In May 1907 the family moved into a small house set idylli-cally on Mendota Court, a few doors down from Lake Mendota.

From the fraternity house docks along the irregular shore it was easy to spot Picnic Point to the west, past the campus. Charles's office on campus was a short walk away.

On May 23, 1908, barely two years after William's arrival, Althea and Charles had a second child, a ten-pound boy, they named John. He soon became the focus of the entire household; Charles developed a special attachment to him. Charles wrote his father that John was "far less of a care" than his brother William had been. "Charles' devotion to John is most touching to see," Althea wrote. She worried that Charles was spoiling the child. William cheerfully accepted his new brother as a companion and playmate. "William has a very generous nature and has never at any time shown the slightest jealousy," Althea wrote her father-in-law. "Considering the attention John gets, it is rather remarkable, but [William] always has the same delight in anything John does and will call our attention to any new thing he does." Charles felt himself "blessed" in a family that was "proving dearer to [him] every day."

Althea often mentioned John in her letters to C. W. When he was a toddler, she wrote how "fearfully knowing" the boy was. She described how quickly he caught on to the idea of opening Christmas gifts, showing each present to everyone before going on to the next.

By the spring of 1910 the Bardeens had again outgrown their home. They bought the stucco next door at 23 Mendota Court, with a living room, dining room, kitchen, large study downstairs, and five bedrooms upstairs. Its glassed-in sunporch and the grape arbor out front were lovely places to sit and read in the summer months. The library was large enough to serve as an occasional space for Dean Bardeen to hold seminars. No one lacked privacy. Even the housemaids had their space in the attic. When a live-in maid was no longer needed, Charles converted the attic into a playroom for himself and the children, complete with woodworking tools, a weight-lifting bench, and a suspended golf ball for practicing strokes.

After John the Bardeens had two more children in fairly rapid succession: Helen on October 13, 1910, then Thomas on April 10, 1912. The pregnancies and births became increasingly difficult for Althea and her health suffered. She stayed in the hospital for over a week after Helen was born, but she maintained an optimistic spirit, writing to C. W. that, although she had been "very ill the first five days," her condition was much improved. "After a necessary operation—I have been recovering by leaps and bounds. I have been sitting up a little more each day, and am beginning to walk a little." After Thomas's birth, "I was obliged to stay in the operating [room] nearly 24 hours before I could be moved," she wrote C. W., "and your letter was better than a tonic, for it made me realize that it was all for a high purpose. I didn't realize up to that time that I had a new child."

While Althea was recuperating in the hospital, Charles was learning the art of raising children. He found that John, then almost four, could be a handful. Althea was entertained by Charles's consternation with John, who until then had been the apple of his father's eye. She remarked in a letter to C. W. that Charles had "always indulged John," and she was "much amused to discover that Charles was obliged to spank John morning and evening before he could be made to obey." She confided to C. W. that she was "glad it happened as I feared he would spoil John's real sweetness of disposition by indulgence."

Charles was not a "huggy-kissy" father, but he was loving and always treated the children with respect. He often invited one of them along on an evening walk to enjoy the outdoors. He was not without humor in dealing with the children. One day the Bardeen boys were helping neighborhood friends annoy the workmen who were laying concrete sidewalks in the Mendota Court neighborhood. Charles appeared and sternly reprimanded the mischievous youngsters. A nearby mother remarked, "Why Dr. Bardeen, I thought you liked children." He quickly retorted, "I like 'em in the abstract, but not in the concrete."

Althea's letters to C. W. suggest that from the day John entered school he stood out in most subjects, demonstrating his "concentrated essence of brain." By age nine, he had been skipped enough to catch up with his two-years-older brother William, who also worked well ahead of his age group. Moving up undoubtedly offered John greater challenges, but he later remarked quizzically that he "just didn't learn to spell."

Before John was five, Charles wrote C. W. that the boy was "showing talent with figures." Through Althea's intervention, John advanced considerably in his mathematics education. Finishing third-grade math when he was eight, he flew through the lessons for fourth and fifth grades the following summer, then mastered the sixth-, seventh-, and eighth-grade material in the fall. In the summer of his eleventh year, he took two mathematics courses in overlapping time periods, "so that he has only half the normal time for each." But as his father wrote to C. W., "He is getting away with them." Charles bragged that John figured out shortcuts for problems that would be difficult for an adult. Althea wrote C. W. that "John has undoubtedly a genius for mathematics."

John's relative youth was a social handicap at school. Although he preferred the advanced course work, he suffered when classmates four and five years older considered him too young for their games. But, he told an interviewer, this left him more time for his studies. After school and on weekends, he found playmates of his own age among the neighborhood children. One of his earlier friends recalled happy times collecting stamps, drawing a map of Lake Mendota, and playing evening games of Run, Sheep, Run and Sardines. John and Bill were both avid stamp collectors and formed their own stamp trading "business," the Four Lakes Stamp Company. In playing with his younger brother and sister, John sometimes used his drawing skills to make paper dolls and furniture for their dollhouse. Althea thought that John's behavior was mature for his age, prompting her to observe that "socially he is reserved, but meets people in a balanced, pleasant way, much like an adult."

John joined his siblings and other neighborhood children in their fascination with science and technology. At about thirteen, he became interested in chemistry through reading a book, *Creative Chemistry*, which described how American organic chemists had learned to create artificial dyes when the German supply of dyes was cut off during World War I. Eager to encourage his son's scientific interest, Charles bought small quantities of organic dyes and other chemicals from the National Stain and Reagent Company in Ohio for John to experiment with in his basement laboratory. With a sheepish grin, John told an interviewer, "I dyed materials, did some experiments injecting dyes in eggs. I was interested in seeing how you get colored chickens."

During the 1920s John and many of his friends were captivated by the magic of radio. They spent countless hours outside listening to the sounds carried by the strange electromagnetic waves detected by their crystal sets. John built his own crystal set and receiver "from dime store wires, oatmeal boxes, little straw suitcases and the crystals." Carefully winding copper wire around an oatmeal box, he made a tuning coil. The whole set, which included a pair of earphones, was stored in an old suitcase, which he would bring out at night hoping to hear voices or music from Chicago. The exercise demanded much patience, for only after a good deal of poking with a wire on his galena crystal would he occasionally hit a "hot spot" and pick up a weak signal. "Some boys even got as far as putting vacuum tubes in their amplifiers, but I never got that far," he recounted.

John Bardeen and his friend John Hames won first place in a Boy Scout semaphore-signaling contest. The two boys were so proficient they were asked to give a demonstration to the Future Farmers of America in Madison. They walked to the top of University Hill and sent them the message, "We are here; come on up."

John's extroverted older brother Bill, who was also present, demonstrated "wig-wag" signaling, which used flags to represent the dots and dashes of Morse code. John would remain close to Bill. The two fell into a pattern in which tongue-tied John would sometimes let Bill carry conversations for him. In later years, John would re-create this pattern with a few close friends and with his wife.

John and Bill entered the combined seventh-eighth grade at "Uni High," Wisconsin's University High School, in the same year—John from third grade and William from fifth.

Established in September 1911, the school had been conceived as a laboratory for training high school instructors and for testing progressive ideas in education. In its philosophy and organization, Uni resembled the Dewey School. The students were accelerated as much as possible to keep lessons challenging. One goal was to "introduce pupils to high school methods and subjects before they reached the 9th grade." The fourth quarter, offered during the summer, allowed students who had missed work or had fallen behind to catch up. It also enabled the brightest students to complete senior high school in only three years.

Althea did her best to arrange course schedules that would allow each boy to pursue his own interests and avoid potentially adversarial competition. They took mathematics, English, and geography together, but she "selected different studies in their other classes to give William a sense of freedom from John."

Meanwhile, Charles William, the Bardeen patriarch, was filtering his experience and ideas down to his grandchildren. For John's

tenth birthday, C. W. sent him *A Little Fifer's War Diary*, an autobiographical memoir about his boyhood experiences during the
Civil War. Coming from a family of staunch abolitionists, C. W.
had viewed the Civil War idealistically, as an opportunity to rectify
the evil of slavery. But he also saw the war as a chance to enjoy an
adventure. Against his mother's wishes, he had left home at fourteen to join the Union army.

The book included entries from the diary that C. W. had kept
as a teenager during the Civil War. He mixed in commentaries
added forty-five years later. Looking back, C. W. characterized himself as given to stretching the truth and as "conceited, boastful,
self-willed, disobedient, saucy, not lazy but always wanting to do
something else than the duty of the moment." This "wholly dis-
agreeable" youngster was present at some of the Civil War's blood-
iest battles, including Fredericksburg, Chancellorsville, Gettysburg,
Spottsylvania, and the Wilderness. But he survived unscathed and
subsequently raised five children. In the front of the book, C. W.
inscribed a reminder of the philosophy on which he had raised his
children:

> My dear John, It is a delight to see you children from time to time
> and watch your development. Your boyhood is very different from
> mine, and your opportunities infinitely greater. Remember that the
> greatest opportunity of all is to serve.

Years later John had copies of the book made for family members. He sent one to his niece in 1975, when she was the same age
that Charles William had been when he left Fitchburg High School
to enlist in the Union Army. His grandfather's career, John told
her, had "always been an inspiration." He found his grandfather's
inscription even "more valid now than when he wrote it."

C. W. continued to be an adventurer in his later years. He some-
times left his family home in Syracuse while he traveled alone in
Europe or Africa. He later wrote of his experiences in his *School
Bulletin*, describing the concerts and plays he attended, such as a
Gilbert and Sullivan production at the Savoy Theater in London, or
his bicycle tour around England and Germany. On a long trip to
Tunisia, he took many interesting photographs of the local people.

C. W. often visited the Bardeens in Madison, endearing himself to
Althea and the children by bringing wonderful gifts, especially at
Christmas. In a letter thanking him for the furs he had sent for her
birthday, Althea wrote, "I have wanted them for years, in fact as long

as I can remember, never having had any." One Christmas when C. W. sent particularly charming toys for his grandchildren, Charles teased Althea for caring "more about them than the children do."

Such sweetness and generosity mingled with the family's serious values, including work service, social duty, and economy. Charles contributed monthly to the American Society for the Relief of French War Orphans, providing a model of social conscience that John would emulate in his adult life. His emerging sense of economy showed up early on. One day during the First World War, Althea wrote to C. W.:

> John opened his bank once and paid his father ten dollars on his war-bond and is nearly ready to open it again and pay him another ten, so that he has nearly paid half of his fifty dollar bond. He has the same sense of economy in his studies as he has in money. He would not fail in a class any more than he would lose fifty dollars.

The Bardeens used money to motivate their children to do chores, offering them an allowance in exchange for fulfilling duties at home. At least for a while they also followed the custom of the choirmaster in "taking off five and ten cents for any shortcomings." Althea suspected that a small monetary enticement offered by the choirmaster at the Episcopal church motivated John's and Bill's continued participation in the junior choir. She concluded that the system of monetary merits and demerits "kept John brave against all odds. The other day he offered with great earnestness and with the wisdom of many years' experience, 'I should think two meals a day would be plenty.' And no one loves food more than John."

John and William took part in many sports, from church league ball teams to competitive swimming and diving. The proximity of the Bardeen home to Lake Mendota made it easy for all the children to learn to swim. In summer months they could be found laughing and swimming in the lake or perhaps diving off the pier down the street. John learned to dive from a twenty-foot-high platform before he was ten and was, as Althea wrote C. W., "doing stunts in fancy diving." The summer of John's first year, Charles bought, with the help of his father, a sailboat with a small motor. Boating jaunts on the lake became a source of great pleasure for him and the children. After a visit to Madison, Charles's sister Bertha "was struck by [his] development as a father," reporting that she did not see how he could be a better one.

Cold weather in winter hardly slowed down outdoor activities.

Ice skating and iceboat sailing on the frozen lake occupied the Bardeen family throughout the winter months. In spring the youngsters dared each other to be first to jump in the icy lake. A fraternity man in one of the houses that lined the lakefront awakened one spring morning to the sound of giggles outside his window. Looking out, he saw three teenagers, John, Helen, and Tom Bardeen, "in their knitted woolen bathing suits upon the cold boulders daring each other to enter the icy water." Eventually all three jumped in. "I shivered just watching them."

Althea played golf with Charles whenever possible. One summer, playing every day throughout the month of August, she succeeded in reducing her average score from 75 to 55. By the summer of 1910 Charles had developed the habit of playing a round of golf "nearly every afternoon." It was for him a time to relax into his thoughts—sometimes with humorous results. A family story features Charles driving home from the golf course and stopping by a favorite drugstore to buy tobacco supplies. He then walked home. Later in the evening he could not find the car and called the police to report it stolen. They soon found the car where Charles had left it, in front of the drugstore.

The couple passed their love of golf along to their children. From an early age John accompanied his parents to the golf course, caddying for his mother when he was still too young to play. He began to learn the game as soon as he was old enough to swing a club. In time John's childhood interest in golf became an adult passion, as it had been for both his father and his grandfather.

John was nine when his father wrote C. W. that Althea had cancer. Charles slipped the news into a short paragraph in the middle of an otherwise casual letter. Falling back upon the physician's trained objectivity, he reported matter-of-factly:

> We discovered a little growth in Althea's breast during the winter and it finally seemed best to have what is called a radical operation, the complete removal of the breast and all surrounding tissue so as to prevent any spread of the tumor. The operation was performed by Dr. Jackson three weeks ago and Althea is now at home again and is as well as could be expected after so radical an operation. It will be some time before she can raise her right arm much, even to write letters. It's hard luck but it's a good thing to get a hold of in time if it must come.

Althea recovered well from the radical surgery, performed in February 1918. But within less than a year the growths reappeared and she underwent a second surgery, followed by X-ray treatments. Having used X rays in his own research, Charles probably understood before anyone else that Althea was losing the battle for her life.

X rays, discovered in 1895, were already in use by 1900 for diagnostic purposes. But it was much longer before medical researchers standardized radiotherapy as a cancer treatment. Charles pulled as many strings as he could in the medical community, but the discussions of his wife's condition with colleagues only underscored the fact that, as yet, medicine could offer little in the way of treatment to cancer patients. The best medical expertise of the time amounted to little more than bold incisions to remove surface and breast cancers and the cautious use of X rays in treatment. Hormonal treatment, chemotherapy, mammograms, pap smears, and fiber optics did not yet exist. Nor did doctors have confidence in the use of radiation to cure common cancers. It was already known that radiation itself caused cancer in some instances.

The years 1918 to 1920 were a roller coaster of hope and despair for the Bardeens. Every time Althea recovered from a surgery and the harsh effects of her X-ray treatments, new nodules would appear or she would become otherwise ill again. At a time when one in eight American women over 45 died of cancer, Althea's chances for survival did not look good.

In March 1919 most of the family became ill during one of the severe influenza epidemics that swept the nation after the First World War. The normally bustling household was unusually still after Althea, William, Helen, and Tom all ended up at the university infirmary. For the first time, John and Charles spent a considerable amount of time together alone. Charles wrote C. W. that he had "seen more of John than usual and finds him a very good companion. He amuses himself evenings, doing problems in algebra and arithmetic and seems to have real talent." The orderly and predictable world of mathematics must have been a haven for John while the rest of the family was ravaged by illness.

The family recovered from the flu, but Althea continued to decline. As the disease bore down more heavily upon her, she must have known she was losing ground. The side effects of the radiation therapy deepened her misery. Still, she gamely kept adding to her life, chairing a newly formed garden club in Madison less than a year before her death.

She began to spend a great deal of time away from home, either in Milwaukee near her primary physician or in Chicago for other treatments. In the hot summer of 1919, Althea's friend Mary Morris visited her in a Milwaukee hospital. Morris wrote to Charles that Althea "did not look ill but very lovely and kept smiling her lovely pointed smile." Mary professed never to have seen "such self control and fortitude as Althea has displayed in this trying time. She filled us all with admiration." In the last year of her life, Althea underwent an almost constant round of frightening and painful surgeries and X-ray treatments. They left her exhausted and ill for weeks. Charles worried that her nausea might be caused by "hysteria and homesickness," though one of her doctors in Chicago felt that it was more probably due to the "rather massive X-ray exposures." Althea's Milwaukee surgeon also denied that the nausea came from some psychological weakness, declaring that he "didn't think she was that kind of a girl and didn't believe it now."

With Althea ill and so often away, the Bardeen household became chaotic. Charles needed a housekeeper but could not find one willing and able to deal with his four children, large house, and invalid wife. The job required nerves of steel.

On April 10, 1920, Althea came home from Chicago, where she had been undergoing radiation therapy. She wanted to be home on Mendota Court, with her family. The day before, Charles had written to his father that she likely had at most a year to live. He did not tell Althea this, nor did her doctors, "although doubtless at times she suspects it." No one in the household could care for her properly. She was given a room at the university infirmary where Charles and the children could easily visit her. For John, the infirmary was "on the way between my high school and home, so that I would stop and see her on the way home from school."

The treatments had not appreciably affected the cancer and had made Althea so ill that she could no longer hold down food. Her strength failed rapidly. C. W. sent her some wine that he had held back in anticipation of Prohibition. She managed to drink a sip. "She said it was delicious," Charles wrote, "and she appreciated very much your sending it." By this time the cancer had invaded her lungs. There was no hope of recovery.

On the evening of April 19, John stopped by to see his mother. "I thought she looked well that day and cheerful," he recalled. "I was shocked to hear the next day that she had passed away. I didn't realize how seriously ill she was." She was forty-seven. John was almost twelve. Charles wrote to his father a few days later, "She put up the bravest possible fight against the greatest possible odds. She had many unusual gifts of which not the least was a magnificent courage that was with her to the end."

Althea's obituary appeared in the *Madison Democrat* the next day, April 21, along with an anonymous tribute emphasizing "her intense feeling for her fellowmen, expressed on all occasions by an ever-ready sacrifice of time and labor to bring to others all that made her life joyous and rich." The article focused on the role that art had played in her life. "She felt in all artistic expressions the pulsations of life and she felt also that the meanest one of us, the least fitted, might with proper guidance be led through expressions in one or the other of the art forms to a higher plane."

Charles found himself in a desperate situation. While mourning his wife he was suddenly overwhelmed with practical responsibilities. His active life as an administrator, teacher, and researcher had relied heavily on Althea to manage the household and raise the children. His work as dean and his continuing medical research had occupied most of his time. A year earlier, on his forty-eighth birthday, he wrote to his father that he had "put in twelve or fifteen hours" to finish a paper on "the proportions of the body to height and weight during growth."

Charles's sister Bertha came to help with the children and household in the days immediately following Althea's death. The visit was a temporary fix. When school let out for the summer, Charles sent the children to various relatives and friends—Tom and Helen to C. W. in Syracuse; John to his Uncle Norman, Charles's brother, in Michigan. "So we got along through the summer that way."

Althea had been a source of love and stability for John. She had helped him navigate in a world that included much older schoolmates and the high expectations of adults. He would have to manage his teenage years without her.

3

To Be an Engineer

That summer, while the children were away, Charles courted his secretary, Ruth Hames. The children already knew Ruth from visits to their father's office and because she was the sister of John's Boy Scout semaphore signaling partner. At twenty-eight, Ruth was more than twenty years younger than Charles. "I have played some golf but have at times neglected it in favor of motor rides with Ruth," Charles wrote C. W. in late August. "Ruth and I expect to be married October 6."

Ruth worked patiently to reduce the chaos of the family. Bill, John, and Helen accepted her presence calmly. But Tom, eight years old when Althea died, resented Ruth and became even more difficult when Ann, a new half-sister, was born less than a year after Charles and Ruth were married. John often acted as Ann's "savior" when Tom's teasing became unbearable.

John's grades at Uni High fell in the year following Althea's death, especially in French. In later years he claimed to have had no talent for languages, but in 1919 Althea had written to C. W. that John "stood at the head of his class" in French. A year after Althea's death, John's French teacher wrote Charles that, although John was "not actually failing now, I have felt all year that he is not in the right class to get the most out of his work."

Bardeen's high school records do not hint at his great future accomplishments. At Uni High he received many more marks of "good" and "fair" than of "excellent," even in some mathematics courses. But the record cards offered no space for teachers' comments, and his permanent record card fails to reflect anywhere that he was often working three to five years ahead of his age group.

Charles believed that Uni High nurtured John's mathematical talent. He explained to C. W. that students there were "allowed to go ahead in any line they take up as fast as they show ability and the hard and fast grade system does not exist." He felt that had John attended public school he would "not only be several years behind where he is now, which in itself would not be so serious, but he would be deadened mentally by routine instead of enjoying healthy mental exercise."

John's seventh-grade mathematics teacher, Walter W. Hart, helped him to recognize his talent for math. "He saw that I was interested so he gave me a lot of extra work and instruction." Hart, who also taught in the University of Wisconsin's Department of Education, had kept alive for over fifty years his famous popular mathematics textbook series, *Mathematics in Daily Use*.

On one occasion Hart noticed that his youngest student's attention had wandered away from the class discussion. John was "in the front seat of the row near the door . . . fully occupied with some personal project." When Hart asked John what he was doing, the boy said he was "solving examples on a page far ahead of that on which the class was occupied." Hart gave John permission to sit in the back of the room and work through the book's problems at his own pace. He "invited others in the class to undertake to 'get in sight of your dust.'" Some students took him up on it. Hart referred to the activity as "legalized inattention." In later years Bardeen would remember Hart with fondness, writing to him in 1962 to "express my deep gratitude to you as a teacher for first exciting my interest in mathematics."

Even with the disruption of Althea's death, John completed all his Uni High course work by age thirteen. But as he was "a little leery about graduating so young," he and Bill decided to attend Madison Central High School for two years, taking additional mathematics, science, and literature courses not offered at Uni. By the time John had turned fifteen and Bill seventeen, the two had completed every course of interest at Madison Central. There was no longer any reason to postpone entering college. In the fall of 1923 they both entered the freshman class at the University of Wisconsin.

John lived at home while trying to be a "college man." Perhaps to appear older, he began to smoke cigarettes, which his father's generation had referred to as "coffin nails."

Sports eased John's social adjustment. Despite his age and relatively small stature, he made the varsity swim team in his sophomore, junior, and senior years. He also lettered on the varsity water polo team. During his junior year, he was mentioned in the *Wisconsin Badger Yearbook* as having swum "some beautiful races in the 200-yard breast stroke to cop first place." The yearbook swim team photo shows a fit young man standing in a relaxed pose by the pool.

In John's sophomore year he joined the Zeta Psi social fraternity. Frowning on fraternity high jinks, Charles refused to pay the membership fee. John raised the money himself by working summers and playing poker or billiards. He discovered he had a talent for billiards and became a three-cushion champion. His unassuming manner probably helped him work up games with unsuspecting students who were surprised to find they had been hustled. He could not earn enough to cover room and board, so he continued to live at home, occasionally taking a meal at the house.

That Zeta Psi was a social, rather than an academic, fraternity may have attracted Bardeen. The university "had little trouble with Zeta Psi on the score of conduct or morale," reported the dean of men. But he had to admit that "they do not stand very well scholastically." John's academic performance was not distinguished. His college transcript shows grades ranging from "poor" to "excellent." His friends did not remember him as a student who spent long hours studying, although he performed better than most of his fraternity brothers.

Walter Osterhoudt, "Dutch" to his friends, belonged to the Pi Kappa Alpha fraternity, whose house sat on the shore of Lake Mendota, down the street from the Bardeens. When Dutch and John were in an engineering course together, Osterhoudt had the impression that John "just floated in and out" of his classes. "I don't think he ever studied." Nor did John always make it to class. "He didn't have to stay on the treadmill like we did." Osterhoudt did recall that some of Bardeen's professors were impressed with his academic work.

Bardeen knew many of the men in the Pi Kappa Alpha fraternity, as he and his siblings had often played in the water and on the dock behind the PiKA house. It was not uncommon to find John there in his college years drinking beer and getting a little rowdy with some of the PiKA brothers, several of whom had at one time or another been "sweet" on John's sister Helen.

Osterhoudt recalled one PiKA adventure that included Bardeen in the summer of 1928. Someone (the PiKAs suspected a rival fraternity) had called the local police to complain about a noisy party at the PiKA house. The police raided the fraternity, carting the revelers down to the city hall, where they were "booked." The offenders were led to the drunk tank, where they were to spend the night. But after they had settled in, a pre-med student discovered that one of the supine old men in the room was not sleeping but was actually dead. The horrified students raised a ruckus and managed to get themselves sent home, albeit with a bad case of the heebie-jeebies.

Another incident during Bardeen's college years reveals something of how he experienced the world in his teens. At seventeen, while he was driving down University Avenue, the vehicle in front of his dropped an axle. John plowed into it. The impact threw him into the windshield, and he sustained some bruises and lacerations that proved more frightening than life threatening. He arrived at the hospital with a bloody face, but it appeared the doctors and nurses had more important things to do than tend to him immediately. Frightened and bruised, he stood up, stomped his foot, and declared, "My father's the dean of medicine and I want service!"

Mathematics remained Bardeen's favorite academic subject. He enjoyed its puzzle-solving aspect. But he was also convinced that he didn't want to "end up being a university professor." An academic career sounded stodgy to him and to his friends, who were mostly children of local businessmen. He decided to major in electrical engineering because he "had heard that that used a lot of mathematics," and he knew that engineers stood a better chance of earning a good living than did mathematicians. But after his challenging studies with Hart, Bardeen was bored by the analytical geometry and other mathematics used in the engineering courses. He began to study calculus on his own.

Bardeen found mentors at the university who recognized and encouraged his mathematical talent. The short, round-faced, and genial Warren Weaver, later head of the Rockefeller Foundation's science program, guided Bardeen in an independent mathematics study and in courses on boundary-value problems and differential equations. In Weaver's courses Bardeen encountered an operational approach to formulating the relationship between mathematical theory and observation. Weaver saw theory as merely an analogy generated by the observed facts, which were more basic. On the first page of his notebook for Weaver's course on electrodynamics, Bardeen wrote, "In a detailed theory of a group of physical phenomena an analogy is exhibited between observed facts and the logical consequences of a self-consistent mathematical structure. The analogy constitutes the theory."

Edward Van Vleck, a mathematics professor at the University of Wisconsin from 1906 until 1929, also mentored Bardeen. This "dignified, formal and reserved" professor was a forbidding figure when riding his bicycle around campus, "with his reddish beard in advance and the coat tails of his cut-away streaming out behind." Having studied in Germany, Van Vleck was a member of the small group of American scholars who had helped to import advanced research in mathematics from Europe. He had a reputation for taking more interest in bright students than average ones and for pushing them to their limit. He influenced mathematics not only through his own scholarship and teaching but also through his leadership of the American Mathematical Society.

Ultimately more important to Bardeen than either Edward Van Vleck or Weaver was Edward's son, the physicist John Van Vleck. Nine years older than Bardeen, "Van" brought modern quantum physics to the University of Wisconsin when he came to teach there in the fall of 1928. "The University was strong in applied mathematics," Bardeen recalled, "but there was little interest in atomic physics until Van Vleck arrived."

Like Bardeen, John Van Vleck grew up in Madison, where he was born in 1899. His Harvard doctoral thesis, a computation of the energy of the helium atom, had been supervised by Harvard's first theoretical physicist, Edwin C. Kemble. One of the early studies of atoms more complex than hydrogen, Van Vleck's thesis was a first step toward understanding heavier atoms, molecules, and later solids. In his early work on the "old quantum theory," Van Vleck had applied pre-1925 notions of the quantum to classical physics problems.

Van Vleck's graduate course with Harvard experimentalist Percy W. Bridgman may have contributed to his subsequent interest in solid structures and magnetism. Van Vleck would later put the fields of paramagnetism and ferromagnetism on a firm foundation with his pivotal text, *The Theory of Electric and Magnetic Susceptibilities* (1932), a careful review of the weaknesses of classical physics theory and both the weaknesses and strengths of quantum mechanics.

At the University of Wisconsin, Van Vleck presented the new quantum physics in a two-semester course, Physics 212, which Bardeen took during the 1928–1929 school year. One of the "earliest of its kind offered in the United States," the course was Bardeen's first serious introduction to quantum mechanics. He found it fascinating.

Developed in Europe during 1925 and 1926, quantum mechanics treated phenomena on the scale of the atom and nucleus, a realm where classical physics breaks down. The revolutionary theory shook the foundations of the classical physics fields of mechanics, electricity, magnetism, optics, and thermodynamics, among others. Old debates, such as whether light is a particle or a wave, dissolved. Light was recognized as something different, a phenomenon having the properties of both a particle *and* a wave. How one made measurements made a great difference to how light was perceived.

The most radical change was the loss of strict causality. Within the new framework, particles and their properties are governed by probability laws. They are specified by a mathematical entity known as a wave function. To determine some physical property— for instance, the mass or speed of an object—one must find the wave function, usually by solving a wave equation. By squaring the amplitude of the wave function, one obtains a measure of the probability that the property in question has any particular value.

Bardeen also took a research course from Van Vleck based on the latter's 1926 textbook, *Quantum Principles and Line Spectra*. Both text and course conveyed the excitement of research at the frontiers of a field undergoing revolution. There was a clear sense of an old and a new world in collision. Among the tools Van Vleck discussed was Niels Bohr's "correspondence principle," a heuristic for connecting the known physics of the old domain with the unknown physics of the new one by requiring that in the overlapping region the physics should agree. Decades later Bardeen would employ an analogous bridging principle in his work on interacting electron gases in metals.

Van Vleck encouraged Bardeen to seriously consider a career in physics. Although he was "intrigued by physics," Bardeen was not yet ready to commit to it. It appeared that "the only opportunities in physics and mathematics were teaching in a university—and I thought that was the last thing in the world I wanted to do!" He did, however, take a year of German to meet an anticipated graduate school requirement, should he change his mind. And he continued to take graduate-level physics courses.

Peter Debye, the eminent Dutch theoretical physicist, spent a semester in Madison. Bardeen took "a very stimulating course from him" covering statistical mechanics and a number of research topics in modern physics. He found Debye an "excellent teacher" who gave dynamic lectures and "had everything very well organized." Bardeen especially appreciated Debye's willingness to present parts of his own current research on dipole moments and the diffraction of X rays and electrons. He thoroughly enjoyed the fact that Debye was up-to-date in quantum mechanics.

Another of Bardeen's physics mentors at Wisconsin was Paul Dirac, then twenty-seven, one of the inventors of quantum mechanics, who visited Madison for six weeks during the summer of 1929. Like Bardeen, Dirac had loved mathematics as a student but had chosen to train as an engineer. When he failed to find an engineering position, he took graduate courses in mathematics at the University of Bristol and then Cambridge. Dirac's course in Madison covered much of the material later published in his classic text, *The Principles of Quantum Mechanics*. Bardeen was impressed by Dirac's "elegant formalism" but felt that "much remained mysterious." He nevertheless earned an A in the course.

Other well-known European theoretical physicists who passed through Madison while Bardeen studied there included Werner Heisenberg and Arnold Sommerfeld. Bardeen heard Sommerfeld speak on electrons in metals shortly after the appearance of the great physicist's pathbreaking quantum theory of metals. "I heard the lectures but I wasn't stimulated at that time to go into that field." Bardeen would later change his mind.

As Bardeen was so much younger than his peers, he did not think twice about taking extra courses to explore other fields outside his course of study. He could afford to take a semester off and did so in the fall of 1926 to extend the summer work he was doing as a requirement for his engineering degree. The job, in the Western

Electric Company's Inspection Development Department, "consisted of developing inspection methods for certain items of interest to the company." Bardeen thought it was interesting work. "Being young I was in no hurry to graduate." He stayed on until Christmas.

By taking extra courses and working at Western Electric, Bardeen delayed his bachelor's degree in electrical engineering by a year, but he was still only twenty when he graduated in June 1928. At that point, as he had already completed some of the course work for the master's degree, he decided to finish the degree before moving on. During 1928–1929 he earned his tuition by serving as a research assistant to Leo J. Peters, who supervised Bardeen's master's thesis. A quiet, pleasant-faced man respected in the field of electrical engineering, Peters squinted at the world through unusually thick glasses, the result of being struck by lightning as a child.

Peters had become interested in the emerging field of electrical prospecting for oil, a branch of geophysical prospecting that related geological conditions with the presence of oil. In electrical prospecting the problem was to relate measured variations of the electrical constants in the earth's crust to minerals below. Geophysical prospecting was then still an art, one Peters hoped to make into a science.

A popular book on the Gulf Oil Corporation's first half century characterized the old-fashioned oil prospector as one who often "had a muddy nose from sticking it into a handful of black earth and sniffing for oil." He was a man who "studied creeks and water holes for seepages and gas bubbles, held lighted matches to the bubbles to see if they would burn and scuffed the ground for paraffin dirt." The new prospector, on the other hand, used state-of-the-art measurements by instruments, such as magnetometers to measure variations in the magnetic field, gravimeters to measure differences in gravity, or seismographs to measure tiny artificial earthquakes set off to explore the materials below the surface. Because the demand for oil was ever increasing, both for automobiles and for other technologies, companies such as Gulf Oil were investing heavily in geophysical prospecting.

For Bardeen's master's thesis, he designed a problem that simulated conditions likely to occur when oil is present. In a long article based on his thesis, coauthored with Peters and published in

1930, Bardeen discussed how "from the geological and other information which is available concerning the region, the conductivity
picture is converted into a picture of subsurface structure or of
mineralization of the region." The method is initiated by sending
an electrical current through the earth at some promising location
or by inducing currents in the particular region by establishing a
varying magnetic field. In the next stage, measurements are made
of electrical quantities at the surface of the earth. These measurements entered calculations based on geological information and
applied classical electromagnetic theory to draw conclusions about
the minerals below the surface.

In the spring of 1929, after Bardeen completed his master's
thesis, he tinkered with the idea of changing his field of study "to
physics, and in particular going to Europe for further study." He
applied for a research studentship in physics at Trinity College in
Cambridge, England. In a glowing recommendation, Peters said that
"in analytical ability Mr. Bardeen surpasses any student that I have
known. He treats difficult problems in a masterly and often in a
unique way." Weaver described Bardeen as "a very independent
young man, doing most excellently the work that interests him
and at times slighting that which does not appeal to him." He added
that the Wisconsin Mathematics Department judged Bardeen to be
"the strongest candidate for such a position we have had in years."
Weaver personally thought there was "a real chance that Mr. Bardeen
may turn out to be a genius."

Van Vleck also supported Bardeen's application to Trinity with
a letter to the physicist Ralph H. Fowler, who had presided over
Dirac's graduate work. "Mr. Bardeen is an exceptional student,
unquestionably one of the two or three best I have ever had. He is
taking my course in quantum mechanics and grasps the subject so
quickly that I feel that he is at times bored because I cover the
ground so slowly, and is never forced really to exert himself in order
to easily lead the class." Fowler forwarded Van Vleck's letter on to
the senior tutor with a handwritten note describing Van Vleck as
"a man of sound judgment and European standards." But Bardeen
did not get the studentship.

He drifted into another year of graduate study. When Peters left
Madison to work in Pittsburgh, Bardeen took a research assistant-
ship for the 1929–1930 term with Edward Bennett, head of the
engineering department at Wisconsin. At Bennett's suggestion,
Bardeen worked on the diffraction of radio-length electromagnetic

waves and the design of antennas. The calculations relied on classical electromagnetism. At first Bardeen was enthusiastic about them because of their predictive power. "The principal aim in pursuing any branch of science is to acquire the power of prophesy with reference to the events treated in that branch," Bardeen had read in Bennett's course notes.

Bennett was struck by Bardeen's "modest acceptance of his own powers" and hoped that he would continue his advanced studies at Wisconsin. But with time Bardeen's interest in antennas waned. When he sought additional mathematics courses, he found that the only advanced courses left for him to take were those taught by Rudolph Langer on the theory of differential equations and analytic functions. Bardeen did not find these subjects particularly interesting. It was time to move on.

In the fall of 1929 the American Telephone and Telegraph Corporation (AT&T) was seeking engineering staff to study wave propagation and antenna design. Thornton Fry, an AT&T recruiter, had heard about Bardeen's antenna work and approached him about a job. But the offer dissolved as the country slid into the Great Depression. By the spring of 1930, AT&T, like many firms, had put a freeze on new hiring.

Another opportunity soon arose when the research laboratories of the Gulf Oil Company in Pittsburgh offered Bardeen a job as a geophysicist. The offer was probably made at the urging of Peters, who had recently moved there. Peters encouraged Bardeen to accept Gulf's offer. The decision was not difficult for John. Not only was Gulf one of the few places in the country still able to hire, he also found the work far more interesting than antenna design. "These were the days when geophysics was just opening up," Bardeen later reflected. "Oil companies were still reasonably prosperous even in depression days. People had to buy gas to run their cars." Years later Bardeen claimed that had he had the luxury to choose between AT&T and Gulf, "I may have chosen that job (at Gulf) anyway."

The rolling hills of Appalachia had a beauty that welcomed Bardeen as he drove to Pittsburgh during the summer of 1930.. The verdant countryside concealed the grinding poverty of farmers who battled with the region's rocky soil. The dangerous working conditions of coal miners and steelworkers were also hidden from view.

As Bardeen entered the city, distinguished by its steel mills and other heavy industry, he noticed a cloud of choking smoke that hung over the buildings. He headed over to Craft Avenue in Oakland, the section that housed the university complex. Then he found the Gulf Research Labs. They were at that time based in a modern, three-story, steel-and-concrete structure, elegantly trimmed around the top-floor windows with bronze. The blocky building was just a year old. Bardeen asked for Peters, who introduced him to a few colleagues.

Bardeen needed a temporary place to live while he looked for an apartment. The process took time, for affordable housing was at a premium. Another new Gulf employee offered John a dormitory bed. "He was tired, dirty, with rumpled clothes and no place to rest. I, with my new B.S. degree from Carnegie Tech, had just started with Gulf and was living in Engelbrecht Hall at Tech. I asked John to go there with me; there were spare beds during the summer."

Bardeen gratefully accepted. Engelbrecht Hall was one of seven dormitories maintained for men by the Carnegie Institute. Its rooms were spartan and small, roughly eight by fourteen feet. Each was furnished with a bed, wardrobe, table, and chairs. And there were bathing and toilet facilities on each floor.

After a brief stay at Engelbrecht, Bardeen moved into a modest red brick apartment building in an east-end, working-class neighborhood. After living there for two years, he would move again into a better building on a hill not far from the University of Pittsburgh and the Carnegie Institute.

According to legend, the Gulf Oil Corporation emerged from a wildcat well in Texas, which came in on January 10, 1901, "with such fury that it wrecked the drilling rig and covered the surrounding countryside with a lake of oil." The driller and two prospectors borrowed $300,000 from the Pittsburgh banking firm of T. Mellon & Sons and formed a partnership, the J. M. Guffey Petroleum Company.

In 1907 the Mellons bought out Guffey and a year later formed the Gulf Oil Corporation. William L. Mellon, the nephew of the banking brothers Andrew W. and Richard B. Mellon, became president of the new Gulf Oil Company. He wanted to create a firm that would integrate "everything from sink drills into land and sea to filling the customer's gas tank at a service station." In 1913 Gulf opened the world's first drive-in service station.

By the 1920s the company also began to invest in research. On the premise that finding and extracting oil was done best "when the experience and knowledge of the geologist are combined with the tools of the physicist," the laboratory began seeking leading scientists. The administration soon recognized that to retain the best scientists it had to offer working conditions competitive with those at universities.

Gulf's hiring policies were successful in attracting the best and the brightest. In 1925 the geologist K. C. Heald came to Gulf from Yale and negotiated a position in which he was paid to conduct basic research. Heald attracted Paul D. Foote, a physicist then at the U.S. Bureau of Standards. Gulf allowed Foote, who eventually became the vice-president of Gulf's research division, to hold a concurrent research position at Pittsburgh's Mellon Institute. Two years later Foote was joined by the geophysicist E. A. Eckhardt and then in 1929 by Peters. Around this nucleus of research scientists—Heald, Foote, Eckhardt, and Peters—grew the Gulf Research Laboratory. Through its aggressive strategy of hiring promising young scientists and engineers, the number of employees in Gulf's research division grew from twenty-eight in 1928 to roughly 1,700 by the early 1950s.

Using constantly improving measuring equipment Gulf's geologists kept records of the geophysical properties of each hole drilled, whether or not the workers struck oil. From this information, Bardeen and other scientists on the staff calculated the likelihood of finding oil in a particular location. Although Bardeen mainly worked on electromagnetic prospecting, and on the interpretation of magnetic surveys, he also studied electrical, gravitational, and seismic methods. He recalled, "It was the early days in geophysics when lots of new ideas were under development and lots of new physics was involved." Soon he was directing a group of more than fifteen employees. For several years he found the problems encountered in his work interesting and challenging.

In the magnetic prospecting work, "We'd get the results in from the field and try and interpret them," Bardeen explained. But he was bothered by the skaky basis of this work. "One may assume that the rocks that give rise to the magnetization are uniformly magnetized and then one can calculate what structure would give rise to the field observed on the surface." He was very well aware that "the assumption of uniform magnetization is by no means valid."

Bardeen also worked on electrical prospecting, in which information about oil deposits came from measurements of resistivity variation. The basic assumption was that "changes of resistivity follow the bedding planes." In their 1932 article in *Physics*, Bardeen and Peters warned that the information gained from such electrical measurements could not be relied on for depths of more than 2,000 feet. Peters described the theory in more detail in a later paper, which won the Society of Exploration Geophysicists' Best Paper Award for 1949.

Bardeen balanced his professional life in Pittsburgh with a relaxed social life limited to inexpensive pastimes, such as bowling. He didn't know at first that, Dutch Osterhoudt, his PiKa friend from Mendota Court, had also been hired by Gulf Labs. Osterhoudt had initially turned down Eckhardt's offer of $150 per month on the grounds that it was not enough. A month later, when Eckhardt renewed his offer, he asked Osterhoudt whether he knew Bardeen, who had just accepted a position at $175 per month. "Do you think you're any better than young Bardeen?" Osterhoudt struck a bargain at $160.

As it turned out, Dutch and John hardly saw each other at work. "Johnny worked in magnetics in a small room at the rear side of the laboratory building. I worked in a small room on the second floor in the seismograph [department]." Soon Osterhoudt was spending most of his time prospecting for oil in faraway places. But the two friends often had a beer together when Osterhoudt came in from the field.

In his third year at Gulf, Bardeen again found himself at a crossroads. He faced up to his growing conviction "that if I wanted to do geophysics in the long run I would have to learn more geology." The subject had never captured his interest as fully as mathematics or theoretical physics. Probably through either Foote or Arthur Ruark, who both held joint positions at Gulf and the University of Pittsburgh, Bardeen learned of a seminar on modern physics held at the university. He decided to attend. "It was on my own time, outside of regular working hours."

Organized by Ruark and Elmer Hutchinson, the seminar of "at most eight or ten people, faculty members from Pitt and Carnegie Tech and a few graduate students" focused on current research problems. Meetings were sometimes led by one of the regular members, occasionally by an out-of-town visitor. Bardeen found the discussions more stimulating than his engineering work at Gulf.

Bardeen soon realized that to change careers and apply himself to mathematics or physics would require more education. He considered only one graduate school—Princeton—"because there was an outstanding mathematics department as well as the Institute for Advanced Study that had just gotten started there a couple of years before." He argued, "I was going on my own money, so I could pick the university where I wanted to go." Late in the fall of 1932, John discussed his plans with his father, Charles. He explained that he had "decided that university life wasn't so bad after all" and that he was interested in studying mathematics. In December Charles wrote a letter to Abraham Flexner, the founding director of the Princeton Institute, and told him he was pleased that his son John would "take a try at higher things" before his "arteries begin to harden." Flexner in turn brought Bardeen's case to the attention of the mathematician Oswald Veblen, one of the first appointees to the new institute. To involve Veblen in Bardeen's application was probably unnecessary, as Bardeen's record and recommendations were outstanding. Moreover, even a university as prestigious as Princeton was underenrolled just then. Bardeen later claimed that he would have stayed on at Gulf had Princeton not accepted his application. It was an unlikely scenario.

Bardeen asked the two Wisconsin professors who had had the most influence on him, John Van Vleck and Warren Weaver, to write to Princeton on his behalf. Van Vleck described Bardeen "as a student of outstanding ability" and one who "easily led the class." He estimated that his former student had a "native ability in mathematical physics comparable with that of many National Research Fellows." Bardeen's "personality is good," Van Vleck wrote, "although he is inclined to be a little reticent."

Weaver, by now at the Rockefeller Foundation, said that as a "general rule" he avoided writing recommendations. He was making an exception for Bardeen because he had "never had a student whom I could recommend to the graduate mathematics group at Princeton with as much enthusiasm or with as little reserve."

He cautioned that it would be difficult to evaluate Bardeen's "actual state of advancement" from the written record. "He is so modest that he often fails to report on things which he has accomplished." Weaver wrote that whenever he questioned Bardeen, it would become apparent that Bardeen had not only solved the problems but had done so "by methods which were surprisingly simple and mature." Weaver said he had "a strong suspicion that [Bardeen] is the most able student with whom I have ever had contact." He also mentioned that Bardeen was "somewhat shy, and not particularly communicative," but said he was nevertheless "attractive socially."

Princeton admitted Bardeen in mathematics for the fall semester of 1933 but without financial support, complicating his decision to leave Gulf. He was not yet twenty-five, but returning to academia was risky. "It was 1933 when jobs were hard to get. I had a good job at Gulf. I didn't know if I would be able to get a good job again if I quit this one to go back to school."

By the spring of 1933 he had concluded that he should not, indeed could not, "resist any longer" following in his father's academic footsteps. Like his mother who had taken a comparable risk at about the same age when she broke from her family to study art, John decided to follow his passion.

"I quit my job not knowing if I'd ever get another." As for money, he could draw on his small inheritance from his grandfather, C. W., who died in August 1924, a week before his seventy-seventh birthday. At the time, C. W. was still actively at work. John also had some additional savings that his frugal lifestyle in Pittsburgh had allowed him to accumulate. So "I could take a gamble."

Peters and Eckhardt hated to lose Bardeen. Gulf Research was then preparing to move into a larger facility in Harmarville, eight miles up the Allegheny River. Bardeen was in their plan. Eckhardt asked Osterhoudt to "drop down and see your friend John Bardeen and try to talk him into staying with us."

Dutch found John adamant. "Bardeen swiveled his chair around and pointed to his small old blackboard." He replied, "I'm tired of sitting here in this little office, staring at the same damn blackboard and the same four walls. I'm going back to school and get my doctorate." He eyed Osterhoudt, "Why don't you come with me?"

But Dutch shook his head. He loved prospecting and would remain at Gulf for another eighteen years.

By the time Bardeen arrived at Princeton, some of the world's leading mathematicians and theoretical physicists would be there or arranging to come. In March 1933, Hitler had come to power and Jewish physicists and mathematicians working in Germany had abruptly lost their positions. A large number emigrated to Great Britain or the United States. A few ended up at Princeton. When Bardeen heard that Einstein had accepted a lifetime appointment at Princeton's new Institute for Advanced Study, he thought perhaps he would study with him!

The night before John left Pittsburgh, he attended a small dinner party in the apartment of a Gulf colleague, Bruce Reline. There he met Jane Maxwell, a slim and articulate twenty-six-year-old woman who was renting a room from Bruce and his wife Mary Margaret. Jane and Mary Margaret had grown up together in their hometown of Washington, Pennsylvania, about twenty-five miles southwest of Pittsburgh. Their families had known each other for years. Jane held a teaching position at the women's college of Carnegie Tech and was also taking graduate courses at the University of Pittsburgh toward her master's degree in zoology. Bruce had told Jane that he hoped she would stay home for dinner that evening. John Bardeen would be coming. "I think you ought to meet him," said Bruce.

The Relines' matchmaking had a number of elements in its favor. Jane's and John's fathers were both medical doctors—hers practiced in Washington, Pennsylvania. Like Althea, Jane's mother, Elizabeth "Bess" Patterson Maxwell, had had a career before marriage as a teacher and school principal. Like Althea, Bess had given up her career to raise her children and help her husband in his office. And like John, Jane had had her share of childhood misfortunes. The eldest of five, she became the caregiver for the younger children when her mother became debilitated by arthritis.

Jane had done well in her studies, despite the extra responsibilities. She had initially studied history at Wellesley College after graduating from the Washington Female Seminary. But before long she wrote home, "I've decided that I'm really interested in science and so the thing to do is to find out something about it." She decided to study medicine but, like John's sister Helen, could not convince her father to allow her to become a physician. She couldn't even convince him to let her attend nursing school, as Helen had done. So Jane moved to Pittsburgh after completing her bachelor's degree and continued studies toward a master's degree in zoology. She supported herself by working at Carnegie Tech, first as a laboratory assistant and then by teaching. She often visited her close-knit family on weekends.

In the Relines' tiny kitchen, John and Jane sat on one side of the rectangular table. Bruce and Mary Margaret were on the other. Jane eyed John surreptitiously. She saw a trim, athletic, twenty-five-year-old man with intense eyes and a gentle smile. That he was brilliant was obvious to her, despite his reticence. She was attracted by his thoughtful demeanor, dry wit, and quiet confidence.

John, in turn, was struck by Jane's warmth and energy. She spoke easily, with detached humor, about her life and views. Something about her eyes, her self-assurance, perhaps a hint of sadness, arrested him. Before leaving, he unobtrusively captured her hand for a quick squeeze, just enough to let her know he was interested in getting better acquainted.

Soon he would be miles away. On the long drive east, his thoughts often returned to the lively biologist he had met at the Relines'. As he crossed the forested peaks of the Appalachians, he planned a trip back to Pittsburgh, maybe over Christmas.

4

A Graduate Student's Paradise

It was dark when Bardeen reached Princeton, but the campus was not deserted. Most of the buildings were still lit on that late summer night. Every now and then a bicycle headlight flickered by as a student returned from late study in the library or from a laboratory.

Driving half a mile west of the main campus, Bardeen pulled up to the Graduate College, where he would live for the next two years. He stopped for a moment to admire the Gothic structure designed by Ralph Adams Cram. Another physics student, Philip Morse, described the elaborate building as "the embodiment of scholasticism." Numerous gargoyles and entryways projected a sense of age and culture. The carillon tower beside the entry gate looked eerie and beautiful in the moonlight.

The next morning Bardeen made his way to Fine Hall, the mathematics building. After registering, he walked next door into the Palmer Physics Laboratory. The department chair, Adams Trowbridge, grabbed Fred Seitz, a fourth-semester graduate student from Stanford, and said, "Here's a new student. Could you take him to lunch?" Seitz readily agreed.

Seitz and Bardeen struck up a friendship that would be life-long. John told Fred about his oil prospecting work. Fred was fascinated. He thought the electrical and magnetic methods that John described for seeking oil sounded more effective than the sonic technique that he knew about. He also thought that Bardeen appeared mature for his age. Seitz decided this was probably because Bardeen had lived on his own for three years while "working in the corporate world." Having taken two years of graduate engineering courses at Wisconsin, Bardeen was more advanced than Seitz had been at the time the latter entered Princeton's program for the spring 1932 semester.

Bardeen and Seitz walked together to the Graduate College and into its formal dining hall, where they sat down at a long table with about half a dozen other students. Seitz knew them all. He reckoned there were only about 150 graduate students in the entire university.

Across the table sat a fellow "who was much worried about passing his preliminary exams. He was in the middle of taking them and in deep trouble," Seitz recalled. Extending over several days, the "prelims" were the major hurdle for a Princeton graduate student, the crucial step before passing to the stage of thesis writing. The ordeal included a rigorous written section and several sessions of intensive oral questioning by a committee. "This fellow looked at John and at me and said, 'Does he know what he's getting into? Don't you think you should tell him?' "

Seitz explained the home rules of the Graduate College. Dinner was served every evening in the great hall following the dean's daily grace, which was said in Latin. He told John that he needed to buy a formal black scholastic gown to wear at dinner, adding that he could probably get one inexpensively from a graduating student. The gowns grew increasingly stained and tattered as they were passed on. "They were rags," recalled Seitz. "We thought of them as bibs."

Bardeen nodded stiffly as he registered Seitz's next instruction: Always wear a tie to class. The Princeton style was clearly very different from the informal one he had witnessed from the docks of Lake Mendota. Bardeen thought Princeton's rituals might take some time to get used to, but they didn't. Like Seitz, he would find the formalities "insignificant." Princeton's rituals would be less graciously accepted by Einstein, who arrived on the scene a month later with his wife, Elsa.

Abraham Flexner had conceived of the Institute for Advanced Study as a haven for scholars. Its purpose was to nourish advanced research and thinking by freeing the most creative academics from their usual pressures. Oswald Veblen, one of the institute's first appointed mathematicians, suggested that Flexner start by building up faculty in the area of mathematics. Accordingly, Flexner hired Hermann Weyl, whose work had helped to put quantum mechanics on a firm mathematical basis. Einstein's subsequent acceptance of a permanent post moved the institute onto the world stage. The physicist Paul Langevin compared Einstein's well-publicized move to Princeton with transferring the Vatican to the New World.

The Einsteins, like the Quakers who had settled Princeton in the 1680s, admired the town's physical setting, with its lush forests and many streams. "One great park with wonderful trees," Elsa Einstein wrote to a European friend. Albert, however, deplored the aura of gentility in the Princeton community, with its endless ceremonies regulating life in that intellectual enclave. He described Princeton as "a quaint and ceremonious village of puny demigods on stilts." He missed the coffeehouses of Berlin, where intellectuals could relax for hours while exchanging deep ideas. Eugene Wigner, a fellow European on the faculty, also lamented Princeton's lack of "coffeehouses in the European sense." Bardeen didn't care about coffeehouses, but he did care about games, such as bowling or bridge, which were readily available in Princeton. Johnny, as he was known to his Princeton friends, bowled regularly with his roommate Cassius Curtis or with his friend Robert Brattain. As in Madison, John parlayed his social encounters during games into collegial connections.

Bardeen felt very comfortable with Bob Brattain, who entered the program in physics at the same time Bardeen entered in math. Raised on a cattle ranch in the state of Washington, Brattain was quite unconcerned about high culture. And, like Bardeen, he also loved games. The two bowled as a team to win the doubles graduate bowling tournament and then faced off against one another for the singles championship.

In bridge they were "enemies." Bardeen's partner, John Vanderslice, another mathematics graduate student (who Brattain referred to as "Slice"), was an internationally ranked chess player. Occasionally Brattain invited Bardeen along to New York City to join in one of the weekend-long bridge marathons that Bob's brother Walter sometimes hosted in his Greenwich Village apartment. "We played until everybody got so sleepy that they went to sleep. Then we'd sleep for a while, get up, eat something and play bridge. We played bridge continuously for the whole weekend."

Bardeen immediately hit it off with Walter Brattain, an outgoing experimental physicist then working in lower Manhattan at Bell Telephone Laboratories, the research and development arm of the American Telephone and Telegraph Corporation (AT&T). Walter was among the few Bell Labs researchers who already recognized that quantum mechanics would be important in solving AT&T's communications problems. When Walter learned that Arnold Sommerfeld, the great European quantum theorist, would be lecturing in Ann Arbor at the 1931 Michigan summer school in theoretical physics, he convinced Bell Labs to let him attend Sommerfeld's course. The lectures covered the new "semi-classical" electron theory of metals that Sommerfeld had developed several years earlier. Brattain recognized the importance of this theory for his own physics problems and for others of interest to Bell Labs. Afterwards, when Brattain returned to Bell Labs, he summarized what he had learned in Michigan in a series of talks he offered to his colleagues. Bardeen soon encountered the same material in his studies at Princeton.

John and Walter could not have known then that fourteen years later they would co-invent a device that would change the world. John did know that he greatly enjoyed socializing with Walter. He loved Walter's good humor and colorful stories about the Brattain family's pioneering heritage. Walter would often tell how, before entering college, he had spent an entire year "herding cattle in the mountains, with a rifle, in my own camp. I only saw my family on occasional weekends and saw practically no other individual outside of my mother and father and brother and sister."

Walter's and John's stylistic differences were already clear. For example, in their bridge playing both men aimed to win, but Brattain played aggressively while Bardeen played thoughtfully. Although different in their styles, their passions, values, and interests aligned. At that time they even worked on the same physics problem. "Walter and I had a common interest," Bardeen later wrote, "the theory of the work function as derived from the quantum theory of metals." A measure of the energy required to remove an electron from the surface of a metal, the work function was a concern of any technology that relied on vacuum tubes.

The two also had in common John Van Vleck, who had introduced both to quantum mechanics. Van Vleck had been on the faculty at the University of Minnesota before moving to Madison.

Walter had taken Van Vleck's quantum theory course at Minnesota in the late 1920s. But at that early stage in their lifelong friendship, John and Walter "didn't really talk that much physics," Bardeen recalled. "It was more of a social acquaintance."

Bardeen's closest colleague and friend at Princeton was Seitz, who after completing his Ph.D. in 1934 stayed at Princeton for an additional year with a Proctor Fellowship. In later years Bardeen and Seitz would often interact professionally. They would work together in the same department at the University of Illinois from 1951 through 1965.

After Seitz, Bardeen's closest physics friend at Princeton was Conyers Herring, who arrived in 1934, when Bardeen was starting his second year. After graduating from the University of Kansas, Herring had spent a year studying astronomy at Caltech. He transferred to Princeton after changing his field to physics. Bardeen and Herring would also remain friends and colleagues. They would overlap in the late 1930s at Harvard and in the 1940s at Bell Laboratories.

Another of Bardeen's friends at Princeton was Walker Bleakney, a young faculty member. Bleakney had attended graduate school in Minnesota with Walter Brattain, who occasionally visited Bleakney at Princeton. They "were part of the ring with John," Seitz recalled. "Their friendship went very deep" and they "were thick as thieves when having fun together." Not only would they bowl and drink beer together but they also "spoke the same Midwestern language."

Bardeen's Princeton friends also included the chemists Henry Eyring and Joseph Hirschfelder. Princeton encouraged communication among students in related fields, such as chemistry and physics, or physics and mathematics. The physics and mathematics students regularly drank afternoon tea together in Fine Hall. "At 4:30," Seitz recalled, "everyone who could walk or go on crutches met in what was called the social room and spent about twenty minutes to a half-hour talking."

Bardeen attended seminars at both the university and the Institute for Advanced Study. The institute, though institutionally separate from the university, was housed in Fine Hall. Bardeen recalled institute seminars that included such greats as Einstein, John von Neumann, Oswald Veblen, and Hermann Weyl. He later described the institute as a model of collaborative work that created "a strong synergistic effect of accomplishing much more than the individuals could be expected to do on their own." The notion of enhancing research through interdisciplinary collaboration appealed to Bardeen throughout his career.

Princeton was then entering an extraordinary era of change. Modernization of the curriculum had brought major additions to the physics faculty, including Edward U. Condon and Howard P. Robertson in 1928. Princeton's strengthening of its mathematics and physics offerings was part of a larger institutional growth in American universities during the 1920s and 1930s. This nation-wide trend transformed American physics, removing the earlier gap between American and Old World scientific centers that had led young theorists of the previous generation, such as J. Robert Oppenheimer or John Slater, to study in Europe. By the time Bardeen entered his graduate program, that gap was no longer noticeable.

Among the European physicists who had taken an interest in building up American physics was the eminent Dutch theoretical physicist Paul Ehrenfest. He had explained to the heads of American departments that hiring Europeans would work much better if positions were offered in pairs, preferably to researchers in close specialties, so that they would be able to speak with one another and feel less isolated in the "wilds" of America. The University of Michigan conducted the "Ehrenfest experiment" successfully in 1927, when Samuel Goudsmit and George Uhlenbeck were simultaneously offered positions in its physics department.

Princeton followed suit in 1930, hiring two Hungarian members of the Berlin circle of physics and mathematics, John von Neumann and Eugene Wigner, who were old friends. Princeton had "reluctantly" taken Wigner on to attract von Neumann, who had been in the class behind Wigner in their Budapest grade school. At Princeton the two friends enjoyed deep conversations during long walks together. The subjects they discussed ranged from popular culture to mathematics.

A year later, when their one-year appointments ended, both Wigner and von Neumann were happy to accept a new five-year contract. The appointments included the condition that they both spend half the year at Princeton together, the other half anywhere else. Wigner accepted an appointment at the Technische Hochschule in Berlin for the spring semesters. But before the five years had ended, the situation in Germany for Jewish physicists had grown so dangerous that Wigner broke his ties with Berlin. Bardeen later wrote that of all his professors at Princeton he was "most stimulated by the two young Hungarians."

Bardeen thoroughly enjoyed his Princeton seminars on quantum mechanics and relativity. One of his favorites was the one that von Neumann offered on "Operator Theory" during the first and second semesters of Bardeen's time at Princeton. The seminar dealt with Hilbert space, an infinite dimensional space of functions. During his second year at Princeton, Bardeen sat in on Paul Dirac's yearlong course on quantum electrodynamics, a more advanced continuation of the course Bardeen had taken from him at Wisconsin. Robertson's course on general relativity and cosmology during the spring of 1935 impressed Bardeen so much that he would use it later as a basis for teaching his own course on relativity. "Robertson's lectures were of supreme elegance," Seitz later reflected.

Bardeen found Robertson easy to talk with. He was "certainly one of those who made the department in terms of social interactions, creating a friendly atmosphere." Bardeen recalled that he was the "sort of person who would not intimidate anyone from asking questions about anything." Herring said that Robertson was "prominent in the beer parties," as well as "very competent in his field of general relativity and cosmology." Both students said the same about Condon, who Herring called "good-natured, very approachable." Condon "always had a good physical way of explaining everything and just was very easy-going." Bardeen, Seitz, and Herring were among the members of a small group of young physicists that gathered informally around Condon, united by their interest in theoretical physics. Their meetings typically began in the afternoon and ended in the evening at the Nassau Tavern, where the discussions continued over beer.

Bardeen attended many special lectures and seminars. One distinguished visitor was Ralph H. Fowler, from Trinity College, to whom John had applied unsuccessfully in 1929 for a research studentship. Others who lectured at Princeton while Bardeen was studying there included Erwin Schrödinger, the inventor of the wave mechanical formulation of quantum mechanics, and Isadore I. Rabi, whose molecular beam experiments had made it possible to measure the radio frequency spectrum of atomic nuclei. Bardeen made an effort to attend these lectures "most of the time, even though the talks were not in areas which affected my work directly," for his goal was to gain "a broader knowledge of what sorts of problems people were interested in." He wanted "to see what major hurdles were being faced"—in short, what it meant to be a great physicist. Although, he admitted, many of the talks were "over my head," he was "getting a little feeling" for their subjects.

The Princeton graduate students discovered that they were quite free to follow their interests. "Only a few courses were offered at the graduate level, so most of the students took what was offered," Bardeen explained. Beyond these, students worked with individual professors to create specialized courses to help advance their careers. There were no official requirements. Routine burdens, such as homework assignments, were "minimal." Seitz called Princeton in that period "a graduate student's paradise."

Bardeen stayed on the fence between physics and math. He thought he had more talent in math but considered physics more interesting. In any case, choosing was unnecessary as the physics and math graduate students took the same courses. The only difference between Bardeen's program in math and Seitz's in physics was in the oral prelims; even their written prelims were identical. Bardeen straddled the two fields as long as he could. Although his Ph.D. was in math, he selected a physics problem for his thesis.

Bardeen "found that there wasn't much opportunity of working with Einstein after all." The great man had done his major work decades ago and was then mainly devoted to relocating German-Jewish refugee physicists. The research he was doing focused on two immensely difficult projects not appropriate for a young researcher: finding the inconsistencies in quantum mechanics and developing a unifying theory for all of physics.

In shopping around for a thesis advisor, Bardeen spoke first with Condon, but he found that all of Condon's suggestions concerned filling gaps in the textbook he was then completing with G. H. Shortley, *The Theory of Atomic Spectra*. That "didn't sound too interesting" to Bardeen. Seitz had had a similar experience with Condon, with whom Seitz had expected to work, given that Condon was known to be "one of the few individuals on the East Coast with a working knowledge of quantum mechanics." Seitz found Condon "all wrapped up" with his textbook. In an effort to be fair, Condon explained to Seitz that the problems of his book were not top-level thesis projects. He helped Seitz arrange to work with

Wigner. According to Seitz, Condon told Wigner, "Atomic theory is a cold fish by now. What we ought to see is if we can make a start on solids." And Wigner had replied, "Yes, that has been on my mind."

Bardeen also spoke with Robertson about working on a problem in relativistic quantum electrodynamics. Exploring a few problems that Robertson suggested, he soon realized that the esoteric field was not ripe for progress. He thought he would be constantly frustrated by "all those infinities" that arise in relativistic quantum field theory, because single electrons can make a virtual transition to an infinite number of states. A decade and a half later, in 1947, Julian Schwinger, Richard Feynman, and Sin-itiro Tomonaga would develop a method for avoiding the infinities using "renormalization" theory. The three would win the 1965 Nobel Prize for their contribution.

The problems that Wigner was addressing, on the application of quantum mechanics to real (rather than ideal) solids, looked more manageable to Bardeen. It also appealed to Bardeen that solid-state work would be of great use in the world.

Wigner thought that his interest in solids had emerged from the curious fact that each material has its unique characteristics. Attempting to explain this interest to an interviewer, Wigner reached for his keys which at first he could not find. "Maybe they're in your overcoat, Eugene," suggested Seitz, who was there. Wigner rushed out into the hall to check his overcoat. He returned after a short time rattling the keys. Then he raised them and let them fall. He explained:

> If I drop my keys, I don't worry at all whether they will be broken. But if I drop a cup or a glass, I worry at once, and usually it is broken. That is a fundamental and noticeable difference. The explanation of these facts uses, to a very large extent, the basic fact that there are fundamental differences in the structure of different solids.

Wigner wanted to know the cause of these differences in structure. His childhood experience of tanning leather in his family's shop had nurtured an interest in real materials. He claimed that he had always been attracted to problems on the borders between disciplines. The problems of solids, such as why the "atoms in crystals are so often located on symmetry axes or in symmetry planes," drew on several disciplines. In studying solids he could benefit from seven fields in which he had prior knowledge: atomic physics, quantum mechanics, chemistry, chemical engineering, molecular structure, elastic vibrations, and group theory.

Bardeen wondered whether Wigner, only six years Bardeen's senior, would be a suitable advisor. He had heard rumors about the rigors Wigner imposed, in particular his expectation that students have a firm grasp of quantum mechanics. That was not a problem for Bardeen, as he *wanted* to gain mastery of that field. He was more concerned with the amount of time Wigner could offer. The latter was still spending only the fall semesters at Princeton, the spring semesters in Berlin. Bardeen also worried whether he and Wigner would be able to communicate effectively. The courtly Wigner was at least as quiet as Bardeen. One Princeton student judged Wigner "too polite for this informal society." In his own office, Wigner would ask for permission to remove his jacket. Wigner's biographer, Andrew Szanton, described a telling interchange he had had with Wigner. When a coughing spell interrupted the conversation, Wigner apologized. "It is my fault," he said, but "not my *conscious* fault."

Bardeen questioned Seitz about his experience with Wigner. Seitz was completely positive. He later called working with Wigner "one of the most remarkable experiences of my life." Not only was Wigner the perfect mentor for Seitz, but the two also became close friends. Having recently arrived at Princeton in 1930, Wigner felt like a "fish out of water." During long walks together, Seitz tried to address Wigner's questions about American customs, while he in turn quizzed Wigner about European politics.

Bardeen decided to throw in his lot with Wigner and never regretted it. He found he needed only occasional meetings to keep his thesis on track. Wigner later told Seitz "that he rarely communicated with Bardeen." But what Bardeen remembered about Wigner was that he always had ways to motivate him with his penetrating questions. Most importantly, Bardeen felt that Wigner instructed him in the art of choosing crucial problems. "He could see what was essential and what the important problems were."

Bardeen also felt that Wigner taught him how to go about attacking problems. The first step was to decompose the problem, either into smaller problems with less scope or into simpler problems that contained the essence of the larger problem. Bardeen said that Wigner stressed reducing to "the simplest possible case, so you can understand that before you go on to something more complicated." In other words, "You reduce a problem to its bare essentials, so that it contains just as much of the physics as necessary. I think that was a good lesson to learn." However, Seitz and other colleagues later pointed out that Bardeen's problem-solving methods differed from Wigner's in at least one important way: While Wigner typically opted for an elegant, "very refined mathematical approach," Bardeen was happy "to bully through" using any method that worked.

In his later years, Bardeen would sometimes note that William Shockley, his Bell Labs group leader and co-Nobel laureate, often used the same approach of simplifying problems to their essentials. But the physicist Philip Anderson, who knew both Shockley and Bardeen, pointed out that while Shockley believed in simplifying problems, he sometimes found it difficult to try alternative approaches if the first attempt failed. If Shockley's "try simplest cases mantra failed to work, that was it." Bardeen, Anderson said, "had, in addition to his brilliance, the persistence and judgment . . . to recognize that when one line failed one had to look deeper."

Wigner was "very encouraging about the attempts I was making towards quantum electrodynamics that didn't go anywhere," Bardeen recalled. When the work was superceded before Bardeen had a chance to publish it, Wigner suggested that Bardeen instead try to calculate the work function of a metal, the energy that must be added to release an electron from its surface. This problem was of great interest to industry because reducing the work function of filaments could save huge amounts of power. Bardeen took the problem up.

In the early 1930s, only two other graduate programs offered training in quantum solid-state theory: one program at MIT, in Massachusetts, was directed by John Slater; the other, at the University of Bristol in England, was led by John E. Lennard-Jones, Nevill Mott, and Harry Jones. As MIT was only a long day's drive from Princeton, Wigner's and Slater's students could, and did, occasionally visit each other for a few days, sometimes for a whole week. Reflecting on the roles of Princeton, MIT, and Bristol in shaping solid-state theory during the 1930s, Bardeen felt that "practically all descendents can be traced back, one way or another, to those three." But Princeton, he smiled, was "certainly the most exciting place."

A year later Herring also made the decision to work on the

theory of solids under Wigner. Wigner later told an interviewer that "Conyers knew more solid-state physics than anybody I ever met." Seitz, Bardeen, and Herring, in that order, Wigner's first three graduate students, were in the first handful of theoretical physicists who would refer to themselves as solid-state physicists. But while Wigner's infectious passion for solid-state physics problems burned out after a few years, Seitz, Bardeen, and Herring all stayed within solid-state theory throughout their long careers.

For the Christmas break in 1933, Bardeen followed through on his plan to drive back to Pittsburgh. As soon as he arrived, he phoned the Relines. Jane was there. Bruce put his hand over the mouthpiece and said, "It's John Bardeen. I think he wants to ask you out but he's forgotten your name." It was a moment she had been hoping for. John asked Jane to a New Year's Eve party. She had an earlier engagement across town. To encourage the romance, Bruce offered to get her back home in time for her date with Bardeen.

Afterwards, John had to rush back to Princeton. He began to seek reasons to visit Pittsburgh. When Peters and Eckhardt invited him to work at Gulf for the summer, he readily agreed. But the time John and Jane spent together was minimal in that summer of 1934, because Jane had already arranged to work at the Woods Hole Marine Biological Research Laboratory on the Massachusetts coast.

When John learned that his younger brother Tom was marrying his longtime sweetheart Janet Smith, he suggested that Tom apply to Gulf for a position. He figured that Tom had a good chance of being hired. Unlike John, Tom had excelled all along as a student, receiving honors at the University of Wisconsin. One story has Tom developing his own derivation on a mathematics final when he did not know the official one. Like John, Tom had also been on the swim team, serving for a time as its captain. He had been president of the Student Athletic Board. Although Tom went on to work toward his Ph.D. in electrical engineering, he was not completely satisfied with the program at Wisconsin. With a master's degree in hand and a strong recommendation from his older brother, Tom secured a post at Gulf Labs. He arrived in time to overlap with John for a few weeks during the summer of 1934.

Tom's quick intelligence made him an instant asset at Gulf. Over the years he took responsibility for much of the company's seismic instrumentation. By the time he retired he had nine patents to his credit. Gulf eventually rewarded him with its senior scientist appointment, a rank equivalent in honor and pay to that of its upper-level executives.

John saw Jane whenever he visited Tom in Pittsburgh. The family in Madison was not fooled by the "visiting Tom" ruse. They also teased John about the letters that arrived in Madison from a mysterious woman friend whenever John was home. John's sister Ann recalled, "He wouldn't say a word about them, and he was very closed-lipped about the whole thing."

When he was in Pittsburgh John also visited his former Gulf buddies, with whom he could relax and sometimes carouse. Seitz recalled a visit that he and Bardeen made in June 1934, after they passed their prelims. Having studied intensively, they were "greatly over-prepared" for the written part, Seitz recalled. He and Bardeen would work until they "thought they were safe." Then Joseph Hirshfelder, a Wigner student in chemistry who was examined with them, "would come around with an exotic question and we all went back to the books."

The oral part proved an ordeal, however. The questions that a committee of professors threw at them were aimed at revealing the limitations of their understanding. Robertson "raised the height of the cross-bar until you tripped," recalled Seitz. His questions highlighted one of Bardeen's weaknesses, his difficulty to verbalize all the steps in his reasoning. In one question, Robertson asked John to explain what "electrodynamics would look like if there were magnetic poles." Bardeen gave the correct answer, but stumbled when asked for the details of his argument. "I was using too much intuition and could not give a convincing argument of the sort he wanted." He passed the prelim, but never could forget the embarrassment of having fumbled an important question.

Afterwards, Bardeen and Seitz piled into Bardeen's car for a vacation in Chicago. On the way they stopped in Pittsburgh, where they were greeted "noisily and as honored guests." They were given rooms in an engineering fraternity house across the street from Carnegie Tech. Seitz later described the "big party for John" thrown by the fraternity on Saturday night:

> Without too much encouragement, he soon became the life and soul of the boisterous gathering. That night I saw John in a rare departure from his usual sober demeanor, a treat that few of his later friends

would experience in quite the way I did. I might add that I was put
quietly to bed long before John was ready to quit.

Bardeen realized that he would have to grasp much of what was
known about the physics of solids. He would have to learn how the
electrons in metals interact with one another and with the crystal
lattice. These were issues that concerned him throughout his
career. Back at Princeton, he retired to the library and looked up
everything he could find on work functions. Wigner, who encouraged all his students to immerse themselves in the literature of
their field, reported "No one in my experience ever became acquainted more quickly with a rather complicated subject."
Bardeen added the habit of spending daily time in the library to his
arsenal of problem-solving tools. Over the years he built up an enormous fund of knowledge about solid-state physics.

He had to decide how far back to go in the literature. Since the
beginnings of civilization, the properties of materials had been
studied by blacksmiths, potters, jewelry makers, and other artisans.
Artists and builders also knew much about the materials they used.
But at the end of the nineteenth century, there was still no framework for answering many basic questions, such as why do metals
and insulators behave so differently in transporting heat and
electricity?

Paradoxically, as the metallurgist and historian of technology
Cyril Stanley Smith once noted, physicists interested in explaining
real materials were forced to leave the domain of real materials for
a period of time. For almost three decades, from 1905 until 1933,
they had puzzled over the more abstract problem of explaining ideal
materials, much simpler models of metals, insulators, and the like.
Here was a historical example of Wigner's principle of approaching
complex problems by considering simpler cases first. By starting
with these hypothetical cases, physicists could get to the heart of
the matter and develop a basic machinery for dealing with the full
range of properties and conditions that characterize real materials.

Bardeen realized that it was the work of these three decades
that he needed to zoom in on. The work had left many problems
unaddressed—a field ripe for progress. He also recognized that he
was standing on the edge of a frontier. Since the birth of the
quantum in 1900, the physicists who studied solids had limited
their work to ideal materials. This restriction had allowed them to

make much progress, especially during the seven years following the invention of quantum mechanics in 1925. The new mechanics had offered the tools for studying atoms, molecules, and solids. But until 1933—until the work that Wigner and Seitz were just then doing—the theory was incapable of dealing with *real* solids, "far more complex things than have been allowed in the domain of respectable physics in the past," as Cyril Smith wrote. Wigner and Condon were among the first to recognize the opportunity to address real problems of real solids, such as the structure, cohesion, plasticity, diffusion, strength, electrical conduction, and magnetism of materials like sodium or copper. For the first time in history, one could go beyond treating such problems of ideal materials and move into the real world. Wigner offered this opportunity to his first graduate students.

The problem of the photoelectric effect, which Einstein had studied in 1905, was in some ways analogous to the problem of the work function. Einstein had computed the energy in the form of light quanta needed to release an electron from a metal's surface. In exploring the photoelectric effect he had paved the way for viewing all forms of radiation in terms of Max Planck's revolutionary quanta, the irreducible bits that comprise electromagnetic energy. It was for solving this problem—not for his theory of relativity—that Einstein won the 1921 Nobel Prize.

Another important model for Bardeen was Einstein's 1907 calculation of the specific heat. Using quantum concepts Einstein had derived the amount of energy needed to raise the temperature of one gram of material by one degree Centigrade. And to understand the crystal structure of a metal, it was necessary to go back only to 1912, when Max von Laue, Walther Friederich, and Paul Knipping demonstrated in their Munich experiments that crystals can diffract X rays. Their work offered the first startling look at the crystal structure inside metals.

Bardeen realized that he needed a thorough understanding of the quantum theory framework on which he and others in his generation would construct the theory of real solids. Statistical methods were essential because solids contain multitudes of atoms—over a thousand billion billion (10^{21})—in every cubic centimeter. By the end of 1926, two different quantum statistics were available: the Fermi-Dirac and the Bose-Einstein statistics. Fermi-Dirac statistics apply to particles, such as electrons, that obey the

Pauli exclusion principle, the electron zoning ordinance that Wolfgang Pauli proposed in 1925 to explain the closure of electron shells in atoms. The principle states that two identical particles in the class later referred to as "fermions" cannot occupy the same quantum state. The other statistics apply to "bosons," particles of radiation, like photons, or X rays, that don't obey the Pauli principle.

In 1926 Pauli began the development of the quantum theory of solids with a quantum-mechanical calculation of one of the phenomena that experimenters were studying, the weak paramagnetism of metals. He used this problem as a test case to address the fundamental question: Which of the two quantum statistics describes matter? Ample data were available on paramagnetism. He performed the calculation both ways, using Fermi-Dirac and Bose-Einstein statistics. He found that Fermi-Dirac statistics worked and Bose-Einstein did not.

This simple triumph pointed to the need for reworking the existing formulation of the theory of metals, then based on classical Maxwell-Boltzmann statistics. Pauli could see the utility of doing this, but he also realized that the calculation would be exceedingly messy. He turned in disgust from the approximations needed to tailor the theory to real phenomena, warning his students against working in what he called the "physics of dirt," an area he cautioned "one shouldn't wallow in." Fortunately, Wigner, Seitz, and Bardeen disagreed with Pauli.

Arnold Sommerfeld, who had been Pauli's professor in Munich and whose Michigan lectures Brattain had heard in 1931, considered solid-state physics beautiful *because* its problems were real. In his classic calculations of 1927, Sommerfeld followed Pauli's program and structured a "semi-classical" quantum theory of solids. Avoiding the full use of the quantum-mechanical machinery based on Schrödinger's wave equation, and employing the Fermi-Dirac statistics only as needed in modifying the classical theory, Sommerfeld was able to address a whole range of problems that had previously been insoluble. Still it remained a puzzle why Sommerfeld's theory should work at all, as it was based on the unreal assumption that electrons in metals move freely. This question would nag Bardeen for years.

A partial answer was offered in the brilliant 1928 doctoral dissertation of Heisenberg's graduate student, Felix Bloch. Bloch discovered that when electrons move through a perfect crystal (in which the atoms are evenly spaced), they behave *as though* they are free particles. On this basis he constructed the conceptual framework known as "band theory." Just as the electrons in atoms are confined to energy levels, the electrons in metals are confined to energy bands (extended levels). Other physicists, including Léon Brillouin and Rudolf Peierls, elaborated on Bloch's theory.

Bardeen especially admired the theoretical bridge that Alan Wilson erected in 1931 between the theoretical works of Bloch and Peierls and the practical problems encountered by experimenters. Wilson's *tour de force* was to assemble the available pieces of the band theory to explain the difference between metals and insulators. His simple answer was: insulators have completely filled bands; metals have partially filled bands. Within the partially filled bands, electrons can move and carry current.

Wilson's work also clarified the ghostlike notion of the "hole," a concept of solids that Peierls first described in 1928. An empty electron state near the top of an otherwise filled energy band, the hole behaves as *though* it were a positively charged particle. The hole in a solid is like the moving empty seat in a large garden party, where guests are seated at various tables. As one guest who had been seated at a certain table with all chairs filled moves over into a previously empty seat at another table, she leaves her former seat vacant. Then when someone else leaves a different chair to move into her seat, he leaves his former seat vacant, and so on. To one observing the scene from a helicopter, it would appear that the empty place is moving. For the case of semiconductors and semimetals, the moving holes can be treated as positive charges, because the electrons are negatively charged.

Wilson also explained the behavior of semiconductors, materials with properties that are partway between those of metals and insulators. Semiconductors were enormously controversial in the early 1930s; some physicists were convinced they did not exist. Wilson pictured them as materials having a gap between their highest filled energy band and their lowest unfilled one. Under ordinary circumstances, electrons cannot jump the gap. Therefore, the material behaves like an insulator. But when energy is added in the form of light or heat, some electrons gain enough energy to traverse the gap. Conduction begins when an electron enters the first unfilled band. In the 1940s experimenters would find that adding impurities ("doping") can enhance conduction by creating shallow levels, in effect offering stepping stones for the electrons.

Bardeen realized that the most dramatic moment in this little-known history was occurring just then—*at Princeton!* Seitz's doctoral thesis of 1933 was the first step in a revolution that would bring the quantum theory of solids to the study of real materials. Until then no one had been able to calculate any *real* band structures. Focusing on sodium, the simplest metal, Wigner and Seitz began by dividing the crystal into small cells associated with individual ions. In each cell they assumed that the electrical potential was spherically symmetrical. This simplifying assumption made the mathematics able to compute such properties of real metals as cohesive force, elastic constants, compressibilities, thermal and electrical conductivities, and various optical properties.

To Bardeen the Wigner-Seitz work "looked like it would open up a new area." At MIT, Slater, who made the Wigner-Seitz method central to his training of graduate students, wrote of the Wigner-Seitz papers, "There were so many approximations that it was hard to accept the numerical results very seriously. However, for the first time they had given a usable method for estimating energy bands in actual crystals."

Within a decade the first crude methods for calculating the properties of metals would lead to better ones. The new framework became an umbrella covering many subfields—including crystallography, electrical conduction, semiconductors, ferroelectricity, and magnetism—that were earlier thought to be independent. Considered together the subfields became known as solid-state physics. Most scientific fields arise through differentiation from other broader fields. Solid-state physics, however, arose by a conglomeration of fields, as the historian Spencer Weart pointed out.

At least a dozen major reviews of the new field of physics were in print or in press by 1933. The most comprehensive was the monumental, almost 300-page article by Hans Bethe and Arnold Sommerfeld in the 1933 *Handbuch der Physik*. Graduate seminars on solid-state physics soon began to be offered at universities. One of the earliest was Wigner's seminar, "Theory of the Solid State," first taught in 1932. He covered a wide variety of topics, including the distinctions between different solids, symmetry groups in relation to the different kinds of crystals, aspects of the growth of crystals, and approximation methods. Seitz prepared the course notes for this seminar in 1934, the year in which he, Bardeen, and about a dozen other students attended.

Bardeen learned that designing the best approximations for a theory of solids was an art in itself. It was not at all clear how to represent the electrons inside solids. Were they to be thought of as tied to individual atoms (or ions)? Or were they better represented as swarms of particles that move without allegiance to any single ion? The true picture had to fall between these extremes. Numerous approximation schemes grew up around these questions.

The effort to deal with the interactions between electrons, or between electrons and ions, set the stage for a new field that became known as "many-body" theory. In many-body problems, the interactions are so crucial to the description that ignoring them misses important observed phenomena. The art in solving such problems involved representing the interactions and deciding which interactions to include in the theory. Such problems would engage Bardeen throughout his career.

In his thesis calculation, Bardeen first needed to make workable approximations for the electron interactions. Starting with a method pioneered by Douglas Hartree and Vladimir Fock, Bardeen wrote down a wave function that described each electron approximately, with its own single-electron wave function. He next modeled the distribution of electrons at the surface of the metal, including higher contributions arising from electron forces that correlated electrons with other electrons. It was a heroic early work in the field. He was ready to write up the thesis by the spring of 1935, publishing it as a joint work with Wigner.

"Wigner actually did most of the work," Bardeen later claimed. But Wigner said that Bardeen "started to work on his own line and has worked practically independently." He added that Bardeen "has not needed exaggerated mathematical equipment but rather the handling of physical ideas and this he has done in an unusually original manner."

As Bardeen began writing up his thesis, he was surprised to learn that Harvard's Society of Fellows, recently established in 1933, was considering him for their third class of junior fellows. It was an extraordinary opportunity. If he could land the prestigious fellowship, he would be able to concentrate entirely on research for three years. Harvard was an outstanding physics center. So was nearby MIT. But first he had to be interviewed by a group of senior fellows of the Harvard Society.

The purpose of this formidable ritual was "to see the man beyond the scholar, if possible a man with a long view." Bardeen was completely tongue-tied during the ordeal. His interviewing committee included the famous philosopher and mathematician Alfred North Whitehead, the internationally known historian Samuel Eliot Morison, and the leading biological chemist Lawrence J. Henderson. "I was placed before this very distinguished group of people who asked me questions. I think I was too scared to hardly say a word."

When Bardeen failed to make much of an impression at the interview, John Van Vleck stepped in. Van Vleck had recently left the University of Wisconsin and was now on Harvard's physics faculty. His intervention succeeded, and Bardeen's fellowship began in the fall of 1935. "I'm sure it was Van Vleck that got me in," Bardeen admitted years later.

As it turned out, Bardeen had the luxury of choosing between two attractive fellowships, for Princeton also offered him one of its prestigious Proctor fellowships for 1935–1936. As the Harvard fellowship paid substantially more ($1,500 a year plus board at Harvard's Lowell House) and was guaranteed for three years, the choice was easy to make. "Three years of job security was no small item at the time." Another consideration was that Harvard's fellowship was in physics, whereas Princeton's was in mathematics. By then "it was obvious that I was going in the direction of physics."

Harvard junior fellows were expected to arrive with doctorate in hand. For Bardeen this proved impossible. In May 1935 he received an urgent request to come home to Madison immediately. His father Charles was jaundiced and very ill. He was not expected to survive long. Two years earlier he had had his thyroid removed because of a cancerous growth. The cancer had returned.

Charles passed away on June 12, a few weeks after John came home. A postmortem discovered a difficult-to-diagnose form of pancreatic cancer that had spread to Charles's liver. He was buried beside Althea, under a boulder from Lake Mendota.

Charles Bardeen's obituaries highlight many features that John came to share with his father. One said that the older Bardeen's "relaxations came through walks or picnics with his family, a game of golf with some old cronies, conversation or dinner with a group of friends." Although Charles had been "endowed with a brilliant mind," he was, like John, "sympathetic with those less gifted." The morning after Dean Bardeen's death, a janitor said he "had lost a good pal."

Later reminiscences of Charles Bardeen expanded the analogy between him and his son John. In 1957, at the dedication of a new medical school building to Charles Bardeen, Harold Bradley described Charles as "a prodigious worker," with "a deep dislike of public acclaim or personal recognition." He referred to him as "a quiet giant, given to few words; modest, self-effacing, shunning praise and public approbation." Paul Clark subsequently wrote in a history of the Wisconsin Medical School that "Dr. Bardeen spoke in a somewhat mumbling fashion, moving his lips only slightly, but he thought with clarity and there was no mumbling in his decisions." He, too, referred to Charles as "a prodigious worker, a patient, tolerant man, simple in his tastes, keenly intellectual . . . reserved and persevering almost to stubbornness." Among Dean Bardeen's lasting educational contributions was his establishment of a preceptorship in which medical students worked closely with one of a dozen highly competent practicing physicians.

In the weeks after his father's funeral John was intensely anxious to finish his thesis. He worked on it whenever possible, but he could not complete the arduous calculations that summer. Not only was it difficult to concentrate, but also Wigner was in Europe and thus unable to approve the thesis before fall. Bardeen did not receive his Ph.D. until January 1936. That Harvard allowed Bardeen to complete his doctorate during his first semester as a junior fellow was perhaps another result of Van Vleck's continuing influence.

At the end of that long and difficult summer of 1935, when Bardeen's extended family of stepmother, siblings, aunts, uncles, and cousins offered him comfort and stability, John set out for Harvard. Despite some uneasiness about not completing his thesis, he was filled with excitement.

5

Many-Body Beginnings

On arriving in Cambridge, Bardeen proceeded immediately to Harvard's Jefferson Laboratory, a massive three-story brick structure completed in 1883. It was the first university physics building in America designed for research as well as instruction. Bardeen climbed its granite staircase and found the department office, where he spent some time on administrative chores.

He next checked into Lowell House, the student dormitory complex where he would live for the next three years. He found Building G in a group of adjoining brick buildings surrounding a courtyard. He climbed to the third floor and, entering his new home, suite 31, noticed that his living room was equipped with a wood-burning fireplace. It was far too hot to even think of using it now. He opened several windows to let a breeze into the stuffy apartment.

He decided to unwind with a short stroll around Cambridge. Walking a block or so, he came to the Charles River. He stopped to watch a group of rowboats gliding past. Then he entered Eliot House, another Harvard residential complex resembling Lowell House. He tried to find the room where the junior fellows would gather once a week for a formal dinner. After this brief tour, John went back home to Lowell House to unpack.

If Princeton was a graduate student's paradise, then Harvard was a postdoc's heaven, especially for a junior member of the Society of Fellows. When the society was established in 1933, "The idea was to give promising young students free time to do whatever they wanted in their research." Bardeen soon realized: "There were no obligations." They could work in any area they found compelling, with access to all of Harvard's libraries, laboratories, and other facilities. Bardeen would flourish there.

The deeper aim of the society was to groom exceptionally promising men (and later women) between twenty and thirty years of age for intellectual nobility. Selected by the senior fellows of the society "for their promise of notable contribution to knowledge and thought," the younger members were expected to advance into positions of leadership in politics, science, literature, or the arts. Interdisciplinary interaction was encouraged, for the society's elders believed in the benefits of study at the borders of disciplines "where growth is still going forward." Seven or eight junior fellows were appointed each year. At the time Bardeen came into the society, about twenty junior fellows were in residence.

Bardeen might have reverted to his usual pattern of dividing his time between physics and games had not the society imposed certain social rituals on its fellows. The junior fellows were expected to eat lunch with each other on most days and to dine formally with the senior fellows every Monday evening at Eliot House. Bardeen came to consider these dinners "a great privilege and inspiration, even though I was mostly a listener." After the traditional drinking of sherry—a time "when the ordinary small talk of academic life is passed"—the group sat down to dinner at a U-shaped table. The high point of most dinners was a discussion with some eminent guest, such as a Supreme Court justice, or a nationally known poet, invited by one of the senior fellows. These guests contributed to the "general broadening influence of the society on its members," Bardeen later reflected.

In time, most of the junior fellows assumed positions of influence. Of the three studying physics in Bardeen's class, one was Ivan Getting, then conducting experiments in cosmic-ray and nuclear physics. He later became vice-president of engineering and research at Raytheon, chairman of the Naval Warfare Panel, and president of the Aerospace Corporation in California. The other was the theorist James Fisk, who would climb the Bell Labs ladder to become its president. Fisk would also serve on many high-ranking scientific committees, including Dwight D. Eisenhower's President's Science Advisory Committee (PSAC), where he was influential in his support of a nuclear test ban. Fisk's term on PSAC would overlap with Bardeen's. In later years Fisk became a Harvard senior fellow.

Bardeen also befriended many junior fellows in fields other than physics. One was David Griggs, a geophysicist who would later became noted for demonstrating the flow of rocks under high-pressure conditions. Bardeen and Griggs both took a strong interest in the famous high-pressure experiments of Harvard physicist Percy W. Bridgman. Other friends were the philosopher Willard Van Orman Quine, who was at the time immersed in mathematical logic, and the behavioral psychologist Burrhus Frederick Skinner. Bardeen's career would benefit many times from his acquaintances with former Harvard junior fellows. For instance, in 1945 Fisk would be instrumental in bringing Bardeen to Bell Labs.

Bardeen's interaction with the other fellows was not confined to intellectual matters. The mathematician Garrett Birkhoff sometimes borrowed Bardeen's car "for the nominal sum of $3 per month in 1937–38 when, unbeknownst to each other, we were both courting our future wives in Wellesley." Birkhoff "used John's car one evening to drive to Jake Wirth's restaurant in Boston for a German meal with three distinguished friends: Heinrich Brüning, the former chancellor of Germany, John von Neumann, and Stan Ulam. The front seat would only hold three of us, so Stan had to sit in the rumble seat."

The junior fellows were not required to teach, but were encouraged to do so if they thought the experience might enhance their training. Bardeen decided to offer a course on general relativity, basing his lectures on notes he took in Robertson's course at Princeton, supplementing them with material from Arthur Eddington's authoritative text, *The Mathematical Theory of Relativity*. Eddington's philosophy of science matched Bardeen's evolving sense of what physics should be. Eddington wrote on the first page of his book, "The physicist is not generally content to believe that the quantity he arrives at is something whose nature is inseparable from the kind of operations which led to it." He explained:

> To find out any physical quantity we perform certain practical operations followed by calculations; the operations are called experiments

or observations according as the conditions are more or less closely under our control. The physical quantity so discovered is primarily the result of the operations and calculations; it is, so to speak, *a manufactured article*—manufactured by our operations.

Bardeen found that he felt at home in a department of physics. He attended seminars and immersed himself in the physics literature. He continued his habit of stopping regularly at the library to read important new papers. At that time, as Bardeen later reflected wistfully, "It was still possible to read all of them." He reread Bethe and Sommerfeld's major 1933 article on the quantum theory of solids, a review of all the major solved and unsolved problems in the theory of solids.

By then Bethe had resettled in the United States. He was among the 100 or so German-Jewish physicists who fled Germany and came to England or America after Hitler came to power. In 1935, after spending two years in England, Bethe emigrated to the United States and accepted a permanent post at Cornell. He became the central figure in Cornell's growing program of nuclear physics—one of the three most rapidly growing subfields of physics, along with solid-state and quantum electrodynamics. Having created the basis for graduate training in solid-state theory with the Bethe-Sommerfeld article (mostly written by him), Bethe now played the same role in nuclear physics, publishing a series of detailed nuclear physics articles in *The Reviews of Modern Physics*. They became known as the "Bethe Bible."

Bardeen was attracted to Bethe's no-frills, practical style of physics. The tables of experimental numbers that Bethe often included in his publications reassured Bardeen that the theory he described was in step with observed data. Bardeen also joined in Bethe's appreciation for the American tendency to take part in team research. Bardeen readily accepted Bethe's invitation to visit Cornell, where the two theorists conferred about applying solid-state theory to nuclear physics in calculating the energy level density of heavy nuclei.

Bardeen also appreciated the "style and approach" of two excellent British texts that appeared in 1936 on the quantum theory of solids: Nevill Mott's and Harry Jones's, *Theory of the Properties of Metals and Alloys* and Alan Wilson's *The Theory of Metals*. The point of view of these texts was compatible with his own. Written to address the needs of experimenters, they stressed the "close correlation between experiment and theory." Not yet in print was Fred Seitz's *The Modern Theory of Solids*, which would appear in 1940. The first comprehensive textbook in modern solid-state physics, Seitz's text became an instant classic. The book would help to establish solid-state physics as a distinct area of research. It was used for decades in physics departments throughout the world.

While in Cambridge, Bardeen wrote a paper with John Van Vleck on a method for approximating the interactions between electrons in solids, the "tight-binding method." "I had more inter-action with Van Vleck than anyone else at Harvard. He was one of the first there who was interested in solid state, and so I'd talk about work with him, but did the work pretty much on my own."

Bardeen often visited MIT where he interacted so much with John Slater and his students that "years later, in his memoirs, Slater remembered me [incorrectly] as a postdoc at MIT," Bardeen said. Slater had been among the first to recognize the great opportunities offered by the Wigner-Seitz method. As the head of MIT's Department of Physics since 1930, Slater encouraged students to compute the band structures of materials such as lithium, copper, and sodium chloride. Although interested in many of the same problems of solids that had excited Wigner, Slater approached them less creatively. He referred to himself as "of the prosaic, matter-of-fact type."

Slater had taken his Ph.D. at Harvard in 1923, before any theorists in physics were on the faculty. Working under Percy Bridgman, he attempted to explain some of his advisor's pioneering measurements on solids under high-pressure conditions. By increasing the pressure from a few thousand to 100,000 kg/cm^2, Bridgman had managed to alter the spatial ordering of the atoms in the crystal lattice. That, in turn, affected the material's electrical, magnetic, thermal, and optical properties, which Bridgman had carefully measured.

Bridgman encouraged Slater to apply quantum mechanics to his data; in particular, to explain the compressibility of alkali halides (materials in the same class as ordinary salt). Having studied physics in the pre-quantum-mechanical era, Bridgman had no working knowledge of the modern theories. Slater did his best to develop an explanation in terms of a balance between the electrical attractions of alkali and halide ions, and postulated repulsive forces of an unknown nature. But he could not account for the source of the repulsive forces. "I was convinced by these facts that the quantum theory of 1923 was not adequate to describe the nature of molecules and solids." After studying for a time in European centers of physics research, he accepted MIT's offer to become the chair of its physics department. Karl Compton, who had successfully built up Princeton's programs in math and physics, was now the president of MIT and was encouraging more growth in these areas.

Conyers Herring spent two years (1937–1939) working in postdoctoral study at MIT under Slater, overlapping with Bardeen in 1937–1938. Herring had met Slater as a student at Princeton and had been impressed by the speed with which Slater produced physics papers. Slater would say, "Well here's something that needs to be calculated," and go on to quickly do so, writing down the equations, and "zip zip zip, he's got a paper." One of the papers that Slater wrote in his short time at Princeton was his classic work on the approximation method known as the Augmented Plane Wave method. Unfortunately, Herring soon recognized that Slater's outlook "was somewhat different from mine." He did not appreciate Slater's emphasis on "grinding out" results.

The mismatch in style between Herring and Slater showed up when Herring encountered difficulty in calculating the energy of a distorted lattice. After some weeks Slater suggested that Herring drop the problem and do "something that you could get finished with." Herring preferred "to just keep butting away at it." His approach, like Bardeen's, demanded perseverance. His continuing attempts to solve the problem of the distorted lattice paid off in his development of a far more powerful approximation method, the widely used orthogonalized plane-wave method. "And that was real pay dirt."

Herring's "strongest intellectual contacts" in Cambridge were with Bardeen, not Slater. Herring recalled "that the things he [Bardeen] was working on were very close to the things I was working on." For example, Bardeen was calculating the wave function for various alkali metals, "and I wanted to do it for beryllium," said Herring. "I used to go up to Harvard every so often and talk with him about some of these band techniques and I learned quite a few things from him."

At MIT, Bardeen also interacted often with Slater's cocky graduate student from California, William Shockley, who was completing his doctoral thesis under Slater on sodium chloride. Bardeen had already met Shockley while he was studying under Wigner at Princeton.

Seitz and Shockley were in those years close friends. The two had taken a memorable drive together across the country in September 1932. Both were traveling from Stanford, California, to their East Coast graduate schools—Seitz to Princeton and Shockley to MIT. Seitz had found Shockley "strongly influenced by the Hollywood culture of the day, fancying himself as a cross between Douglas Fairbanks, Sr. and Bulldog Drummond, with perhaps a dash of Ronald Coleman." Seitz felt uneasy when he realized that Shockley had stowed a loaded pistol in the glove compartment of his 1929 DeSoto convertible. "I was handy with a rifle at that time, but still looked askance at traveling thousands of miles in the company of a loaded pistol."

One night during the trip Shockley startled Seitz by firing his pistol at a group of howling coyotes. Having encountered torrential rains, they had "parked the car off the highway at a rise" and were "wandering through the desert." The next morning a gas station attendant warned them "two desperadoes were loose in the area." He assured them "the local police had been alerted."

Bardeen and Shockley often interacted during their year of overlap in Cambridge, 1935–1936. "We went to joint seminars. I talked about his work with him." Bardeen enjoyed Shockley's quick intelligence, imagination, and self-confidence. The two had many common interests and spoke at length about calculating wave functions for alkali halides. They also discussed Shockley's current work on surface states, which Bardeen would draw on a decade later in his research that led to the transistor.

Shockley's surface states work followed up on studies conducted several years earlier by the Soviet physicist Igor Tamm. Cutting off periodic structures at the surface of a sample, Tamm showed, gives rise to "dangling bonds" and causes the formation of surface states. Shockley showed that surface states also arise when energy bands cross.

Shockley's thesis calculation required many hours of number crunching on office calculators. Because it was easy to introduce errors during this process, Shockley devised checks along the way, including a novel "empty lattice test," to see if the approximations provided agreement for the case of free electrons. It was a "case in which we knew what the exact answers would be," Shockley explained, an example of bridging unknown with known physics.

In March 1936, during Bardeen's second semester at Harvard, Shockley received an enviable job offer. The director of physical research at Bell Telephone Laboratories, Mervin J. Kelly, had received his physics doctorate under the noted University of Chicago physicist Robert Millikan. Kelly was one of the first leaders of industrial research to recognize the promise that solid-state physics held for industry. He learned about Shockley on a recruiting visit to MIT. When he then heard that Shockley was preparing to accept a physics instructorship at Yale, Kelly quickly countered with an offer of a research position at Bell Labs having better terms. Kelly assured Shockley that Bell Labs would offer him the freedom to conduct basic research in solid-state physics. Shockley was working at Bell Labs by July 1936.

Herring recalled that Shockley left behind a "residue of stories" at MIT. In one Shockley had cleverly rewired an elevator "so that when you pushed for floor one you'd go to floor four and when you pushed floor three you'd go to the basement. And then he took secret observations to see how long it took people to get used to it."

Bridgman was perhaps the greatest influence on Bardeen at Harvard. The common intellectual "ancestor" of Van Vleck, Shockley, and Slater (the physics "grandfather" of Shockley through Slater, and of Van Vleck through Edwin Kemble), Bridgman impressed on his physics progeny his strong conviction that observation, not theory, was of primary importance in physics.

Bridgman was widely known as a solitary researcher who spent far more time in his lab gathering data than he spent in the classroom. On weekends and in summers, Percy (Peter to his friends) would immerse himself in solitary gardening at his New Hampshire home or in solitary hikes in the mountains. Although usually courteous and helpful, he was known to be "fierce in his inner disdain of sloppy or wishful thinking." Some of Bridgman's colleagues noted that he "indulged in no elaboration that would not improve the measurements." The economy with which Bridgman worked in his laboratory probably impressed the frugal Bardeen.

On the foundations of Bridgman's work, wrote Bridgman's biographer, Maila Walter, "the new field of high pressure physics was established." Bridgman had invented an important tool when he was a graduate student that allowed him to apply high pressures to solids. In subsequent decades he improved the apparatus and made countless precise measurements of matter under extremely high pressures.

In the 1920s Bridgman had hoped that his student John Slater could explain some of his experimental results using the new quantum physics, but the quantum theory was not yet sufficiently developed. A decade later, when Bardeen appeared on the scene, Bridgman showed him some of his data on alkali halides. Bardeen was excited to have the opportunity to work with the unsorted data of this master experimenter who, Bardeen recognized, was "at the height of his productivity."

Bardeen's teamwork with Bridgman established a model for Bardeen of highly productive collaborative work between a theorist and an experimentalist. He would in the years ahead create similar collaborations whenever possible. The two often discussed details of measurement techniques and experimental strategies. Bridgman suspected that through pressure alone he could bring about a phase change in cesium at about 22 kilobars. Before attempting this feat, however, he wanted to be armed with a better theory. He discussed with Bardeen why he expected the high pressure to pack the ions more closely and cause the cesium to transform from its "body-centered" structure to the more densely packed "face-centered" structure. Bridgman suspected that the change was related, somehow, to the atomic forces involved, but he did not know enough quantum theory to explore this hunch. Here was Bardeen's chance to stand up in the physics community he had recently adopted and address a problem posed by one of its most prominent senior members.

Bardeen constructed a semiempirical theory that allowed him to predict the exact pressure at which cesium would change from a body-centered to a face-centered structure. When Bridgman tried the experiment, he found the transition at just the pressure Bardeen had predicted. "This was one of the first cases of being able to calculate a structure change before it was found experimentally," Bardeen boasted. "Bridgman was very impressed with John," claimed the Harvard physicist Edward Purcell. "For the first time one could calculate things that Bridgman was gathering data for."

In another calculation, Bardeen used atomic spectral data to extend his earlier theory of the work function and compute the binding energy of lithium and sodium as a function of both volume and pressure. "The alkali metals were of particular interest to me," Bardeen explained, "because there are large changes in compressibility over the 45 kbar range which should be calculable from a first-principles Wigner-Seitz calculation." He readily engaged in the long calculations. "Many days were spent working with a desk calculator on a calculation that could be done in minutes with modern computers." For the case of sodium, the results "checked very well with experiments over the entire pressure range." Unfortunately, there were small but significant departures (about 15 percent in compressibility) for the case of lithium. The problem, Herring discovered somewhat later, came from using the preliminary, rather than the final, values for the electric field, which Seitz had inadvertently published. "John never forgave me," joked Seitz.

None of Bardeen's colleagues recalled Bardeen ever mentioning the philosophy of science for which Bridgman is remembered today. "Operationalism" was Bridgman's attempt to explain how physicists acquire scientific knowledge and achieve clear thinking. Maila Walter interprets Bridgman's search for scientific meaning as a way of coping with the new quantum physics, for which he "was intellectually and emotionally unprepared." Bridgman wrote copiously on operationalism during the 1920s and 1930s; Bardeen could not have remained untouched by the grounded convictions of the man who developed this philosophy.

"The concept is synonymous with the corresponding set of operations," Bridgman wrote in 1927 in his celebrated monograph, *The Logic of Modern Physics*. He considered the operations by which measurements are made the true basis of scientific knowledge; physics concepts were merely "penumbra" surrounding the operations. Bridgman believed that studying Einstein's route to relativity would help students understand how important new concepts could and should be formulated. He explained that Einstein's writings reveal how he formulated the theory of special relativity starting with the operations basic to the concepts of time, length, and simultaneity.

In his later years, Bridgman also emphasized the importance of what a theory does—how it helps physicists make sense of the world and contribute to social good. The emphasis on usefulness resonated with the progressive values of physicists nurtured in the early years of the twentieth century.

Bridgman had another, less tangible but possibly as important, influence on Bardeen. Like four of Bardeen's other teachers—Peter Debye, Paul Dirac, Eugene Wigner, and John Van Vleck—Bridgman was a future Nobel laureate. In 1946 Bridgman became the first physicist on the Harvard faculty to win a Nobel Prize, received for the invention of an apparatus for producing high pressures and for his pioneering experiments establishing the field of high-pressure physics. (Dirac's Nobel Prize in 1933, shared with Erwin Schrödinger, had been for fundamental work on the theory of quantum mechanics; Debye's prize in 1936 was for chemistry. Seven years after Bardeen first became a Nobel laureate in 1956, Wigner would share his Nobel Prize with Maria Goeppert Mayer and J. Hans D. Jensen for the application of group theory to the atomic nucleus. And in 1977 Van Vleck would share the Nobel Prize with Philip Anderson and Nevill Mott for their studies of the electronic structure of magnetic and disordered systems.)

The Nobel Prize connections between certain scientist mentors and their students have been examined by the sociologist Harriet Zuckerman in her classic study of Nobel Prize–winning scientists. She suggests that the scientists who are destined to become Nobel laureates and their future Nobel laureate mentors recognize each other intuitively. There is a reciprocal relationship between them. The mentors "selectively recruit" students of Nobel caliber, and the students "self-select" to work with these mentors, often years before either receive their awards. According to this interpretation, in studying with physicists who were "Nobel-bound," Bardeen was "being socialized" for a "position in the aristocracy of science."

At Harvard, Bardeen continued to concentrate on the type of problems he had begun to address in his thesis calculation—"many-body" problems, in which the interactions between electrons, or between electrons and the lattice, play a significant role. He was often frustrated because physics was not yet equipped with adequate tools for treating these interactions. Indeed, he could not solve most of the problems he struggled with at Harvard. He worked on them anyway. John's brother Bill had once described this attitude of his younger brother in the context of their playing football as youngsters. "John just hangs on and won't let go."

One of the problems Bardeen worked on at Harvard concerned the sharpness of the "Fermi surface," a two-dimensional surface drawn in momentum space. This construction bounds the physical states occupied by the electrons in a metal. By the 1950s, Fermi surface studies would become an active topic of research, but in the 1930s few physicists recognized the importance of this geometrical surface. An exception was Bethe, who discussed the Fermi surface in his 1933 *Handbuch* article with Sommerfeld. "It was clear to me that [these Fermi surfaces] would be important," Bethe later said, "and that it made a great difference whether they were nearly a sphere or were some other interesting surface."

One of the mysteries of the Fermi surface was the fact that states inside are fully occupied, whereas those outside are completely empty in the ground state. This puzzling property known as the "sharpness" of the Fermi surface had been observed at MIT in 1934, in Henry O'Bryan and Herbert Skinner's measurements of the low-energy X-ray emission spectrum of light metals. Many-body effects, such as electron exchange or correlation interactions and particle motion due to finite temperatures, were expected to destroy the sharpness, but they did not. Bardeen tried to explain this result, but he was unsuccessful. His contribution only reformulated the problem as a question of explaining why the scattering lifetimes of electrons at the Fermi surface grow longer as the temperature falls.

The sharpness question was eventually answered in 1951 by Viktor Weisskopf, who showed that the increase in lifetimes of particles close to the Fermi surface was a consequence of the Pauli exclusion principle. Not until 1957 did Joaquin Luttinger, Walter Kohn, and Arkady Migdal more fully explain the sharpness of the Fermi surface within the context of Lev Landau's Fermi liquid theory.

Another many-body problem on which Bardeen worked unsuccessfully concerned nuclear energy levels. Collaborating with Eugene Feenberg, who had taken his doctorate under Kemble in 1933 and was now a postdoctoral research associate at Princeton, Bardeen attempted to move beyond his earlier discussions with Bethe on employing many-body methods in computing the density of the energy levels in heavy nuclei such as uranium. He also tried to extend earlier work by Wigner on isotopic spin to a study of the effects of symmetry on nuclear energy levels, but this too "was too elaborate a calculation for its time."

Similarly, Bardeen tried—but failed—to explain the "exchange forces" that arise as a consequence of the symmetry of the quantum-mechanical wave function when two particles exchange position. These forces seemed to increase the calculated velocity of an electron at the Fermi surface all the way to infinity, an obviously anomalous effect of the mathematics employed. He suspected that the difficulty arose from an incorrect formulation of the long-range part of the electron interaction. But in 1936 Bardeen could not yet understand the deeper meaning, namely that electron–electron interactions are electrically "screened"; the electrons in such a system attract positive charge to them and travel with a cloud of positive charge around them. The result is that the charge (of electrons plus their positive cloud) appears neutral. This calculation also did not result in a full publication (only an abstract), but the work prepared Bardeen for his later research.

Bardeen also struggled at Harvard with how to treat the electrical forces between displaced ions inside crystals in the presence of electrons that screen the ions. His simplifying assumption—that the unscreened potential of an ion moves along with ions in the crystal—was equivalent to the widely used approximation method known as the "random phase approximation" (RPA). Again he made no breakthrough on the problem itself.

Why then did Bardeen often describe his time at Harvard as "the most influential years of my life"? He was probably recognizing the critical training effect of his work in those years, a training that derived from his persistent efforts to overcome challenging problems of physics. In struggling with problems too difficult to solve, he was honing his skills as a theoretical physicist.

Explaining superconductivity was another of the problems that Bardeen grappled with unsuccessfully at Harvard. In 1911 the great Dutch experimentalist Heike Kammerlingh Onnes had observed that at very low temperatures, certain metals and alloys lose *all* their electrical resistance. Over the next four and a half decades nearly every theorist working in physics sought to explain this puzzling effect. The work resembled a search for the Holy Grail.

In the late 1920s and early 1930s, following the development of the quantum theory of solids, theoretical physicists engaged in a flurry of work that produced many theories of superconductivity. They all failed. In his 1933 *Handbuch* article with Sommerfeld, Bethe lamented that superconductivity still resisted quantum-mechanical treatment: "Only a number of hypotheses exist, which until now have in no way been worked out and whose validity cannot therefore be verified." Nevertheless, Bethe remained confident that "despite lack of success up until now, we may assert that superconductivity will be solved on the basis of our present-day quantum mechanical knowledge."

All the theories of superconductivity put forth at that time were based on novel assumptions about the electronic structure of superconductors. For instance, a theory of Felix Bloch, Lev Landau, and Yakov Frenkel assumed that the ground state of a superconductor bears a "spontaneous current," suggesting that this is why the energy is at a minimum in this state. Bloch's calculations, however, could not find any current in the minimum-energy state, a result that came to be known as Bloch's first theorem on superconductivity. He soon added a second theorem: "Every theory of superconductivity can be disproved!" This widely quoted second theorem reflected the frustration of the many who failed to explain superconductivity in the 1930s.

Another conception was put forth by Niels Bohr and Ralph Kronig, who imagined the electrons in a superconductor joined in a chain or lattice. Walter Elsasser based his theory on the assumption of relativistic electrons, while Richard Schachenmeier assumed exchange forces between conduction electrons and bound electrons. Léon Brillouin associated superconductivity with electrons trapped in metastable states. All these theories were hampered by incomplete data and inadequate theoretical tools. As a postdoc just sensing his prowess, Bardeen could not resist pitting himself against the many who had failed to crack the riddle of superconductivity.

Developing a "feel" for what was known experimentally was to be another of Bardeen's characteristic first steps in approaching a new physics problem. In his efforts to learn what experimenters already had observed about superconductivity in the late 1930s, Bardeen read a short monograph recently published by the Cambridge physicist David Shoenberg. Its message echoed Bethe's: "Although wave-mechanics has found a satisfactory qualitative explanation for most metallic phenomena, superconductivity has remained as anomalous as ever from the theoretical point of view." Recent experiments, Shoenberg wrote, "have not materially changed this position," but "they have at least made possible a coherent statement of what it is that the theory has to explain." Fifteen years later Bardeen would recommend Shoenberg's book, by then in its second edition, to his postdoc Leon Cooper.

In 1933 a breakthrough in superconductivity occurred. The London brothers, Fritz and Heinz, published a speculative theory that had remarkable predictive power, although it did not make

use of the full machinery of quantum mechanics. At the time
Bardeen was among the few who recognized the importance of this
theory. The Londons had fled Hitler's Germany. Settling temporarily in Oxford, they collaborated in formulating a theory of superconductivity designed to explain the surprising result, reported in
1933 by Walther Meissner and Robert Ochsenfeld, that superconductors expel magnetic fields. This phenomenon implied that the
transition between the normal and superconducting state of a metal
is reversible and can thus be described by thermodynamics.

The Londons devised two equations relating the superconductor
to the electric and magnetic fields. The vanishing of the resistance
observed by Meissner and Ochsenfeld followed from the second
equation. What the Londons realized, and expressed in their second
equation, is that what is proportional to the electric field in a superconductor is not the current, as in the normal case, but the
change in the current with time. If there is no electric field, the
change in current will be zero, and an existing current will flow
forever.

Solving the "London equations" together with Maxwell's equations for the electric and magnetic fields yielded the observed
experimental results, for instance, that the magnetic field decays
exponentially as it enters the superconductor, with a penetration
depth between 10^{-6} and 10^{-5} cm. Moreover, the current flow in the
penetration layer near the surface shields the interior from the
external field, the so-called "Meissner effect." The Londons'
assumption that the normal current (which exists along with the
supercurrent and satisfies the usual Ohm's law of resistance) is
short-circuited by the superconducting current under steady-state
conditions explained how superconductors respond to electromagnetic waves.

Attempting to place this theory into a quantum-mechanical
framework, the Londons suggested that there is "rigidity" of these
ground-state wave functions in the presence of a magnetic field. In
other words, if one applies weak fields, there is no effect on the
wave function. But if one applies a sufficiently large magnetic field,
the system ceases to be superconducting. An analogy would be
pounding on a table. If one pounds gently, nothing happens because
the table is rigid. But if one applies great force, the table will break.

Why should the wave function be rigid? The Londons said it
was so because there exists an energy gap in the electronic structure

of the solid, a finite interval between the energy of the quantum-mechanical ground state and the low-lying excited states. Because of this gap it takes a great deal of energy to reach the lowest excited state. Bardeen sensed that this notion of the gap was the key to explaining superconductivity.

Physics was not the only challenge Bardeen faced while at Harvard. Although he thought that his romance with Jane was flourishing, she disagreed. Nearing thirty, she wanted to know whether she and John were going to be a permanent couple. Unless they could see more of one another, how could they decide? "Every time I saw him I liked him better," but "I got small doses." It frustrated her that John avoided all talk of marriage. "In those days, it was thought that you should support your wife if you got married," Bardeen later explained. He could not yet do so.

Jane made a move. She again arranged to work at Woods Hole in Massachusetts. While studying the reproductive patterns of wasps there for her master's thesis, she had many opportunities to see John during the summers of 1936 and 1937. From Boston she wrote home, "Dinner at the Wayside Inn Thursday night, the Pops concert at Symphony Hall Friday night, dinner at the Blue Ship on T Wharf and a drive last night. . . . John and I are having a grand time." After that weekend, they drove to Pennsylvania to see Jane's family.

As John continued to avoid commitment, she took another step. "Having decided that absence had not produced the result I wanted, I decided to try presence." In August 1937 she wrote home that she was "resigning from Tech" and taking a position as a teacher of biology at the Dana Hall School in Wellesley, Massachusetts. Jane's family knew that John was nearby in Cambridge. "But his proximity has not been the influence you may think it is in bringing me to move East," she protested. She claimed she had "left that factor out, since I still am not sure he would make me a good husband." She rationalized that she had done all she could with regard to her career at Carnegie Tech and that in Massachusetts she would be able to take courses toward her doctorate at Radcliffe or Wellesley. "I can't pass up all chance for getting ahead in my own profession while waiting for some man to discover what a good wife and mother I could be." Her family was not fooled.

During her job interview at Dana Hall, the schoolmistress, Miss Helen Cooke, asked Jane whether she had any "plans." Jane prevaricated that she had none. She saw no need to volunteer her hopes of a future with John, for it was frowned on for a married woman to work outside the home. By November Jane's Dana Hall colleagues were showing "great interest in 'that nice young man from Cambridge,' who comes to see me occasionally and frequently telephones when I am out."

Jane still expressed reservations about John. "He really is fond of me and always wants to be with me when he seeks entertainment, but he works more than he plays and Physics is his first love." Almost in the same breath, she resolved to be "more philosophical about the situation," because "I do not see how there can ever be any other man for me."

6

Academic Life

John shocked his Harvard colleagues when he brought Jane to the Fellows Society picnic in May 1938, and the more so when he mumbled that he guessed he would take his wife to Minnesota. Few of his colleagues were even aware that he had a serious love interest. Miss Waldo at Dana Hall excused Jane from two classes so that she could attend the Fellows Society picnic. "Isn't that sporting of her?" Jane wrote home. "She highly approves of my plans and was so amused when I broke the news." Even Miss Helen Cooke, the schoolmistress "gave her blessing" but "was less enthusiastic."

Jane had enjoyed teaching biology to bright young women and was pleased to learn some months later from a friend that "all of her Dana Hall children had passed their college boards with 'good,' 'honor,' or 'high honor.'" Even those who had tried her patience had done well. "It made me feel good that my struggles with them had turned out successfully." Miss Cooke and Miss Waldo "were very complimentary about my success this year and said they wished, for Dana's sake, that I were staying."

John's new willingness to commit to Jane stemmed from his successful negotiations with the University of Minnesota. He had been offered a teaching position there starting in the fall of 1938. The land grant college in Minneapolis was modernizing its instructional program. Physics was a targeted area. John Tate, dean of the

College of Liberal Arts and Sciences, managing editor of the *Physical Review*, and director of Walter Brattain's Ph.D. thesis, had recently put Minnesota "on the map" of nuclear physics by adding J. William Buchta, John H. Williams, and Alfred O. C. Nier to its physics faculty. He hoped to do the same for solid-state physics.

Now a full professor at Harvard, John Van Vleck again wrote a letter of recommendation on Bardeen's behalf. Van Vleck had special connections with Minnesota, having taught there from 1926 to 1928. He sent Buchta a copy of the letter he had written to Johns Hopkins about several young physicists. Of Bardeen he wrote that Harvard thought "so highly of his work that we would like very much to keep him here but the financial situation is such that this is probably not possible."

Late in March, Minnesota offered Bardeen an assistant professorship. The salary, $2,600 a year, was less than he had made at Gulf five years earlier, but Bardeen knew he was lucky to have any offer at all. The academic pay scale could not be compared with the industrial one. Most universities had *no* theoretical physicists on their faculty. He accepted the offer.

Minnesota also offered a position to Alfred Nier, who had overlapped with Bardeen at Harvard from 1936 to 1938. Nier recalled Bardeen being irked when he found out that Nier had started at Minnesota with a higher salary than he. The reason was that Nier was well known at Minnesota, having done his graduate work there. When two physics posts opened for the fall of 1938, one was immediately extended to Nier. And when Harvard offered Nier another year of support, Minnesota countered with an offer a little higher than the usual starting salary.

Wedding plans now took center stage for John and Jane. They agreed to keep expenses down by scheduling the event directly after one of John's professional engagements. In July he had agreed to lecture on the physics of metals at a summer school hosted by the University of Pittsburgh. It made sense to have the wedding in Washington, Pennsylvania, where Jane's family lived. "The date will probably be the 18th," Jane wrote her mother late in May, "unless John can make it by the 16th. That is the day his lectures are over." They planned a driving vacation for their honeymoon. Jane wrote her sister, "Johnny called tonight to say that he has just bought a new car, a 1937 Ford sedan."

A crucial issue was where to hold the ceremony. "My mother would be pleased," Jane began, "if we had this wedding in a church." John replied, "You can be married in the church if you want to, but not to me." Choking back a laugh, she recognized that negotiation on this point would be fruitless.

By getting married in the parlor of Jane's family home, John and Jane were following a tradition set by Jane's parents when they were wed in 1906. J. R. and Bess Maxwell then settled in Washington, Pennsylvania. Jane was born a year later, followed by two more girls, Betty and Sue, and two boys, Jim and Sam. Dr. Maxwell saw patients in his downstairs office at home for fourteen years. When he moved his practice to the McGugin Building, plenty of space became available at home for celebrations, such as Jane's wedding. To please his new mother-in-law, John conceded to having the Reverend Dr. Lippencott, the father of Jane's friend Mary Margaret Reline, perform the service.

Only a few friends and family witnessed the wedding on Monday, July 18, 1938. It was a drizzly Appalachian summer morning. Most of John's immediate family found a way to be there. Ruth and Ann drove from Madison, Tom and Janet from Pittsburgh. Several of John's paternal relatives came from Syracuse, New York. All of Jane's immediate family attended the ceremony. Mary Margaret and Bruce Reline were there to celebrate the long-awaited outcome of their matchmaking.

Rejecting the "traditional white and veil," Jane wore a simple white dress with a fringed jacket. John's teenaged sister Ann admired the way the fringes shimmied when Jane moved. Jane's aunt Margaret took the only snapshots of the event.

After the ceremony in the parlor, everyone gathered in the dining and family rooms, where tables had been set for lunch. Then Jane ran upstairs to change into traveling clothes. As the couple drove off in their new Ford, John cringed at the clattering of the tin cans that Jane's younger brothers had tied to their car. "That didn't please John a bit," Jane recalled. He removed them as soon as they were out of sight.

Then the new Mr. and Mrs. Bardeen headed northwest to Madison so that John could show his boyhood home to his bride. "We drove as far as Canton, Ohio, the first night and picked out the best hotel in town." As they drove out the next morning, "who should drive up along side us but Ruth, Ann, and their two friends,"

Jane wrote. "We stopped for a chat and complained about the lack of privacy on our honeymoon!" Jane was getting a feel for John's family. She was happy to be part of it.

Next they had "some lousy luck with a tire," a blowout and no proper tools to fix it. Jane wrote home, "We borrowed a jack from a farmhouse and then had a hard time to get it under the car." Jane began to suspect that she would become the family's "fixer." In later years, John's frustration with mechanical things often brought outbursts of anger, not at people but at the tools that failed him. "John was pretty angry, as you can imagine," Jane reported. "Finally we had to flag down another Ford in order to get a wheel wrench." Eventually they reached a town "where we bought a tire and had the spare repaired. Then we shopped every Ford agency we came to trying to buy the missing tools."

They finally reached Madison about 4:30 P.M. on Wednesday— just an hour and a half ahead of Ruth and Ann, who had also spent the night in Warsaw a few blocks from John and Jane's hotel.

The next day John took Jane for a swim in the lake. They unpacked the Ford and "stowed the stuff away in Ruth's cellar and attic." Ruth offered them their selection of family photos. Jane found some "adorable baby pictures of John."

Ruth was having trouble getting household chores done without a maid, so the younger generation "pitched in and helped with the work." With no time to send out their laundry, Jane did it herself with Ruth's help "and took a lot of kidding about washing and darning on my honeymoon. John was secretly much impressed and so proud of me that I didn't mind the work."

On Saturday night John drove Jane around the Wisconsin campus, and they went out for a beer at a student hangout with John's brother Tom and his wife Janet, who had arrived from Pittsburgh earlier that day with their daughter Judy.

Sunday morning the newlyweds drove to Green Bay to visit John's brother "Willie" and his wife Charlotte. "They killed the fatted calf for us and showed us the town." With no trace of resentment at spending her honeymoon on family rounds, Jane's letters reflect only the pleasure with which she discovered that John valued family as much as she did. It was John who insisted that Jane write frequently to her own family, as promised. "He even threatened to write this letter himself."

After Green Bay they drove to Minneapolis, their new home.

"All the people we have come into contact with here are very pleasant," Jane wrote. "I think we are going to enjoy living here." Ann Hustrulid, "another bride in the department," helped them look for an apartment. The market was tight, but they found a small, reasonably priced flat at 701 Seventh Street S.E., a short walk from the university. They arranged for it to be ready by the time they returned from their honeymoon.

From their spacious room at the Curtis Hotel, Jane wrote after eight days of marriage that she had "found more happiness being with John than my most rosy dreams led me to expect. Every day I discover new and lovable things about him and feel that I am a lucky girl to be his wife."

Heading west Jane marveled at the spectacular views in Yellowstone Park. In Montana they visited one of John's geologist friends from Gulf. At Glacier National Park they hiked fourteen miles and made a pact to return for a longer vacation. At Lake Louise they again hiked fourteen miles. To save money, they made their own breakfasts. Jane wrote to her mother that John made "very good coffee, believe it or not."

Occasionally they rested, with Jane writing thank-you notes and John working on a physics paper. But John resisted making too many stops. When Jane asked if they could get out of the car to have a better look at a hill covered with wildflowers, he replied, "Can't you see them from the car?"

John was anxious to get to Vancouver and see his sister Helen, who was in a hospital there. She had contracted tuberculosis in nursing school at Yale, where she had met and married Donald Beach, then a medical student. Helen and Donald had been living with their young daughter Glenis, now age two, in Langley Prairie, British Columbia, until Helen became too ill to stay at home.

The Bardeens visited with the Beaches for a few days. John fished with Don; the new sisters-in-law got along "famously." But Jane intentionally made herself "scarce," sensing that John and Helen needed time to "be free to talk about anything they wanted." Helen, who could no longer work, asked John to help get her affairs in order, so that in the event of her death Glenis would be cared for.

It had been almost a year since Helen had been near Glenis, who had not been allowed to see her mother in the hospital. It was therefore a special treat for Glenis and Helen when John, Jane, and Don took them for a two-hour drive in the car. "It is a desperate fight for her," Jane wrote of Helen, "but it is encouraging to know that people do recover from T.B. May she be one of those!"

After saying good-bye to Helen and Don, the Bardeens headed down the West Coast to San Francisco and Yosemite National Park. They walked "about four or five miles through downtown Frisco and along the water front" and had "great fun" on the trolley cars. Jane wrote to her family about the splendor of the redwood forests and their stay at a "charming place on the edge of a grove." She and John had become "pretty fair hikers by now, although I have a tendency to lag behind on upgrades. John sometimes has to give me a tow." She apologized for writing fewer letters than usual. "We go so hard all day and John wants me to be with him all the time so that when night comes I haven't time or energy left for writing."

The newlyweds made one last stop in Madison before heading to their new home in Minneapolis. They had driven more than 8,000 miles and had visited twenty-four states and two Canadian provinces. In Madison, John and Jane "rooted in the cellar" and found some "Chinese dishes which belonged to John's mother." They took them and the family photographs along with them. They carefully packed the dishes, photos, and wedding gifts into the car for the trip to Minnesota. Several wedding gifts caught up with them in Madison, including an elegant lace dinner cloth from Eugene Wigner. They stayed an extra day in Madison when they learned to John's delight that Wigner was in town.

Their apartment was ready when they returned to Minneapolis on September 7, 1938. Unfortunately the few household goods they had shipped from Pittsburgh had not yet arrived. Cashing a check for several hundred dollars, a wedding gift from Jane's father, they bought a few furnishings, including the walnut bedroom set they would use for the rest of their lives. For another twenty dollars they bought six chairs and a secondhand dining table on which John did "a swell job of sanding and staining." John's dark red drapes from Lowell House, which had been altered to fit the new windows, "went fine" with Jane's rug. Grateful that they didn't have to "go into John's capital," they had, with the help of wedding gifts, "paid cash for everything."

Soon they realized they would also need a crib. Both were eager to start a family, but it was a surprise when Jane became pregnant during their honeymoon. John was thirty, Jane thirty-one. Jane had already "suspected something" by the time they got back from their trip, as she had begun to be plagued with morning sickness. "John, like a dear, brings me tea with sugar before I get up, gets his own breakfast, and thus permits me to take my time about getting started in the morning."

Mindful of appearances, Jane instructed her doctor, "Don't let this baby come before spring." "People are bound to count on their fingers," she confided to her sister, "since the baby will arrive well in advance of our wedding anniversary." By December, feeling the pinch of her expanding belly, she bought a few items of maternity clothing. A month later John teased that she "was going to produce a litter." Her doctor disagreed, "being more used than my fond husband to the disastrous effects of pregnancy on the figure."

Jane was deeply touched when her new Minnesota friends threw a surprise baby shower for her in late February. "The most amazing present," handmade by the husbands of two of her friends, was "a large wooden drawer in which to keep baby clothes." Jane explained to her mother that it was "an exact replica of the book 'Tiny Garments' which Betty sent me." Her friends had it "painted blue except for the sides which are natural color wood with the grain suggesting the pages. The drawer front is curved to resemble the back of the book and on the top Lynn had copied in black the title of the book and the picture of the baby that decorated the front cover of the real book."

Before leaving Dana Hall, Jane had asked Miss Cooke for a letter of recommendation, should she want to seek another teaching position. Having a child ruled that possibility out for her. Like most women of her generation, Jane considered a career incompatible with raising a family. When a former professor from the University of Pittsburgh urged her to continue her genetics studies at Minnesota, she quipped that she was working on "a genetics project of my own." Following in the footsteps of Althea and her own mother, Jane soon assumed all the household and family responsibilities, abandoning her own teaching and research so that her husband could concentrate on his.

The Bardeens entered Minnesota's academic circle easily. Jane often hosted parties at home. "I'm really enjoying cooking and don't mind spending a lot of time on it," she wrote to her mother, "but I certainly don't get much satisfaction out of cleaning."

John continued to have trouble making small talk. Writing to her mother about a new couple they had met, Jane expressed her hope that "we'll continue to know them better because John really opened up and talked, which is unusual for him especially with people he doesn't know well." She began to assume the role that had earlier been played by John's brother Bill, and later by Dutch Osterhoudt, of smoothing over some of the awkward gaps in John's conversations.

Jane had many friends among the wives of John's colleagues. The closest were Vera Williams, Ann Hustrulid, and Florence Rumbaugh. The Rumbaughs and the Bardeens were a frequent foursome, sharing companionship, humor, and support. Florence and Jane commiserated when both their husbands contracted the flu, "exchanging ideas by telephone on care and behavior of sick husbands." Jane reported, "John hates to take medicine, but he *will* stay in bed, which is something."

Again, John cultivated friendships through sports and other games. Both the Bardeens enjoyed bridge, then popular among academic couples. John played billiards and golf with other physicists, sometimes during lunch breaks. One year he won the faculty billiards championship.

John would usually play softball when someone worked up a game. At the physics department's spring picnic, faculty and graduate students traditionally faced off against one another. Nier bragged that with Bardeen, Williams, and Rumbaugh on the team, the faculty always won, which "ruined" the event. On one occasion the game broke off after a faculty-to-student score of "25 to 3 or something like that, and we chose sides to make it more fun."

The annual Physics against Chemistry game fared no better. In the late 1930s, physics chair Jay Buchta revived an earlier tradition of a friendly softball game between the two departments at their annual joint picnic. He tossed the gauntlet with an insulting note signed by everyone on the physics faculty. Chemistry accepted with an equally insulting letter sealed in a glass tube of pungent hydrogen sulfide. Only one member of the chemistry faculty, however, was skilled enough to make its softball team. Physics was short on skillful graduate student players. In the end, the physics faculty ended up playing against the chemistry graduate students. "We beat them, of course," Nier recalled smugly more than fifty years later.

Bardeen was soon recognized as one of the department's star athletes. Opponents who might have been lulled into complacency by his average build and mild manner did not take long to notice the superb control with which he wielded the tools of any game that he played. The control extended to his tongue. Not only could he refrain from saying anything until he knew precisely what he wanted to say, but he could also stop and start abruptly as circumstances dictated. Nier claimed that Bardeen could be quietly grumbling on the golf course after a bad shot, and immediately stifle his tirade when another golfer stepped up. Bardeen would then resume complaining where he had left off when the other's shot was completed.

Teaching a full program was a new experience for Bardeen, one that at first absorbed most of his time. He taught both undergraduate and graduate courses, initially finding both a bit intimidating. He had previously taught only his small Harvard seminar on relativity. "Since you're the teacher," he said to Jane, "tell me what I ought to be doing." She suggested, "Tell them what you know." He learned that like his father he was far better teaching one on one than lecturing to a group. He worked with several graduate students but was not at Minnesota long enough to help them complete their doctorates.

At the end of his first academic year at Minnesota, Bardeen felt lucky "to have the opportunity to teach during the summer and get a little extra money on the side." But the heavy teaching load, especially for the beginning physics course, drained him. "John had a hard week," Jane wrote her mother in August. "The sudden plunge into teaching got him so on edge he couldn't sleep." Worried that he was wearing himself out, Jane planned a family drive to Madison. "John certainly needs to get clear away from Physics and teaching for a while. This summer school work is very exhausting." To sooth his nerves, she indulged his sweet tooth with "a batch of blackberry jam."

Jane soon learned that "vacations are for concentrating on research." They were "about the only time he can get caught up with papers which have to be written," she explained to her sister Sue. Money continued to be tight, especially when both John and Jane needed expensive dental work in the fall of 1938—just when they were setting up their household and saving for the baby.

John's first project at Minnesota was to write up the paper he had delivered in Pittsburgh before the wedding. He began with an overview of outstanding problems in the physics of metals before quantum mechanics. "These difficulties have been almost completely removed by modern quantum theory as applied to the problem by Sommerfeld, Houston, Bloch, Mott, and others." Laying out the experimental picture concerning electrical conductivity he wrote, "Perhaps the most important thing an adequate theory of conductivity must explain is the remarkable difference in conductivity between metals and insulators." The early theories by Paul Drude and Hendrik Antoon Lorentz, he wrote, "did not attempt to give an explanation of these facts in any detail; they merely attempted to give the mechanism of conductivity."

Bardeen's paper discussed Arnold Sommerfeld's semiclassical theory of conductivity, mentioning William Houston's estimate of the electrons' mean free path, computed on the basis of Sommerfeld's theory. He also described Felix Bloch's theory and its remarkable consequence that "a perfect periodic lattice will have no resistance." Outlining the difference between metals and insulators according to the band theory, he explained why "the 'hole' in the otherwise filled band acts like a particle of positive charge and positive mass in the region of negative curvature, and like a particle of positive charge and negative mass in the region of positive curvature near the bottom of the band."

Moving into a more detailed discussion of conductivity and resistance, he appealed to works by his mentors Peter Debye and Percy Bridgman. Finally, he reported on problems of current interest in the field, such as the resistance of ferromagnetic metals and the conductivity of alloys. Disclaiming any attempt at completeness, he pointed out "some phenomena for which the Bloch theory has not yet provided even a qualitative understanding." On his list were superconductivity and the curious observation by W. J. de Haas and his colleagues in Leiden of a minimum in the experimental plot of electrical resistance versus temperature for gold at liquid helium temperatures. With great insight he pointed out, "It may be necessary in some cases to go beyond the Bloch picture to explain things which depend to a large extent on electron interaction, or on the cooperative action of a large number of electrons." His conclusion summarized his own planned program of research in many-body theory, which he launched into after sending off the article.

"Any labor pains yet?" was "John's first question when he wakens and when he comes home in the evenings," Jane wrote on May 4, 1939. It's "going to be a big, fat girl," he had teased her throughout the spring. "But of course he really wants it to be a boy." Jane assured her mother they would telegram immediately when the baby came, "unless Johnny gets too excited over becoming a parent." She made backup plans, asking a friend to call her mother if John forgot. While waiting, John painted furniture. He "is doing a fine job on the baby's basket, but the bassinette still requires some fixing—John will probably be doing that the night before I return from the hospital!"

James Maxwell Bardeen, named after Jane's brother, arrived on May 9, 1939—four days late according to the doctor, but a little early according to Jane's own calculations. Both John and Jane delighted in Jimmy. They were playful parents. "He was so much fun last night," Jane wrote when Jimmy was six months old. "He was lying on his tummy on the couch with his head and shoulders raised up, supported by his arms. He started swaying from side to side and rolling his head. I imitated him and then he would repeat. He taught John the same trick. John calls it 'playing bear.'" A year and a half later, when Jimmy could string words together, John set out to teach him about Newton's laws. The little boy's babble about bodies persisting in their state of motion was a hit at department gatherings. John's enjoyment of children extended beyond his own family. He frequently ended up playing with the youngsters at social gatherings.

Even after Jimmy's birth Jane and John continued to enjoy a newlywed relationship. For Jane's thirty-third birthday, "Johnny and I went gallivanting," she wrote her mother. They had "dinner downtown," then saw a show of Pinnochio "in celebration of the day." Some months earlier she had baked a Valentine cake for John "decorated with small red cinnamon candy hearts." When Jimmy was seven months old, Jane wrote to her family, "The kind of love that grows as you live with and really learn to know a man is so much more satisfying than the romantic fluff girls dream about."

John's unusual reserve and his intense focus on physics nonetheless created a gap in their relationship, one that was often painful to Jane. Her feelings of isolation from John's most ardent passions, especially physics and later golf, did not dissolve as the marriage matured. In Minnesota, she briefly tried to play golf in an effort to be closer to John, but she soon gave it up in favor of attending to home and family. Physics would remain John's "first love" throughout their marriage. But in his ventures off into new scientific territory, Jane would serve as his anchor, as his family always had.

By the summer of 1940 it was clear the Bardeens needed more space. Jane wanted a yard for Jimmy. They rented a small, three-bedroom frame house at 171 Arthur Avenue South East. "The house is not large but we are very comfortable," she wrote to her mother in August. An open living and dining area and a small eat-in kitchen comprised most of the first floor. "The closets are wonderful— almost big enough to be rooms!" They treated themselves to a few conveniences. "John went downtown and bought me a swell Thor washing machine." And "you knew about our new Frigidaire."

They had intended to buy rather than rent their next house, but the looming possibility that the United States would enter the war against Germany made them hesitate. "We've decided that we had better wait and see what happens before we sink all our capital into real estate and go into debt besides," Jane wrote to her mother. "If Germany wins, as she may well do, there will undoubtedly be economic upheavals in this country." Although short of money, the family made do. "Both of us feel that living within the income is more important than being lavish with our gifts," Jane wrote home at Christmastime in 1939.

Jane was willing to "try to be economical" so that they could send John to his physics conferences at the family's expense. The American Physical Society meeting "really is an important occasion since it provides useful contacts and keeps him and his talents in the public eye." When John flew to Washington in March 1940, Jane said he was "as excited as a small boy would have been" to be on his first airplane. He reported back to Jane, then shivering in the below-zero temperatures of a lingering Minneapolis winter, that the weather had been "perfect all the way."

The Bardeens starting combining family vacations with John's professional trips. In late spring of 1940, he drove to Pittsburgh to attend a meeting of the American Physical Society, while Jane and Jimmy traveled ahead by train to visit with family in Washington, Pennsylvania. They accompanied John on the long drive back to Minneapolis.

If Bardeen could not attend a meeting, a colleague would sometimes present his paper, as Nier did during the Christmas break of 1940. Bardeen appreciated Nier's help in getting his paper to the Columbus, Ohio, meeting of the American Association for the Advancement of Science, but he wished that his department would do more to support faculty travel to conferences.

Nier was one of many nuclear physicists then studying uranium fission, a phenomenon that opened the possibility of building an atomic bomb. Those who knew about fission worried that Nazi scientists might build such a bomb. To do so, as things were understood at the end of 1940, one had to first separate a substantial amount of the rare, highly fissionable, uranium isotope of mass 235, from the more abundant isotope, ^{238}U, found in natural uranium. With only three units of difference in atomic mass, the separation of these isotopes was extremely difficult.

One of several possible methods for separating the isotopes was thermal diffusion. Nier asked Bardeen to help him develop a theory for the diffusion process. Nier also worked extensively on an electromagnetic isotope separation method, after Enrico Fermi asked Nier to prepare some ^{235}U to use in his research at Columbia University. Fermi suggested that Nier obtain the ^{235}U using the mass spectrometer that Nier had developed to separate isotopes. Nier was successful. He wrote, "On Friday afternoon, February 29, I pasted the little samples [collected on nickel foil] on the margin of a handwritten letter and delivered them to the Minneapolis Post Office." Two days later he was awakened "by a long-distance telephone call from John Dunning," who had bombarded Nier's ^{235}U samples with neutrons from the Columbia cyclotron. Dunning "clearly showed that ^{235}U was responsible for the slow neutron fission of uranium."

An article about Nier's triumph was featured on May 5, 1940, in the Minneapolis *Star-Journal*. Jane sent a clipping to her mother, who had met Nier on one of her visits to Minnesota. "Looks like a Hollywood version of the earnest young scientist, doesn't it? The work he has done is extremely important."

In 1940 and 1941, the United States military began to prepare for the nation's nearly inevitable involvement in the war. Calls to military research laboratories disrupted the families of many American physicists. Faculty who were not called into service assumed extra duties. "John is teaching Will Wetzel's geophysics course and so is very busy here," Jane wrote to her mother in late 1940. Around the same time, Lynn Rumbaugh was called to the Naval Ordnance Laboratory (NOL) in Washington, D.C. He left Florence to tend to their three children, with a fourth on the way. Having worked earlier at the Department of Terrestrial Magnetism, Rumbaugh "knew a lot about magnetism," according to Bardeen, and so "was called on an urgent basis to help us do something about the German magnetic mines." The Bardeens invited Florence to dinner on her birthday. Lynn phoned from Washington while she was there to wish her a happy birthday. He had written her that "John had been slated to go down there too, but the Navy decided they had better not take three men from the same department." John and Jane were relieved, but they also knew it was but a temporary reprieve.

Jane dreaded a separation from John. "It is a completely new experience for me to be alone," she had written earlier when he was away at a conference. "Our only other separation was when Jimmy came." She feared that she "would be less courageous than Florence if he should be called 'to the wars.'" Although they welcomed the money that a summer of defense work might bring, Jane wrote to her mother that she "hate[d] to think of being a widow for the whole summer."

John's sister Helen caused additional worry. In late 1939, to avoid the dampness of the coast, she moved with Donald and Glenis to McBride in the Canadian Rockies. With Helen so often in hospitals, her marriage deteriorated. A few months later, Donald asked for a divorce, relating that he had fallen in love with a nurse at work "All of us are heartsick at the trouble Helen has had," Jane wrote.

Jane and John considered taking care of four-and-a-half-year-old Glenis for a while. They discussed the matter with John's stepmother Ruth. Donald clearly could not cope with his ailing wife, small daughter, growing medical practice, and now his new love. They doubted that Helen "will ever be well again—and even if the T.B. is arrested she will certainly not be able to earn hers and Glenis's living." They also discussed the possible consequences of Helen's death. John agreed that he and Jane should offer to take Glenis, but that if Helen died, Glenis "should live with her father."

After a flurry of correspondence in December, Helen sent a telegram to John and Jane one Thursday evening in January 1941 to say that "Glenis would arrive Friday morning on the night plane from Seattle." Jane met the plane, as John "had a first hour class he couldn't cut on such short notice." She found her new charge to be "the most unconcerned small person, just put her hand in mine and came home with me. Has all the poise of a woman of the world and takes flying alone hundreds of miles right in her stride."

But the child's adjustment was not smooth. "Being an only child," Jane noted, "she and Jimmy (who has also had that handicap) don't get along together in complete harmony." Jane sent Glenis "to the University nursery school so that she will have playmates of her own age and can be helped by persons skilled in child care." Jane confided to her father that "she has tried to see just how much she could get away with, but I think she accepts me as boss now. She obviously needs a lot of loving and I try to give her all I can without turning Jimmy into a green-eyed monster. Her coming, of course, has meant a rather violent change in his way of life, particularly in his emotional self."

Whether or not Glenis and Jimmy ever got past their initial jealousy is not recorded in Jane's letters, but in later years Glenis recalled the time with Uncle John and Aunt Jane as a period of serenity in an otherwise troubled childhood. She remembered laughter and the sounds of Beethoven, Mozart, and Brahms wafting from John's phonograph by the living room window.

In February, less than a month after Glenis arrived, a new perturbation added to the worries of war and family. "I shall not see the doctor for a month or so, but I am unmistakably pregnant," Jane wrote to her mother in early 1941. "I think John would like another boy so he could be pals with Jimmy, but it makes no difference to me."

Bardeen worked on his physics through all the family turmoil. After completing his review of electrical conductivity, he focused his attention on superconductivity. He hoped to explain the London theory on first principles. Heisenberg's uncertainty principle—that there is a specified limit to the accuracy with which two "conjugate variables," such as position and momentum, can be measured at the same time—suggested that if he could explain the observed diamagnetism of superconductors, an account of the infinite conductivity would also emerge.

He suspected that Walther Meissner and Robert Ochsenfeld's demonstration that superconductors expel magnetic fields was a manifestation of the fact that the electrons inside superconductors move in much larger orbits than anyone had thus far realized. He tried to construct a theory, assuming that the only electrons he would have to include were those whose momentum differed little from electrons at the edge of the Fermi surface. He hypothesized that the diamagnetism of superconductors arose from a surface current involving electrons with large orbits traveling in the neighborhood of the Fermi surface. If the electrons at the surface ran around in such a way as to create a field in the opposite direction to the applied field, electrons inside would not feel the applied field.

Building on this intuitive picture, Bardeen assumed that when circumstances are favorable, small displacements of the ions inside superconductors cause the electrons to gain an amount of energy that more than compensates for the energy spent on ionic displacement. The disparity between these energies, he argued, would cause gaps to form in the electronic structure near the Fermi surface. Drawing on the London theory, he hoped to explain superconductivity in terms of these gaps. Unfortunately, when he tried to formulate this theory in three dimensions, the numbers were off by more than a factor of ten.

Bardeen sent a version of his new theory to Seitz, who in turn showed it to Wolfgang Pauli when both were in Ann Arbor during the summer of 1941 at the annual Michigan Summer School in Theoretical Physics. Pauli "shook his head in disapproval," according to Seitz, who wrote back to his friend, "You haven't made the grade with Pauli." Once again Bardeen did not commit his calculation to print, choosing to publish only an abstract.

Bardeen would have worked further on the theory, but in March 1941 he was, as expected, called by the Naval Ordnance Laboratory. By the summer of 1941 he was working in Washington, D.C. But as he studied war-related problems, "the concept of somehow getting a small energy gap at the Fermi surface remained in the back of my mind." With characteristic tenacity, Bardeen held on to that nagging idea.

7

Engineering for National Defense

The news that crackled across the radio was not good in June 1941. Bardeen listened as he drove across Wisconsin, Illinois, Indiana, Ohio, and the Appalachian Mountains on his way to Washington, D.C. He was not happy to be heading off to the nation's capital, leaving Jane and Jimmy behind, but he considered it his patriotic duty. More than a year earlier Jane had written to her mother, "John and I both feel it may be only a short time until this country is involved—heaven forbid—but we certainly couldn't go on living without protest in a world dominated by Nazis and Communists."

Americans had looked on in fascinated horror as Britain took a brutal pounding from Germany's *Luftwaffe* in 1940. By July, Hitler had conquered Poland, Belgium, the Netherlands, and France. The only European country still standing firm was Britain. Led by Winston Churchill, the new prime minister, Britain pledged to resist Hitler no matter the consequences. Addressing the nation on May 19, 1940, he proclaimed, "Now one bond unites us all—to wage war until victory is won, and never to surrender ourselves to servitude and shame, whatever the cost and agony must be." Week after week, month after month, German planes bombed British cities, killing thousands, displacing thousands more, and wreaking havoc on the British economy. But there would be no surrender.

British scientists set aside their own research to develop technologies useful in war. British radar allowed intelligence to track enemy planes and mount an offensive, thereby minimizing damage. With its excellent communications system, the British Royal Air Force (RAF) could send its Spitfire and Hurricane fighter planes at just the right moment to confront German bombers in the air. The RAF destroyed more than twice the number of German planes that it lost itself. The Battle of Britain was a decisive air victory for Churchill's forces, but the island nation's resources were more limited than Germany's. Then in June 1941 Hitler suddenly gave Britain some breathing space by turning his attention to the east. He unleashed a massive invasion of the Soviet Union.

United States President Franklin Delano Roosevelt became increasingly certain that the U.S. would have to become involved in the war. He encouraged a shift in public opinion to support interventionism and accelerated the steps that had already begun toward U.S. rearmament. Although many scientists had been isolationists, by the time the Nazi Blitzkrieg overran Poland in September 1939, the belief that Hitler had to be stopped was widespread. The German-Jewish scientists who had fled to the United States or Great Britain encouraged military preparedness. Einstein, a socialist and a pacifist, declared publicly that the democracies ought to prepare for German aggression. American aid to the Allied forces, reduced now to Great Britain and the French resistance, flowed freely but informally until March 1941, when the arrangement was formalized by the Lend-Lease program.

Federal support for laboratories engaged in military research entered a phase of unprecedented expansion. Vannevar Bush, an MIT professor of electrical engineering, became a crucial link between scientists and the government. On becoming president of the Carnegie Institute of Washington in 1939, he took over the chair of the National Advisory Committee for Aeronautics (NACA). Based on the NACA model, he and other scientists developed a plan for organizing research within the armed forces and for mobilizing civilian science in the service of the military. Bush gained Roosevelt's authorization for the National Defense Research Committee (NDRC), established in June 1940 to coordinate research with the needs of the army and navy. Through the new organization, Roosevelt released emergency funds for research, but the amounts proved inadequate.

In May 1941 Bush secured a commitment from Roosevelt to

create a more powerful organization funded directly by Congress. The Office of Scientific Research and Development (OSRD) could award contracts to universities and industrial laboratories, as well as coordinate their activities with military labs. Bush took charge of the OSRD, while James Conant became chair of the NDRC. Out of these committees grew the government network that coordinated America's wartime research.

The Naval Ordnance Laboratory (NOL), into which Bardeen landed, was part of this network. In 1929 the Bureau of Ordnance's Mine Laboratory, a small experimental unit founded in 1919 and housed in the Gun Factory of the Washington Navy Yard, was combined with the Experimental Ammunition Unit to form the Naval Ordnance Laboratory. Not until the Second World War, however, did the NOL become a prominent research facility.

By 1939 the navy was desperate for new technology to defend against the magnetic firing mechanisms of German mines and torpedoes, which were devastating supply lines across the Atlantic. This need obliged the Bureau of Ordnance to expand the NOL, which, in January 1942, became an entity separate from the Gun Factory. Bardeen's call to Washington was part of this expansion. Lynn Rumbaugh had been brought to the NOL by Ralph Bennett, a craggy engineer with whom Rumbaugh had once worked at the Carnegie Institute in Pittsburgh. Bennett was at the NOL because he had volunteered as a reserve naval officer after examining a British ship whose magnetic field had been "degaussed" (neutralized). He believed he could improve the degaussing technology. The process involved surrounding a ship with a circular network of coiled cables. High currents generated on board would pass through the cables, turning the coiled system into an electromagnet. By appropriately selecting the current, one could make the ship's magnetic field either so strong that mines would explode too far away to cause damage or so small that even nearby mines would not explode.

In the summer of 1940 the navy called Bennett to Washington for a six-week assignment, which stretched to seven years, during which, as technical director, he transformed the NOL into a first-rate laboratory. Focused not only on its original agenda of degaussing ships but also on more than a hundred related projects, the NOL aggressively recruited scientists and engineers against stiff competition from other laboratories. Bennett recruited Rumbaugh in the fall of 1940. The following spring Bennett and Rumbaugh persuaded Bardeen to join.

The physics chair at Minnesota, Jay Buchta was at first reluctant to release Bardeen, because like many American university physics departments, Minnesota's was struggling to cover the classes of its drafted professors. Bennett and Rumbaugh persuaded Buchta to spare Bardeen for the summer—if Bardeen wanted to go.

At first Bardeen resisted leaving his family and his academic life for the grueling work of a military laboratory. Nor had he forgotten how bored and frustrated he had become during his final months at Gulf. That Jane was expecting a child also worried him. He suspected that if he worked at the NOL for the summer, he would end up staying for at least a year, leaving Jane with the entire burden of Jimmy, Glenis, and the new baby.

Salary was another drawback. By hiring civilian scientists for a daily wage on personal service contracts, the navy could recruit technical help more quickly than it could through civil service channels. But the $16 per day contract rate that Bennett initially offered Bardeen did not cover the cost of uprooting the family. When Buchta wrote Bennett that Bardeen was due for a pay increase in the fall, the navy increased its offer to $17 per day.

Bardeen finally realized that it could only be a matter of time before he would be drawn into some sort of war work. And he had just the right background for the work at the NOL. In terms of the science, many of the problems being investigated there, including magnetic, pressure, and gravity changes, were related to the prospecting work he had done at Gulf. He decided to go. As he had feared, his summer position grew into a one-year leave of absence and eventually into four years of research for the navy.

Jane stayed behind with Jimmy, then not yet two, to arrange the family's relocation. Dragging Glenis to the nation's crowded and chaotic capital did not appear to be in the child's best interest. After consulting with the rest of the family, Jane sent her to stay with John's brother Bill and his wife Charlotte in Green Bay.

When Jane told her doctor she would be moving to Washington, D.C., he said, "You can't have this baby there—it's not a fit place." The U.S. Public Health Service had identified the District of Columbia as deficient in health facilities even before the war. Now the influx of wartime workers critically overburdened the city's hospitals and physicians. John and Jane decided that she should have the baby in Washington, Pennsylvania, where her father practiced medicine and where there was a good hospital. Six months pregnant, Jane packed up their household. With Jimmy she boarded the train to Pittsburgh. They were met by Jane's family, who brought them to her parents' home. On September 15, 1941, Jane telephoned John to announce the birth of his second healthy son. "Good," he muttered. "I always liked having a brother." They named him William, after John's brother, and called him Billy. By November Jane was ready to travel with the two boys. This time the train ride felt completely different, for they were traveling to be *with* John.

Until then, Bardeen had fended for himself in the District of Columbia, where thousands of men and women had swarmed to contribute to the war effort. Living conditions varied, but no one could avoid the ubiquitous rationing, from gasoline and food to housing. Special two-hour time changes saved lighting expenses. Automobiles crawled along on congested roads, and public transportation was inadequate. After a stay in a tiny apartment at 2910 Brandywine, John found a somewhat larger two-bedroom apartment at 3813 W Street S.E. It was part of Fairfax Village, one of many small apartment developments that sprouted up in the city to accommodate the influx of war workers.

In an uncharacteristic act of domesticity, John went about furnishing the apartment after he received the shipment of household goods Jane had sent from Minnesota. Jane appreciated his effort to arrange furniture, dishes, and linens. When she and the boys arrived, they could just move in. The modest apartment felt even smaller when Jane's sister Betty took up residence with the family after landing a job at *National Geographic* magazine. She slept on a cot in the boys' bedroom; the boys slept in bunk beds after Billy outgrew his crib.

Built on a reclaimed swamp, Washington, D.C. oozes a sticky, enervating heat in summer. Recalling his office at the NOL, Bardeen called the Navy Yard the "hottest place in Washington." Technical professionals worked without air-conditioning in crowded labs and offices. If one had the good fortune to work near an open window, the view was most likely a surreal scene of weapons research. Nearly every day cannons fired test rounds that rattled the windows. Bardeen kept his office windows shut to avoid the fumes from a paint shop directly below, rendering his workplace stifling in the summer.

The NOL was generally considered one of the more agreeable

military laboratories. R. H. Park, chief physicist at the time Bardeen
arrived, singled out the NOL as the one laboratory within the
Bureau of Ordnance that was free of such "objectionable conditions" as poor access to information, "invidious distinctions"
between the higher officers and civilian workers, and "poor esprit
de corps." The NOL also suffered less interference from outside
agencies than did other labs.

William Anspacher, an engineer in Bardeen's group, shared a
cramped office with John and four or five other physicists. Two
desks away, in a different group, but also in the "unaired loft" of
the Gun Factory, sat the physicist Charles Kittel, with whom
Bardeen struck up a friendship. Kittel and Bardeen would overlap
again later at Bell Labs.

Bardeen sat at a corner desk facing the wall, often with his feet
propped nonchalantly on his desk. Tables crowded the tiny room.
Anspacher wondered why Bardeen never complained, because as a
supervisor he might have requested more space. Anspacher could
not imagine how Bardeen managed to get so much work done. A
casual observer "would swear he was sound asleep," yet he was
always alert when he needed to be. To Kittel it appeared that
"everything John's group worked on was done quietly, correctly,
and with a strong imagination."

The mathematician David Gilbarg, who worked a few buildings away, stopped off regularly in Bardeen's office to discuss torpedo design. Gilbarg and his team were building torpedoes about a
foot long, scaled to one-twentieth the size of a working torpedo.
They tested the scale models by measuring the impact against a
wall or by taking high-speed motion pictures of the action in a
huge tank of water designed to simulate field conditions.

Bardeen's analytical and pragmatic approach to problems made
him a valued member of the NOL staff. He did not hesitate to probe
if something perplexed him. He also developed a reputation for
answering questions at whatever level the inquirer could understand. According to Anspacher, if Bardeen found that his explanation was pitched above someone's head, he would reformulate his
answer in progressively simpler terms until his meaning became
clear. "There was no intellectual snobbery about him at all."

The members of Bardeen's group worked on whatever projects
required their expertise. For example, they analyzed the feasibility
of different mine designs. In one design they studied during the

summer of 1941, the mine detonated when it detected gravity changes associated with a ship passing above it. The group found the design useless in actual field conditions because the available LaCoste-Romberg gravity meter was not sensitive enough to detect the feeble gravitational fields of ships. A more promising design, which Bardeen analyzed during 1942, involved pressure-sensitive detonation.

He also analyzed the target areas of submarines—the projection on a vertical plane of the regions where a submarine's magnetic field is greater than or equal to the mine's magnetic sensitivity—as well as the firing areas for particular mines and exploder mechanisms. He directed the preparation of a catalogue of ships' magnetic fields for use in naval planning. His report in August 1942 offered a way to classify the magnetic signatures of various ships, with an eye toward setting mines targeted at particular classes of enemy ships.

Bardeen often consulted with experts at other institutions. For example, during March 1942 oceanography problems brought him to New York to speak with scientists at the New York Academy of Sciences and then to Princeton to interview theoretical physicist John Wheeler, who was being considered for recruitment at the NOL. The following month Bardeen traveled to New York to confer about oceanography with scientists at the American Museum of Natural History. His visits to naval stations took him to Newport, Rhode Island; Norfolk and Cape Charles, Virginia; Seattle, Washington; and Mayport, Florida. In May and July 1943 his work on torpedoes brought him to the Westinghouse plant in Sharon, Pennsylvania, where torpedoes were manufactured, and to stations in Seattle, where torpedoes were tested.

That summer he also visited Einstein and "spent a very enjoyable afternoon with him in his study on the second floor of his modest frame house in Princeton." Einstein had come to the navy with an idea for an induction coil placed inside a torpedo. When a warhead passed under a steel ship, current would be induced in the coil, indicating the presence of the ship. Bardeen reported, "We had thought of the idea but did not pursue it because of the attenuation of the signal by the steel covering of the warhead."

Bardeen proceeded to ask Einstein about this problem. Einstein "suggested using a plastic warhead or a plastic window. He had carried out calculations which indicated that the methods should be feasible." Bardeen was impressed that Einstein "had thought deeply about the problem and had answers to all my questions." The NOL's interest in the induction coil mechanism was revived "largely as a result of this conversation." However, a production model for Einstein's torpedo could not be built in time for it to have wartime use.

These forays offered Bardeen interesting breaks from his routine, but they could not quell his uneasy feeling that he was just marking time at the NOL. He was weary of working on projects that were not his own and chafed to return to fundamental physics. On top of that, the chaotic and inflexible nature of the military bureaucracy rattled his nerves. He much preferred working in groups smaller than those typically found at the NOL. And he liked focusing on only one or two problems at a time. As a supervisor he was forced to manage dozens of people working on many different projects.

Bardeen's assignments and titles changed several times during his four years at the NOL. When he first arrived, he headed a group working on degaussing methods under Rumbaugh. Next he was put in charge of the Field Analyst section of the Mine Counter-measures Unit, where he and his group of twenty-one men analyzed data on ships' fields. In the fall of 1942 the Research Division underwent a major overhaul, in which the focus of its Mine Counter-measures Unit (renamed the Mine Research Unit) tightened to deal only with underwater phenomena and degaussing. Bardeen headed one of five sections in Mine Research, with fifty-one men under his direction.

The work in the Mine Research Unit was monotonous for a highly trained scientist. The unit also faced personnel problems because its technical people were constantly siphoned off into Mine Development. Adding to the frustration, the results of the Mine Research Unit typically did not interest any naval officer in a position to develop them. Many disgruntled employees left at the earliest opportunity.

Bardeen also struggled with the inaccessibility of naval officers who could authorize research programs. As the layers of bureaucracy thickened with time, Bardeen complained to Jane that he found it difficult to accomplish anything because every detail had to pass through a tedious process of approval. "If he wanted to send work, it had to go up the chain, stopping at every place. Then it had to come back down again. Took forever."

In a further reorganization that fall, the Mine Research Unit became simply the Research Division, with two large subdivisions—Engineering and Influence Field Measurements—formed to oversee the work of the smaller units. Bardeen became head of the Influence Field Measurements Division and was now responsible for ninety-three people working under his supervision in Acoustic, Magnetic, Graphical Analysis, and Detection units. He planned and directed the NOL's research program on influence field measurements. Teams in his division collected and interpreted data for applications in devices such as mine firing units, torpedo exploders, and depth charges.

The navy's policy of compartmentalization was one reason for Bardeen's unhappiness at the NOL. The navy considered it less of a security risk if each person worked without knowledge of other parts of the same problem, or of the project as a whole. Parceling out pieces of a project to different researchers meant that each would work on an isolated fragment of a larger problem. From the scientists' viewpoint, this policy destroyed effective collaboration and resulted in false starts, repetition, and a good deal of misunderstanding. Not only did compartmentalization make the work more difficult, it reduced much of it to sheer tedium.

The situation improved slightly in December 1942, when the NOL shifted its strategy from blind problem solving to "project orientation." A project manager was assigned to each new incoming project. Then an administrative group would put together a team whose members would be the most qualified to work on the job as a whole, regardless of their positions within the hierarchy.

Bardeen's security clearance required him to keep technical details from anyone without an official "need to know," even Jane. Not discussing his work at home was not a hardship for John. Like most of his colleagues, he welcomed the relief from shoptalk during his all-too-brief time at home. Many of the government scientists and engineers worked from early morning to late evening, six or seven days a week. Bardeen averaged fifty-three hours per week while at the NOL.

John's frustrations were amplified when he suddenly gave up smoking, not for health reasons but because of the difficulty of obtaining cigarettes, which were rationed during the war. He had been in the habit of smoking a pack per day. As Jane refused to let John smoke any of her cigarettes, he began rolling his own using an inefficient rolling device. The trouble was not worth its limp and leaky products, so he stopped smoking altogether.

At one point Bardeen had the chance to leave the NOL and work for the Manhattan Project. In the fall of 1943 both he and Fred Seitz were invited to join Eugene Wigner's nuclear reactor group at the University of Chicago. Seitz recalled, "Eugene got worried about whether the reactors would hold up under the intense radiation bombardment they had in the graphite reactors." John and Fred discussed their invitation on the train to Chicago. Although tempted, Bardeen eventually decided to turn it down. "John felt he was tied down to Washington," recalled Seitz. He remembered John mentioning, "kids in school." Seitz "was free," however, and spent the next two years working with Wigner on the plutonium reactor.

Bardeen thoroughly disliked organizing and supervising the dozens of prima donnas gathered at the NOL. Most of those in his group were accustomed to following their own lines of inquiry. Bardeen hated making them work on problems that were not of their choosing. He could be blunt when necessary, but he did not enjoy it. Years later William Whitmore sent Bardeen some "memories of the day at NOL in 1942 when you asked me if I knew anything about hydrodynamics." Whitmore had replied to Bardeen, "Not particularly." Bardeen then handed him the standard textbook, Sir Horace Lamb's *Hydrodynamics* (1895), "and said 'start learning.'"

A letter from Bardeen to the chief of the Bureau of Ordnance shows that Bardeen tried at least once to break out of his bureaucratic straightjacket and terminate his contract. He was not successful. His restlessness had increased so much that by the summer of 1944 his superiors noticed he was taking less initiative in his assignments.

Because his work was unfulfilling, Bardeen's recreation time meant more to him than ever. In good weather, he tried to hit the links with golf-playing colleagues. On winter evenings, he bowled with the NOL team. Bardeen was "one of the NOL's low-score golfers and high-score bowlers." He won a medal for having the highest average on the NOL bowling league.

Bardeen's greatest pleasure came from his family. He often sat with Jimmy or Billy on his lap, reading to them, or catching up on his own reading. Billy memorized one book that he especially loved about train engines. "John had to go through it very methodically," Jane recalled. If he missed even a word, "he'd get scolded."

Not all the family news was happy. In December 1942 the Bardeens learned that Helen had passed away. As executor, John administered her small estate to preserve it for Glenis. Bill and Charlotte sent Glenis home to her father in Canada. On another family front, unbeknownst to John, his younger brother was developing a drinking problem. Tom often visited John while on business travel to Washington, D.C. Security regulations prevented them from discussing their war-related work with each other; John learned later that Tom had modified Gulf's airborne magnetometer to help detect enemy submarines.

In the fall of 1943 Jane discovered that a third child was coming. This time everyone hoped for a girl. Jane's brother Jim, a medical officer stationed in India, wrote, "I am thinking of you constantly and hoping it may be a girl." In the spring, taking both boys, Jane traveled again to Washington, Pennsylvania, for the birth. With her family nearby she could rest assured that someone would look after the boys while she was in the hospital. She also wanted the best possible medical care, for at thirty-seven she knew she was at a greater risk of having pregnancy-related problems.

"Is she pretty?" John asked when Jane phoned on April 25, 1944, to announce their new daughter. "Not very," said Jane. The baby had come without much fuss, but her head had been bruised during the delivery. She quickly recovered. They named her Elizabeth Ann and soon were calling her Betsy. John wrote a note to Billy, admonishing him to "help your mommy take care of Betsy." Billy had been sick, and John tried to jolly him up with news of one of his D.C. playmates. Skipper had gotten "his hair cut short so that it sticks straight up in the air." John explained to Billy that his baby sister would take a while to grow into a playmate for him, but before long she would "learn to smile and roll over and sit up," and that soon she would be able to "walk and talk like other children. . . . Aren't you lucky to have such a nice baby sister and such a nice big brother?" He told Billy that he would "be glad when you and mommy and Jimmy and Betsy are all home again, and you can do tricks with me."

On the tenth of June, John traveled to Pennsylvania "to help bring the family back." Once they were all at the apartment, Betsy was installed in a crib in her parents' bedroom. John took special delight in his new daughter. Jane suspected that had she had a third boy, she would have been "invited to have another child."

As the war drew to a close in the spring of 1945, Bardeen planned his return to academic life. In May he wrote to Jay Buchta, who was still the chair of Minnesota's physics department, and proposed returning with a salary higher than the $3,200 he had been earning when he left. "With my present family responsibilities, I don't see how I could get by on that salary." Arguing that "my experience here will be of great benefit to my teaching," he asked for a raise and explained that "my value to the University will be at least as great as if I had stayed on and obtained the normal salary increases." He also mentioned that he had been approached by Bell Labs regarding a position in solid-state research.

Bardeen was bitterly disappointed by Buchta's response. Although the physics department enthusiastically supported his request for a raise, university administrators could only be persuaded to raise his salary to $4,000. It was considerably higher than his former Minnesota salary, but not comparable to what Bell was offering. According to Al Nier the university did not yet recognize the importance that solid-state physics would soon have in the nation's economy.

Certain industries were now willing to invest heavily in basic science, which had received a boost of confidence during World War II because of the dramatic success of radar and atomic weapons. Physicists of Bardeen's caliber were in great demand. Solid-state physics was growing rapidly. The field mushroomed after the war with the advent of new research materials, like pure samples of silicon and germanium, or pure mercury isotopes, and new research technologies, such as nuclear magnetic resonance and the helium liquefier developed by Samuel Collins. New sources of funding were attracting many workers to solid-state physics. The field grew so rapidly that in 1947 the American Physical Society added a Division of Solid State Physics.

Bardeen described to Buchta the Bell Labs offer that "Bill Shockley and Jim Fisk have been trying to interest me in":

> The plans are to set up a section of about 50 people to work on problems of the solid state, with emphasis on electronic processes in crystals (semi-conductors, varistors, thermistors, etc.), magnet-

ics, piezo-electricity, and related problems. The research is to be of a fundamental nature, the practical applications being carried out by other groups. It sounds very attractive, but I am doubtful whether really fundamental work can be carried out in an industrial laboratory.

Mervin J. Kelly, the new executive vice-president, who in 1936 had hired Shockley from MIT, had created the new solid-state department partly in response to the wartime study of crystal detectors. He had noticed that cutting-edge research on semiconductors was being exchanged freely throughout a network of research institutions that included potential competitors, such as General Electric, Sperry, Westinghouse, Sylvania, and DuPont. "All of this art has been made available to a large sector of the radio industry," he wrote in 1943. He recognized that Bell Labs would soon have to confront strong competitors armed with state-of-the-art electronics knowledge.

Kelly modeled his new solid-state department on wartime labs such as Los Alamos Scientific Laboratory, which Kelly had visited toward the end of the war. The semiconductor subgroup to which Bardeen would be assigned was a small unit of the solid-state department. Kelly wanted each of its subgroups to function as a multidisciplinary unit. Bardeen later told an interviewer about Kelly's vision. "He thought that getting all these people together working would add something more than the individuals could do and I think that turned out to be true." Bardeen would be the second theorist with Shockley in the semiconductor subgroup. Shockley, the head of the subgroup as well as co-head of the larger solid-state department, needed a theorist who understood surfaces. Bardeen's doctoral thesis on metallic work functions made him one of the world's experts in this area. The team would also include two experimental physicists, Walter Brattain, John's old bridge enemy, and Gerald Pearson; a chemist, Robert Gibney; a circuits expert, Hilbert Moore; and two technicians, Philip Foy and Thomas Griffith.

Bardeen learned more about the position when he visited Bell Labs to explore the opportunity on May 19, 1945. Kelly made him a verbal offer, emphasizing that the focus of Bardeen's work would be on basic research, although there would always be an eye toward practical applications of potential importance to the communications industry. With no teaching load, Bardeen would have twice as much time to devote to basic physics research as at Minnesota.

Bardeen's formal offer described his position as "Member of Technical Staff associated with our Physical Research Department at our Murray Hill laboratory." His official charge was "to carry on research work in the physics of the solid state." The salary—"$550 per month for our normal five-day week," $6,600 a year—was over 50 percent more than what Minnesota could offer. The letter also stated that the offer was not intended to interfere with war work but rather was meant as a postwar offer. Bardeen was given three months to decide whether he wanted to accept.

Six weeks later Bardeen had his decision. "It was a difficult choice to make," he wrote Buchta on June 11, "because I enjoyed my work and associations at Minnesota, and I like living in a University community." But "BTL appears to offer better opportunity for professional development." For one thing, the Bell Labs job was in solid-state physics. He said he expected physics at the university to be centered on nuclear physics, a field in which he would be "at a considerable disadvantage competing with those who have been fortunate enough to have worked on nuclear physics during the war." Bell was also offering ample research support and a generous salary. Remembering how he had had to struggle just to attend an occasional conference during his time at Minnesota, he told Buchta, "It might be a good idea for Minnesota, recognizing its relative isolation, to encourage (financially) attendance at meetings."

Bardeen then crafted a carefully worded memorandum to the officer-in-charge of the NOL, Captain W. G. Schindler, asking permission to leave the NOL. He explained that a release from military projects "which have little chance of being of any benefit in the present war" would allow him to apply his talents to research that would "be of greater long-term benefit." Playing on the concern that Germany and Russia might become competitive with the United States in the area of communications technology, he hinted that his research on semiconductors, which "are becoming of increasing importance in both standard frequency circuits and radar," would be more important for the country's national security than his present work at the NOL. "The Navy, being a large consumer of electronic equipment, should be one of the chief beneficiaries of the research program" that he was joining at Bell Labs. He characterized Bell as one of the most important centers in the world for such research, being "far better equipped than any other research organization for fundamental work in these aspects of solid state physics."

Noting that he was one of only five or six theoretical physicists in the United States who were well trained in solid-state theory and that "no more will be trained for a long period of time" because of the war's interruption in the education of physicists, he speculated that one reason for Bell's generous offer was the "very limited supply of qualified personnel" available for solid-state research. He also expressed his concern that his four-year hiatus from research might already have dulled his ability to make first-rate contributions to the new field. "Because of the long hours of work," he had "not even been able to keep abreast of current developments." Bardeen wrote Schindler that he "strongly felt that any unnecessary delay in his [Bardeen's] return to his own field of work should be avoided."

Bardeen also used the opportunity of his resignation to make a gentle plea that the government respect its scientific resources. Because of the war, "the armed services have control of scientific talent in this country." He said he hoped "this control will be used wisely, and that they will not jeopardize the scientific growth of the country in pursuit of narrow goals."

Rumbaugh, who was still Bardeen's direct supervisor, rephrased his friend's argument in a subsequent memo to Schindler on July 3. He outlined the bureaucratic steps toward Bardeen's release—from the transfer of his duties to the completion of final reports for the NOL historical officer. Rumbaugh commented that "Dr. Bardeen's decision to resign is especially regretted in view of his international reputation as a versatile theoretical physicist and in consideration of his many outstanding contributions over a period of four years at the NOL in acoustics, magnetism, oceanography, degaussing, simulation, mine sweeping, torpedo research, and other important phases of underwater ordnance."

Don Marlowe, assistant technical director of the NOL, "really worked on" Bardeen to persuade him to stay, but Bardeen would not be moved. Though sincerely sorry to see him go, Captain Schindler approved Bardeen's request. "I consider Dr. Bardeen's request [to resign] to be sound and reasonable and hope that he will find his new work pleasant, productive and most satisfactory. I feel that the Laboratory is losing one of its best men but that the Navy will continue to benefit from his work wherever he may be."

In November 1945, Captain Schindler recommended Bardeen for the Distinguished Civilian Service Award for "outstanding contributions to a large majority of the programs in which the Laboratory was involved," noting his valuable study of pressures and acoustic fields in mine development, and judging his "exceptionally keen" analysis of the influence of magnetic fields to be of "inestimable value in the design of magnetic mines and magnetic mine sweepers." Schindler added, "It would be hard to overestimate the value of the contributions that Dr. Bardeen has made to the Research Program of the Naval Ordnance Laboratory, since there are few problems that have not benefited either directly or indirectly from his keen insight, sound judgment, and exceptional scientific ability." James Forrestal, the secretary of the navy, did not, however, approve the award on the grounds that Bardeen's "accomplishments are not considered sufficiently outstanding to warrant the Navy's highest civilian award." Forrestal offered instead the slightly less prestigious Meritorious Civilian Service Award. Bardeen was already established at Bell Labs by the time he learned of the honor.

8

The Transistor

On Tuesday, October 9, 1945, the Bardeens closed the door of their Fairfax Village apartment for the last time. They had shipped ahead most of their belongings. Still their Ford was packed full. It was 11:00 A.M when they finally drove off. They were elated to be done with the war and their transient lifestyle.

They were driving to Summit, New Jersey, a prosperous residential town known during the Revolutionary War as Turkey Hill. Summit offered excellent schools, parks, and convenient transportation; it was a stop on the Hudson Tubes, the subway to New York City. From Summit one could easily catch a bus to Murray Hill, where Bell Laboratories had built its new research campus.

The Bardeens' house in Summit was their first major financial investment. A comfortable Dutch colonial at 5 Primrose Place, it featured a sun porch, cellar, and detached garage. The children would be able to romp and play in the yard, which had swing sets and "monkey bars." Betsy was now an active toddler, Billy a mischievous four-year-old, and Jimmy a "solemn" seven-year-old. Jane planned beautiful landscaping for the house. Later she would report with excitement when the azaleas bloomed and bemoan their fate when they were damaged in a storm.

Bad luck dogged the family throughout their move. "The nightmare really started Monday," Jane wrote to her mother. The moving company had assured Jane "as late as 11 A.M. Monday that they couldn't come until Tuesday." So the family continued to pack carefully and methodically. But an hour and a half later the movers phoned, "when we were in the midst of lunch, with much packing remaining to be done." They said that "they would be right out." Jane worked herself to a frazzle to meet the shortened schedule, and "after a wild three hours it was all done and we were left with an empty house." The revised plan also included meeting the moving van at the new house in Summit at 8:30 A.M. on Wednesday. The family members spent Monday night with various friends. After Betsy fell asleep, Jane "crept back to the house to finish packing the remainder."

The next day the nightmare continued. As they drove northeast on Route 1, the Ford developed a flat tire. They "found 35 miles an hour very slow going." Tired and frustrated after their exhausting drive, they stopped for the night at Princeton Junction, about 50 miles southwest of Summit. On Wednesday morning the family "got up at 6:00 A.M. and drove to Summit before breakfast," arriving on time. But the moving van did not arrive until midday. Worse than that, "we found the house very dirty." The previous owner had "not kept his part of the agreement to put the house in good condition before turning it over to us." His belongings, even some furniture, were "still scattered everywhere."

The sale had not yet closed, and the Bardeens were not sure they could legally take possession of the house. The seller, who Jane described as "a stinker and a complete heel," was embroiled in a bitter divorce. His wife had left to stay with family in Michigan after "he hit her and broke her jaw last week." They consulted a lawyer, who advised the family to move in, even though they did not yet have clear title.

Bringing the estranged husband and wife together to sign the necessary papers would take time. Removing their belongings and cleaning up their mess took elbow grease. The first evening Jane and John "scoured the bathroom for several hours." Their scrubbing was made more difficult by a lack of hot water; the ancient furnace was on its last legs. "Baths are infrequent and water must be heated on the electric stove for dishwashing and cleaning." Bill Shockley dropped by to help, but mainly distracted them with his storytelling.

At first it was difficult to eat meals at home, for the family had no refrigerator. Nor did they have all their furniture. The pieces

Jane had stored in Minnesota were still waiting to be shipped. To top it off, Jimmy, Billy, and John caught terrible colds and needed nursing, while Betsy clung to her mother all the time. Jane wrote stoically that the world would probably "look brighter next time I write."

Ten days later she could report that she had found a woman to help with the cleaning. Although the house still needed considerable repair, Jane was optimistic that once they had finished "shoveling out the dirt" it would be "attractive and comfortable to live in." John helped with the housework, but he was not adept. Jane grew exasperated when he repeatedly "let the fire go out in the furnace."

Soon, however, life in Summit would become easy and rich for the Bardeens. A few minutes' walk brought the children to Brayton Primary School, where Jim entered first grade and Bill attended preschool. On warm summer days the children could swim and enjoy frozen treats at the local pool. The family enjoyed walking along the trails of the Watchung Mountain Reservation, half a block from Primrose Place.

The house proved very comfortable. It always seemed filled with young voices and laughter. The children were the center of attention. John was a "comfortable presence" for them. Not inclined to fuss over his children, he made it a point to be at their special events. At Betsy's fourth birthday party, he "tried to get some color films of the children" playing with their bubble makers out in the sun.

John's half-sister Ann was a regular visitor in Summit. She marveled at her brother's power to concentrate on his journals or the history books he enjoyed while the children played noisily around him, often climbing in and out of his lap. Betsy later said that she and her friends never felt inhibited by her father's presence or by the fact that he was reading. They did not hesitate to interrupt him. He sat through it all, apparently unperturbed.

Ann was completing her medical internship in Jersey City, about 30 miles from Summit. Charles Bardeen, Ann's (and John's) father, used to boast of the number of women who graduated from Wisconsin's four-year medical college, pointing out that out of 25 students in the first graduating class, six were women. But he discouraged the women in his own family from studying medicine. "He felt that medicine 'hardened' women," said Ann. "He thought women would have an expensive medical education and then get married and have a family and not really use it," John later explained. When Helen expressed interest in studying medicine, Charles diverted her into nursing. But Charles passed away before Ann entered college and she felt free to study medicine. In 1945 she graduated from the medical school Charles had created; she then served a residency there in anesthesiology. Afterwards she accepted the internship in Jersey City. Eventually she would follow her father's example and teach at the University of Wisconsin.

In New Jersey, John had many opportunities to keep up with his sports. He often played golf with his old friend Walter Brattain, and he found time for squash. Joining a Bell Labs bowling league, he bowled often enough to make it worthwhile for him to invest some Christmas gift money on a custom bowling ball.

As in Minnesota, bridge was one of the ways in which John and Jane entered Summit's social community. In their early days there, they belonged to a bridge club that included Brattain and his wife Keren. Jane found Walter's competitive attitude so stressful, however, that she used the fact that she and John caught colds one January as an excuse to quit the club.

Jane enjoyed Keren, with whom she had much in common. Like Jane, Keren had a graduate science training. She had completed her doctorate in physical chemistry at the University of Minnesota, where Walter took his Ph.D. in physics. Bardeen later described the Brattains as "a good match," in which "her calm demeanor" served as an antidote to Walter's "somewhat volatile temperament."

The Brattains were attracted to a simple and wholesome way of life. Keren dressed simply, and "her cooking would have been toward corn muffins," Louise Herring recalled. In their home, the decorations included "a scale and a scythe and a few things like that from western living."

About a year after Bardeen came to Bell Labs, Conyers Herring arrived there with his bride, Louise Preusch, a Barnard College graduate with a mathematics major and physics minor. Herring was assigned to the Physical Electronics Division of the Physical Research Department. Bardeen and Herring continued the warm professional relationship they had established at Princeton and Harvard. Outside work, Bardeen spent less time with Herring than he did with Brattain because Herring's sport was tennis.

In their first years at Bell, the Bardeens occasionally socialized with Bill Shockley and his wife Jean, a warm and gracious woman who was well loved in the community. Fred Seitz vividly recalled an incident in 1938 that illustrated how Jean tried "in a gentle way to hold the reigns on Bill." Shockley had grown a beard, which Jean detested. Just before a visit to Bell Labs by the distinguished British physicist Nevill Mott, Jean asked Bill to please shave off his beard. When he refused, "she grew two big pigtails for the event and tied pink bows on the ends." When Bill expressed displeasure with the pigtails and bows, Jean said, " 'I'll get rid of them if you get rid of your beard,' which he did."

John and Jane saw the Shockleys less often as Jane grew less tolerant of Bill's boastful displays. She was offended by his habit of bad-mouthing Jean in public. Jane began avoiding gatherings where she expected to see Shockley. She recalled a time when Shockley told a group that Jean's inferior genes were the reason their children were less talented than he. Years later Shockley told an interviewer, "In terms of my own capacities, my children represent a very significant regression. My first wife—their mother—had not as high an academic-achievement standing as I had."

Small talk remained difficult for John. At social gatherings he often said nothing at all. Being in a conversation with Bardeen was often more difficult for the wives of physicists, because of the lack of a professional connection. "If you said, 'Aren't we having a nice winter season?' he would say 'Yes.' That would be the end." But while Bardeen's social partner would often be at a loss, "he didn't act as if he was uncomfortable."

Bardeen continued the habit he had acquired at the Naval Ordnance Laboratory of separating his private and professional worlds. In wartime the compartmentalization had been dictated by national security. Now it was self-imposed. He told Jane he didn't want to risk revealing proprietary information, but she knew he just preferred to keep his spheres separate. Jane understood that "he wanted some part of himself for himself." He was nearly always home in the evenings and on weekends, and he made time for family outings and visits. He rarely phoned colleagues from home. Jane also recalled that "when he'd get this far away look in his eyes you knew very well that he was somewhere else." He had drifted into his other world. "Well, it spared him a lot of wasted time," she muttered.

Like his father, John was happy to let his wife rule the family

sphere. Jane did not object. She understood how centrally important family and home were to John and to his work. She was genuinely pleased when he said he was glad to "be doing some physics
again after four years of abstinence." But she could not help feeling
excluded from his "first love," especially when he was embroiled
in a problem. John's low whispery voice plus Jane's continuing
struggle with hearing loss added to her growing sense of isolation.
In later years she resorted to reading lips to deduce what others
said on the basis of context. She was so adept that most people
never knew she was almost completely deaf.

On workdays John usually left Jane the family car for shuttling
the children to their activities and running errands. He made his
own way to Murray Hill, either by bus or in a carpool, where he
enjoyed catching up on local news. Pulling up to the entrance of
the Murray Hill facility on Mountain Avenue was the daily ritual
by which he shifted between his two spheres.

In the first decades after its incorporation on January 1, 1925, the
Bell Telephone Laboratories had occupied an elegant, twelve-story
brick and sandstone building at 463 West Street in lower Manhattan.
The building flanked the western edge of Greenwich Village, a
section of the city populated by artists, musicians, and writers.
Within walking distance of galleries, theaters, coffee shops, and
ethnic restaurants, Bell Labs scientists and engineers could feel the
pulse of the great city. Those whose labs overlooked the Hudson
River could watch steamboats and barges at work, or the crowded
ferries transporting commuters between New York and Hoboken,
New Jersey. Many commuters stood in line on the ferry to grab a
doughnut and steaming cup of coffee in an effort to save a few
minutes of their busy mornings. In the early years, the West Street
building was a spacious home for Bell Labs. But even before the
mid-1930s, the facility began to outgrow its accommodations there.
The dust and grime of the city, and even its noise, became serious
problems, especially for those working on electronics. Many
projects expanded into nearby buildings, such as the Nabisco Building, which had been designed for "baking purposes."

A few executives began to dream of a "country lab experiment,"
a place outside the city aimed at supporting creativity. The idea
emerged to build a facility in New Jersey reminiscent of Edison's

Menlo Park. This was, however, the era of the Depression, and a lack of funds postponed the expensive experiment.

The idea to move the laboratory into a cleaner and larger space reemerged when the Second World War brought new funds and a much larger staff to Bell Labs. In 1941 the massive structure in Murray Hill known as "Building 1" was hastily completed in time to house the wartime exodus from the cramped West Street building. The new mazelike array had been designed for several thousand scientists, engineers, and technicians. Many of its interior walls were movable, so that spaces could be repositioned to fit the needs of particular projects. To the physicist and writer Jeremy Bernstein, the functional space appeared as "a gigantic technological warren within which, at least at first sight, everything resembles everything else." Initially Bernstein experienced the place as so menacing that he worried "in the process of going from one laboratory to another, I would take the wrong turn and never find my way out."

On Bardeen's first day of work at Bell Labs on Monday, October 15, 1945, he showed up with the head cold he had caught during the family's move to Summit. He spent the morning at West Street completing paperwork and having a routine medical checkup. Then he passed through the regular initiation ritual for all new employees, which included selling his patent rights to the company for one brand new dollar bill. "I really feel that this is only fair," Bardeen later told a reporter. "People can cooperate without worrying who is going to get the patent rights and this promotes a much freer exchange of information."

In the afternoon Bardeen found the small fourth-floor office to which he had been assigned in the B wing of Building 1. Although the Murray Hill facility was much roomier than West Street, there was still a premium on space. Many temporary workers were on site, completing the construction or continuing wartime projects. Bardeen's two office mates, the experimentalists Walter Brattain and Gerald Pearson, had been studying semiconductors for over a decade. Sharing an office with experimentalists would have bothered some theorists, but Bardeen welcomed the opportunity to discuss practical matters with them regularly.

Bardeen and Brattain would become close partners at work. Bardeen valued Brattain's integrity and physics know-how. He also thoroughly enjoyed Brattain's unrestrained and colorful use of language, reminiscent of his brother Bill, or of Dutch Osterhoudt. The outgoing Brattain, in turn, appreciated Bardeen's modesty, thoughtfulness, and brilliance. He did not mind his reticence. Brattain's graduate school friend Walker Bleakney had cautioned him about Bardeen. "You'll find that he doesn't open his mouth very often to say anything, but when he does, you listen!"

It took Bardeen little time to become fully immersed in the problems of semiconductors. He approached them in both ways Wigner had trained him to begin any problem: breaking down larger problems, and reading all he could about what others had done before him. Of particular interest to Bardeen was what happens at the interface between two different semiconductors, or between a semiconductor and a metal. There was a large body of literature on this "surface" question. To answer well required the application of quantum mechanics.

Some of the relevant physics was being discussed by the members of a weekly Bell Labs study group that had started to meet in the early 1930s. In those years, employees used the time freed up during Depression-era "layoff days" to educate themselves about the new quantum theory of solids. The group, which now included Bardeen, Shockley, Robert Gibney, Pearson, and Brattain, studied all the important papers on semiconductors, including those written up as military reports during the war.

A great deal had been learned about semiconductors since Alan Wilson's pivotal work in the early 1930s. Silicon and germanium had become the focus, partly because they are elements rather than compounds, making them the two simplest semiconductors. During the war they were used as rectifiers in radar detectors to convert high-frequency alternating signals into direct-current signals. The conversion to lower frequencies was necessary so that the signals could be amplified using simple equipment.

Much of the wartime research had focused on improving the properties of the so-called "rectifying interface," the tiny point of contact between a thin metal wire called a "cat whisker" and a slab of semiconductor, either silicon or germanium. Electrically, this metal–semiconductor interface played the same role in an electronic circuit as a vacuum tube diode, but it could function at higher frequencies. The research also dealt with "forming" the metal whiskers, producing pure semiconductor crystals, adding impurities ("doping") to enhance conduction, manufacturing good crystals of silicon and germanium, and gaining an understanding of the physics involved. The physics of the central device in the radar detector, the crystal rectifier, turned out to be virtually the same as that of the first transistor.

Bardeen reread the important prewar papers on semiconductor rectification by Nevill Mott in England, Walter Schottky in Germany, and Boris Davydov in the Soviet Union. He also read as much of the wartime literature that he could find. Because Bell Labs had been one of the major contractors to the wartime MIT Radiation Laboratory (Rad Lab), the Labs had on hand an almost complete set of the wartime reports on rectifiers. They included classic papers on rectification by Hans Bethe, who had been on the staff of the Rad Lab before moving to Los Alamos, and reports on germanium by members of Karl Lark-Horowitz's group at Purdue. Using a specially developed superpure germanium, the Purdue group had built a rectifier having exceptionally low conductivity in the "back direction," in which current flow was lower. Also included in the collection were papers on silicon by Fred Seitz at DuPont or at the University of Pennsylvania. The next time John saw Fred, he said, "Now I know what you did during the war."

The essential difference between a triode and a diode vacuum tube is the single crucial element known as the "grid" (see Figure 8-1). In a diode, a heated, negatively biased filament emits electrons that travel to the positive plate. Because the negative electrons can flow only in one direction (i.e., from filament to plate), current can pass only one way. The diode functions as a one-way valve.

The triode can function as a valve too, but it has in addition the possibility of amplifying signals. An element called the grid is placed between the filament and the plate in such a way that electrons must pass through it. Much as a faucet controls the flow of water, the grid controls the flow of electrons by virtue of its voltage. When its voltage is negative with respect to the filament, the electrons' acceleration from the filament is retarded. If the grid is positive with respect to the filament, it accelerates the electrons. In a circuit, this device can cause a signal applied to the grid to become amplified in a circuit through the plate.

Around 1910, AT&T needed a technology that could amplify telephone signals. Theodore N. Vail, AT&T's president, had made the decision to build a transcontinental telephone line connecting the East and West coasts of the United States. Diode tubes, already

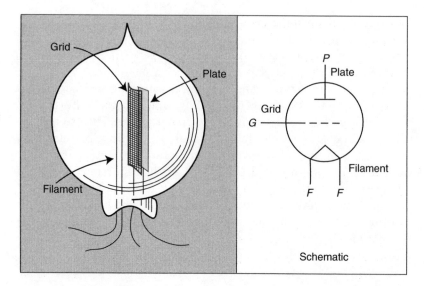

FIGURE 8-1 The triode tube.

in use in telephone switching circuits, could not amplify signals. Triode tubes could in principle do so, but they had only recently been invented, in 1906. When one of the triode's inventors, Lee de Forest, demonstrated his device (which he called the "audion") at AT&T, Harold Arnold, a physicist on the staff, realized that it held the answer to the company's amplification problem. Modifying de Forest's audion, Arnold designed the "repeater" needed for the transcontinental line. Arnold's research group, the first such group at AT&T, subsequently evolved into Bell Telephone Laboratories.

Unfortunately, the very feature that allowed the vacuum tube to function both as a diode and an amplifier—its heated filament— was also its fatal flaw. The filament's work function (the energy needed to remove an electron) had to be overcome so that electrons could be emitted. It was the same problem that Bardeen and Brattain were studying at the time they first met in the 1930s. Thermionic emission of electrons consumes a huge amount of power, especially when many tubes are present in the system. Moreover, the filaments in tubes eventually burn out. For these reasons and others (including the cost, bulkiness, and fragility of glass tubes), Bell Labs looked to semiconductors to replace vacuum tube diodes and triodes.

What made it natural to pursue a semiconductor amplifier was the magic of the interface between a semiconductor and a metal or between two different semiconductors. In a circuit this interface behaves in the same way that the crystal rectifier behaves in a radar receiver: like a valve, it allows current to flow much more easily in one direction than the other. The result of a one-way barrier at the surface, this rectification property of semiconductor interfaces had been discovered during the nineteenth century, but it was not yet understood. It had made possible the "cat's whisker" radio detector, which Bardeen, Brattain, and Shockley had played with as teenagers.

The question arose: If adding a grid to a diode tube can turn the diode into a triode tube capable of amplifying signals, can one add something analogous to a semiconductor diode and create a semiconductor amplifier? And if this were possible, what would function as the grid? Brattain and Shockley were among many researchers who confronted this question in the 1930s. Mervin Kelly, who had been head of the Vacuum Tube Department at Bell Labs from 1928 to 1936, was very familiar with the problem. He encouraged Shockley to pursue it in the late 1930s.

In the late 1930s, Shockley shared a laboratory with Dean Wooldridge (later the "W" of the firm TRW), another physicist that Kelly hired in 1936. Wooldridge vividly recalled a crude experiment in which Shockley tried to convert a copper oxide diode (used as a "click reducer" in the telephone system) into a triode. Arranging two wires so they barely contacted the green oxide on either side of a rusty porch screen and adjusting the voltage applied, Shockley expected the jagged screen to function like the grid of a vacuum tube. The screen, according to Wooldridge, had been "out in the elements for years and years." Shockley was "orders of magnitude away from anything that would work," yet he had before him in the late 1930s "the three elements of a transistor."

Bardeen listened with interest to such stories about early semiconductor experiments, especially as told by Brattain, who would enliven them with his colorful commentary. Brattain loved to tell how one day Shockley came to see him in his lab and asked him to build a workable version of this crude copper oxide amplifier.

The seasoned experimentalist knew it was a hopeless project because there was no physical technique for inserting a screen into a barrier region only ten-thousandths of a centimeter thick. "I laughed at him," said Brattain. "I was quite sure it wouldn't work. But I said, 'Bill, it's so damned important that if you'll tell me how you want it made, and if it's possible, we'll make it that way. We'll try it.'" Brattain was right, of course. "The result was nil. I mean, there was no evidence of anything."

Bardeen also heard stories about Russell Ohl's interesting work just before and during World War II on silicon detectors. A feisty gnome of a man, Ohl was a radio engineer working in Bell's radio lab in Holmdel, New Jersey. By the mid-1920s most radio buffs had replaced their crystal sets with vacuum tube sets, which could better separate stations whose frequencies were close to one another. A few engineers, however, including Ohl, realized that the old crystal sets were actually far more sensitive at high frequencies.

Experimenting with old crystal sets found in secondhand radio shops, Ohl set out to determine which semiconductors work best at the higher frequencies. Trying more than 100 materials, he found that silicon worked best. The trouble was that the individual silicon samples worked erratically, rectifying only when the whisker contacted "hot spots" on their surface. The magnitude of the effect, and even its direction, changed unpredictably.

In one test conducted in 1940, Ohl noticed that a particular silicon sample showed an enormous photovoltaic effect whenever he would shine a flashlight on it. In further study of this mysterious effect, he identified a junction within the body of the sample composed of two kinds of silicon, now called p- and n-type silicon. Further study of this junction might have resulted in an amplifier, but the war diverted Ohl to other problems.

The work that led to the transistor began on October 22, 1945. Shockley asked Bardeen to look over a design that he had sketched in his notebook six months earlier for a silicon "field effect" amplifier. An electric field was applied perpendicular to a thin slab of silicon; the field drew charges in the slab to its surface. In a thin sample, Shockley argued, the field would cause a substantial change in the available charge carriers. In this design, the *field* would play the role of the grid.

But the design did not work in actual practice, and Shockley could not explain why. Drawing on the best available theories about semiconductor rectification, those of Mott and Schottky, Shockley had calculated that for the predicted effect *not* to be observed, the changes in conductivity would have to be less than 1/1,500 of the predicted amount! Stumped, he showed the calculation to Bardeen, who found nothing wrong with it.

Bardeen recognized that the theory must be wrong, or at least incomplete. To explain how was just the kind of puzzle he loved to attack. The problem was both challenging and practically important. In addition, his competitive spirit was aroused, for the problem had stumped another excellent theorist. He worked on it for the next five months.

Bardeen realized that, as in the work function problem, the electrons in this system are more mobile than the ions. This implied that at the surface of the semiconductor the electron wave function (a measurement of the probability of finding an electron there) would extend slightly beyond that of the ions. The result was a small excess of negative charge on the surface, rendering the total charge neutral. By March 1946, drawing on earlier surface-state work by Shockley and Igor Tamm, Bardeen had developed a new theory predicting that a substantial number of the electrons would be trapped in surface states. Unable to contribute to the conduction, the trapped electrons substantially decreased the change in conductivity caused by the applied field. That was why Shockley's field effect design failed.

But were the surface states real? And if so, how did they behave? To progress further, these questions had to be addressed. The solid-state group divided up tasks: Brattain studied surface properties such as the contact potential; Pearson looked at bulk properties such as the mobility of holes and electrons; and Gibney contributed his knowledge of the physical chemistry of surfaces. Bardeen and Shockley followed the work of all members, offering suggestions and conceptualizing the work. "It was probably one of the greatest research teams ever pulled together on a problem," said Brattain:

> I cannot overemphasize the rapport of this group. We would meet together to discuss important steps almost on the spur of the moment of an afternoon. We would discuss things freely. I think many of us had ideas in these discussion groups, one person's remarks suggesting an idea to another. We went to the heart of many things during the existence of this group, and always when we got to the place where something needed to be done, experimental or theo-

retical, there was never any question as to who was the appropriate man in the group to do it.

To better understand the surface states, Bardeen proposed an experiment in which the temperature was lowered enough so that the electrons in the surface states would be "frozen." That way, he argued, the field might be able to get through. Pearson tried the experiment, which involved using liquid nitrogen to cool the semiconductor. When he applied 500 volts, he observed a field effect!

It was a triumph for Bardeen's theory of surface states, but the effect was fifty times smaller than predicted. Bardeen deduced that charge carriers in thin films are less mobile than they are in bulk material. This was a second reason why Shockley's experiment had failed. A successful design would need to get around both obstacles. The group tried many related experiments, but for the next eighteen months they made little progress.

During the remainder of 1946 and throughout 1947, Shockley's engagement with the field effect studies flagged, for he developed new interests during the summer of 1947, when he and Bardeen took an extended tour of European laboratories. It was to be the last time Bardeen and Shockley would share an extended collegial exchange.

The two started out late in June 1947, stopping first in London. Bardeen noted that the city still exhibited "much bomb damage." Food was "in short supply" and suitable lodging was "often difficult to find." Shockley was stimulated by a conference they attended during the first week of July in Bristol on recent work in the new field of "dislocations," the study of departures from regularity in crystals that take the form of linear displacements of the crystal lattice. The fact that the dislocations can propagate through a crystal so captured Shockley's interest that on his return to Bell Labs in August he refocused his research around this phenomenon.

On a Sunday off during the conference, Bardeen brought Shockley along on a visit to Muriel Kittel and her family in Cheltenham. During the war, the Bardeens had become friends with Muriel and Charles Kittel, when both men had worked at the NOL. Kittel had recently accepted a position at Bell Labs. "I told [Muriel]," John wrote Jane, "that you were expecting to assist them any way you could when they get in." Muriel wrote to Jane years later that when "John and Bill turned up at my parents' house to tell me of their pleasure at Charles's appointment at Bell Labs," her parents were surprised and "tremendously touched by such kindness."

While Shockley was learning about dislocations, Bardeen caught up with new research on superconductivity. He discussed the phenomenon on visits in Holland, to the universities in Delft, Amsterdam, Eindhoven, and Leiden. He had productive meetings at Philips Laboratories in Eindhoven with Hendrik G. B. Casimir, Jan Verwey, and others. Unfortunately, there was just one room available in the small hotel Bardeen and Shockley felt lucky to find in Eindhoven. As the room "had only one double-bed," the two men slept "with our heads at opposite ends."

Bardeen wrote to Jane regularly during the two-month-long trip. The separation was almost as long as the one they had experienced during the war. John wrote from Eindhoven shortly before their wedding anniversary. "Sorry we can't be together on the 18th. I will be thinking of you. We will have to have a real celebration on our tenth next year."

It was drizzling when Bardeen and Shockley returned to their room at Queens Gate in London on Sunday, July 19. Despite the rain Shockley made plans to go rock climbing with a British friend and then to spend a few days visiting Edmund Stoner at the University of Leeds to talk about magnetism. While Shockley was gone, Bardeen went to the theater and saw *1066 and All That*, a musical based on British history.

The next Monday Bardeen took the train to Cambridge. As it rattled along, he finished a letter to Jane that he had started on the previous day. He indulged in some early morning grumpiness. "One of my shoes is disintegrating," he wrote. "We have used very little of the food I brought along. I could better have used the space for extra shoes and more bath towels. The latter are seldom furnished in the hotels." On a more positive note, "Bill says I got a raise starting July 1. He is not sure of the amount, but it may be as much as $60 or $70 a month."

In Cambridge, Bardeen visited the Mond Lab, where he spoke with David Shoenberg, the author of one of Bardeen's favorite texts on superconductivity. Shoenberg's graduate student, A. Brian Pippard, was also present. At Shoenberg's suggestion Pippard had studied Heinz London's prewar experiments on the penetration depth of magnetic fields in superconductors. He could draw now on microwave techniques to measure the surface impedance of metals at high frequencies, for during World War II Pippard had worked on microwaves at the British Royal Radar Establishment at Malvern.

Pippard realized that the microwave region was a promising one for studying superconductivity, because microwave energies are intermediate in frequency (between zero, at which super-conductors do not dissipate energy, and the optical frequencies at which they dissipate normally). The London theory, with its radical assumption of the long-range ordering between electrons, had pre-dicted the size of the penetration depth of magnetic fields. If Pippard could measure this depth experimentally, it would be a powerful check of the London theory. During his visit, Bardeen also had the opportunity to discuss superconductivity directly with Heinz London himself.

They decided to go to Paris next. Bardeen wrote Jane that he hoped "to get to Paris next Sunday [July 26] but we don't have any reservations yet." In fact, he confessed, "We don't have any for the Paris–Zurich leg either, but are hoping for the best." After they arrived in France—on "the hottest day in Paris' history"—they visited Louis Néel, one of the experimentalists whose work had put the study of magnetism on a sure footing. Néel "took the not unreasonable attitude that, when in France, you speak French." So on that hot day, the two sat with Néel at an outdoor cafe and "spoke to him in our high school French."

Their last stop was Switzerland, where Bill and John swam in Lake Zürich before visiting Wolfgang Pauli, who in the 1930s had called solid-state physics a "physics of dirt." They followed the visit to Pauli with a few days of vacation in the Alps. Bardeen loved the mountains. "As you must have gathered from the postcards," he wrote Jane, "I like Switzerland very much." In fact, "If I were coming again strictly for vacation, I think I would skip the other countries and come straight to Switzerland." He told Jane that Shockley "bought quite a lot of stuff here in Zürich for climbing in the high mountains," including "nailed boots, windbreaker jacket, and various kinds of mountaineering equipment. The only thing I got is a pair of sport oxfords with corrugated rubber soles for hiking." He justified the expense saying that he could "use them later for golf or all around wear."

Bardeen wrote that he had "learned a lot during the trip, and have picked up some information that may be useful to the Lab. Whether or not it's enough to pay for the trip is hard to judge." In any case, it had been "very hard work—much harder than you can imagine without doing it—but it's also been a grand experience." He added, writing on August 1, "In another week or so I will be home—and glad to be there."

The "magic month" that culminated in the transistor began in the middle of November 1947, three months after Bardeen and Shockley returned from Europe. Brattain had encountered an apparently innocent problem during the course of one of his experiments. Droplets of water condensing on the apparatus were causing a spurious effect. In an effort to avoid the cumbersome two-week-long job of pumping out all the water, he attempted a quick fix. "I'm a lazy physicist," he later reported. "I like to do things in the easiest way." He immersed the system in various liquids including distilled water and was "completely flabbergasted" to find that the photovoltaic effect he was studying increased whenever the liquid was an electrolyte. Soon Brattain was demonstrating these things "to anybody in the group that would listen."

Bardeen suggested that the mobile ions in the electrolytes might be creating a large enough electric field to overcome the surface states. Then Gibney said, "Wait a minute. You've got a potential on there, haven't you?" "Yes," Brattain responded. "Let's vary this thing just a bit," suggested Gibney.

When Brattain and Gibney did so (see Figure 8-2), they noticed that when they used either water or an electrolyte, a layer of positive charge would form on one surface and a layer of negative charge on the other. "We could vary the photo emf [electromagnetic force] from anywhere to a very large value to zero to change its sign." Suddenly the team realized that they might be able to build a field effect amplifier after all! Brattain and Gibney described their findings in a patent disclosure dated November 20. The major remaining problem concerned how to project the change caused by the varying potential onto a second circuit.

By now Shockley was beginning to behave somewhat strangely with respect to the group. He kept to himself more than before. Although he continued to offer suggestions, he saved most of his energy for his own work on dislocations and on the flow of electrons through alkali and silver halides. In the privacy of his home, he began to sketch designs for a different approach to the solid-state amplifier. He told no one else in the group about them.

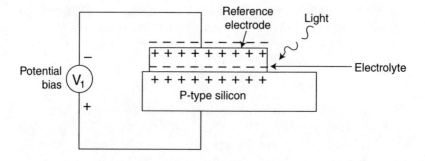

FIGURE 8-2 Gibney's suggestion to vary the potential.

In contrast Bardeen engaged himself fully with Brattain and Gibney's new results. His mind played constantly with the possibilities. On Friday morning, November 21, he walked into Brattain's office and "suggested a geometry for making an actual amplifier." Brattain responded, "Come on John, let's go out in the laboratory and make it."

The two had begun to work together regularly in Brattain's lab. Bardeen loved to peer over Brattain's shoulder and watch him prepare his experiments. Sometimes Bardeen would offer Brattain a hand in routine tasks, such as recording measurements or holding a piece of apparatus in place while Brattain soldered it. Bardeen typically would probe Brattain for clarification on questions of technique or material. In pondering the data Bardeen often made interpretive suggestions based on his deep understanding of the physics. Brattain valued that input. The interplay between the two was reminiscent of Bardeen's work with Bridgman, but it was more fun because of Brattain's sense of play and adventure. The Bell Labs environment supported the collaboration with countless resources hard to come by in an academic lab—technicians, materials, equipment, and even patent attorneys.

Bardeen's proposal of November 21 (see Figure 8-3), although not yet an amplifier, contained most of the elements of the first transistor. Bardeen had modified the basic structure of Brattain and Gibney's promising experiment, using results from the wartime radar program. As Brattain later recalled the design:

The geometry was essentially one of taking a point contact, some-

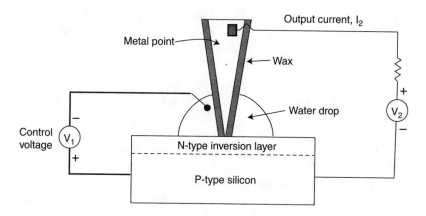

FIGURE 8-3 Bardeen's proposed field effect amplifier, November 21, 1947.

how insulating its surface, and putting the point down on the semi-conductor surface, and then surrounding it with a drop of an electro-lyte to which we made contact with another metal, thus hoping to modulate the flow of current from the point to the semiconductor by a fine electric field through the electrolyte.

Bardeen suggested that for the electrolyte they try a drop of distilled water, and for the semiconductor, a slab of p-type silicon with an n-type "inversion layer" (a thin region that forms in cer-tain circumstances near the surface of a semiconductor). He was well aware of the phenomenon of the inversion layer after his surface-states work in 1946. Indeed, he held a patent on the idea. He hoped to use the fact that in such a layer most carriers of charge are opposite in sign to those in the bulk material. (Thus if holes carry most of the current in bulk samples, electrons would carry most of the current in the inversion layer, and vice versa.) In addi-tion, the wartime research had shown that much larger changes in conductivity occur in bulk samples than in thin films. Thus by using an inversion layer contiguous with the bulk material, instead of a thin film deposited on its surface, one could perhaps get around not only the difficulty of depositing a very thin layer of semi-conductor, but also that of the low mobility of charge in thin films.

For the contact, Bardeen proposed using a sharp tungsten point. Brattain suggested insulating the point from the water using a thin layer of wax. The big advantage of such a "point-contact" design was that there was "a simple way in which you could do it," Bardeen explained. It was "the sort of experiment you could set up and do in a day" because the technology of making point contacts had been refined during the war.

Brattain and Bardeen knew they were getting close when they tested Bardeen's design. It worked to about 10 cycles, giving amplification of current and power, but not of voltage. Brattain said, "I told my driving group that night, going home, that I felt that I had taken part in the most important experiment I had ever taken part in my life." As Bell Labs had not yet released the news, "the next evening I had to swear them to secrecy."

Over the next month Bardeen and Brattain worked steadily on the semiconductor amplifier problem, on occasion even on weekends. They tried to change only one or a small number of features at a time to keep track of their progress as they explored countless variations of materials and geometry. They tried gold instead of tungsten for the electrodes; Duco lacquer in place of wax for the insulation; and water, tetramethyl ammonium hydroxide, and a series of gels for the electrolytes. They achieved their best results using an electrolyte that was a gooey mixture they called "gu," made of glycol borate and glycol bori-borate. The viscous fluid, taken from a lead battery, worked better than water, which tended to evaporate. For the geometry they settled on a configuration in which the electrode was a gold ring set down on the gu, with the sharp metal electrode passing through. They tried using a drop of gu on a junction of p- and n-type silicon and pressing down two points instead of one on the silicon slab.

The most critical change in the design resulted from a suggestion from Bardeen on Monday, December 8. During a lunch-time discussion with Brattain and Shockley, he said he thought they should try germanium instead of silicon. In particular, Bardeen thought that by using the special "high-back-voltage germanium" doped with tin that researchers at Purdue had developed during the war, they might improve the amplification significantly because of its high resistance to currents flowing in the "back" direction. "John Bardeen was great at coming up with approximate guesses of this kind and making the right guess," Brattain recalled.

Brattain happened to have a piece of the special germanium in his laboratory, so he and Bardeen tried the experiment that very afternoon. Brattain pushed the gold point into the germanium through a drop of the gu, to which they applied a few volts. They were startled to measure both a voltage amplification of two and a dazzling power amplification of 330! They also noticed a mysterious change in the direction of the current. "This is the opposite of what one might expect," Bardeen noted.

Might *positive* carriers—holes, rather than electrons—be responsible for the transport? Brattain recorded, "Bardeen suggests that the surface field is so strong that one is actually getting P-type [i.e., positive] conduction near the surface." It was a strange idea, but it fit their data. Could they be producing an inversion layer electrically? Could negative ions in the electrolyte be inducing a layer of positive charge consisting of holes just beneath the surface? Partly through accident and partly through good experimentation, the two had managed to increase the population of holes at the surface. The negative voltage on the tungsten point was indeed driving the electrons away. Increased by the electric field from the droplet, the hole population had raised the conductivity, giving the observed power gain.

Gradually they realized that the holes, the ghostlike particles that had intrigued Bardeen ever since he learned of their existence, were functioning as the grid. These holes, alluded to by Peierls in 1929 and well described by Heisenberg and Wilson in 1931, were the key actors in a new effect they would come to recognize over the next several days.

Two days later, on December 10, Bardeen and Brattain repeated the experiment using a specially prepared germanium sample. They saw an even greater power gain—a dramatic factor of 6,000! But as the frequency response remained poor, the device could not yet be used to amplify the range of the human voice.

They soon realized that the sluggish frequency response was caused by the ions moving too slowly in the gu. They decided to dispense with the electrolyte and instead take advantage of an effect that Brattain happened to notice earlier that day. When he had applied a steady electric field to the glycol borate, the electrolyte had etched the surface of the germanium and caused a green oxide film (similar to rust) to form. "I can remember the green color under the glycoborate," Brattain recalled. "So, seeing this film, we thought, 'Ah. This oxide film must be insulating. If it is, we can form the film and put metal electrodes right on top of the film, get this field effect without the electrolyte and get [power gain at] the higher frequencies.'" Because the film was rigid, unlike the gu, the frequency response should improve. Because it was thin, the electric field should be higher. Thus, they argued, the hole population should be substantial enough to achieve a higher power gain.

For the next experiment, performed on December 11, 1947 (see Figure 8-4), they used the oxide film in place of the electrolyte. Brattain reversed the currents using an n-type rather than a p-type sample. To serve as the voltage plate, Gibney carefully evaporated a gold spot on the oxide film, leaving a hole in the center of the spot for the metal point to pass through to the germanium. When an electrical discharge spoiled the hole, Brattain compromised: he separated the two neighboring circuits by placing the point near the edge of the gold. Once again all the elements for observing a field effect seemed to be present.

But the experiment did not work as planned. When on December 15 they applied a positive voltage to the gold spot and a negative one to a point next to the spot, Brattain saw modulations of both the output current and voltage. There was no power gain, but the voltage *doubled*! It worked the same at high frequencies. "This voltage amplification was independent of frequency [from] 10 to 10,000 cycles," Brattain wrote. Curiously, the modulation occurred only when the gold was positively biased: "I got an effect of the opposite sign," Brattain recalled.

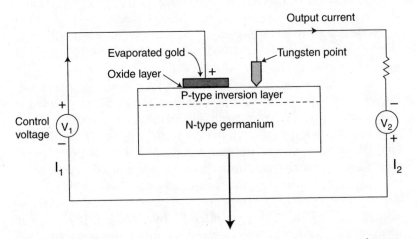

FIGURE 8-4 Bardeen's and Brattain's field effect model of December 11, 1947.

"We knew that something different was happening," said Bardeen. It became clear that "holes were flowing into the germanium surface from the gold spot and that the holes introduced in this way flowed into the point contact to enhance the reverse current." This, Bardeen recalled, "was the first indication of what later was called the transistor effect." Based on the holes introduced into the germanium, the device became known as the "bi-polar" transistor. It worked on a completely different principle than the field effect device that Shockley had sketched more than two years earlier.

"What we didn't know then," Brattain explained, recalling that classic experiment, "was that the oxide that was formed was soluble in water." They had not expected that "when we washed the gu off, we washed the oxide off too." This simple property of the material being used—the solubility of the oxide that forms on germanium—would allow Bardeen and Brattain, but not Shockley, to patent the first transistor.

Had they not switched from silicon to germanium, the oxide would not have washed off. Had they stayed with silicon, they might eventually have arrived at a field effect transistor similar to the MOS-FET (metal oxide–silicon field effect transistor), in common use today. Bardeen would later joke that his and Brattain's invention of the point-contact transistor slowed the quest for the field effect transistor for several years. But as Nick Holonyak later remarked, that "really could not have occurred without all the semiconductor work the point-contact transistor set into motion."

Did Bardeen and Brattain fully understand the behavior of the holes in these historic experiments in December 1947? Did they, for example, realize that the holes flow not only along the surface but into the bulk of the semiconductor as well? These questions have become the focus of a recent debate among a small group of engineers and historians. The available documentation cannot completely resolve the issue.

One party in the debate argues that Bardeen's training as an engineer closed his eyes to the possibility of seeing this "hole injection." We consider this implausible. Bardeen was, after all, fully trained as a theoretical physicist, with a Ph.D. from Princeton and three years of postdoctoral study at Harvard. Moreover, the clue about the flow of holes into the bulk had been published almost a decade earlier, in an abstruse mathematical paper written by Boris

Davydov. Bardeen probably read this paper, for he was in the habit of reading all he could of the relevant physics literature. Even had he not grasped all the implications of Davydov's work, he would have recognized its relevance. Bardeen was among the few physicists who at that time had the proper training to understand Davydov's work.

A more reasonable interpretation, supported by much circumstantial evidence, has been put forth by Holonyak, who has argued that Bardeen understood how the holes behave either immediately or very soon after the invention of the transistor. Characteristically, Bardeen would have been cautious about publicizing his conjecture until there was adequate proof that holes flow not only in the surface but also in the bulk of semiconductors.

The debate seems to boil down to how long it took for Bardeen and Brattain to realize that holes flow into the bulk. "There was a period in which one was not sure whether these holes were flowing in the space-charge layer or whether this was radial injection into the crystal," Brattain told an interviewer in 1964. A year earlier Bardeen had explained in an interview that "our first thought was that the holes were being introduced just in this surface layer here and that the field of the collector pulled them in, so that the action is mainly confined just to this surface layer. We did a number of experiments to try to find out just what was going on here." What these experiments indicated, Bardeen said, "was that you're introducing excess carriers into the surface to lower the conductivity in the neighborhood of the point." Throughout this period of uncertainty, Shockley was conducting his own studies. These "suggested that he could inject holes through a p-n junction into the material."

In any case, this issue of whether the holes flow into the bulk was soon cleared up by the definitive experiments of John Shive (discussed in Chapter 9). By mid-February 1948, Shive had demonstrated "that holes could flow through the bulk, in conformity with Shockley's ideas."

Bardeen and Brattain spent a few days trying to improve the gain in their experiment. "The observed effect" was not very large and no real power amplification was seen. Bardeen suggested "that if we really were introducing carriers into the surface and wanted to get a real large interaction, you have to get the electrodes extremely close together, within a thousandth of an inch or so." He therefore proposed a geometry (shown in Figure 8-5) that would

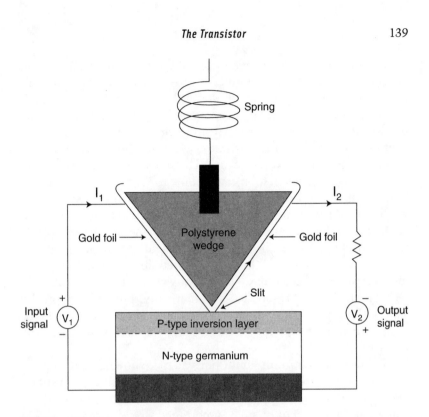

Spring

I_1 I_2

Gold foil ⟶   Polystyrene   ⟵ Gold foil
                    wedge

Slit

+
Input
signal V_1 P-type inversion layer V_2 Output
         −   +   signal

N-type germanium

FIGURE 8-5 The first transistor, December 16, 1947.

allow the holes to flow closer to the input signal. The wire and gold
spot would be replaced by two metallic line contacts set down on
the germanium with a separation between them of only a few thou-
sandths of a centimeter. (At this stage they believed that line contacts would give a stronger effect than point contacts. Later they
realized that point contacts would do if enough current flowed in
them.)

Following the suggestion of Bardeen's, Brattain created the two
line contacts and separated them by about 0.004 centimeters by
wrapping a piece of prewar gold foil around the apex of a polystyrene triangle. "I slit carefully with the razor until the circuit
opened" and filled the cut with wax. Using a spring, he pushed the
triangle down on the germanium. By wiggling it "just right," he
succeeded in making contact with both points. "I could make one
point an emitter and the other point a collector," Brattain recalled.

This particular experiment worked the first time they tried it, on Tuesday afternoon, December 16, 1947. In one of their first trials, at 1,000 cycles per second, they achieved a power gain of 1.3 and a voltage gain of 15. "I had an amplifier with the order of magnitude of 100 amplification clear up to the audio range," Brattain boasted. "It would sometimes stop working, but I could always wiggle it and make it work again." The transistor was born.

That evening, when John came home to Primrose Place, he told Jane about the results. She recalled, "It was one of those days he would drive in and put his car in the garage, which was in the back of the lot, and then come in the back door." As usual, "the surest way to find me was in the kitchen," she said. Wading through the "kids all over the floor," John murmured almost inaudibly, "We discovered something today." Jane hardly glanced up from the sink. "That's great," she responded automatically. But as he typically said nothing about his work, she knew it had been a special day. Some time later she found out "that the something was the transistor."

Bardeen and Brattain could not yet risk showing their invention to Bell executives. Brattain recalled that "at each level of supervision. . . . there was hesitation about informing the next level for fear an announcement of such importance would turn out to be a fluke and that their faces would be red for prematurely claiming something so important." They spent another week verifying everything.

By Tuesday afternoon, December 23, the group was ready to announce the discovery. They gathered nervously for the demonstration, held in one of the executive offices at Murray Hill. Brattain had arranged the apparatus so that some observers could speak over the input circuit while others could hear the output signal over earphones, or see it displayed on the screen of an oscilloscope. He described what happened in his notebook on the following day, Christmas Eve:

> The circuit was actually spoken over and by switching the device in and out a distinct gain in speech level could be heard and seen on the scope presentation with no noticeable change in quality. By measurements at a fixed frequency in, it was determined that this power gain was the order of a factor of 18 or greater.

Bardeen also wrote about the demonstration somewhat later in the day. By then he and Brattain were seeing "voltage gains up to about 100 and power gains up to about 40." What most interested Bardeen at this point was the mechanism by which the holes were entering the semiconductor, spreading out in the inversion layer and affecting the flow of current. In an effort to explain how the holes in the inversion layer might be causing the amplification, he drew an analogy with the familiar system of the triode:

> When A [the gold electrode] is positive, holes are emitted into the semi-conductor. These spread out into the thin P-type layer. Those which come in the vicinity of B [the tungsten point] are attracted and enter the electrode. Thus A acts as a cathode and B as a plate in the analogous vacuum tube circuit. The ground corresponds to the grid, so the action is similar to that of a grounded grid tube. The signal is introduced between A (the cathode) and ground (grounded grid). The output is between B (the plate) and ground. The signs of the potentials are reversed from tthose in a vacuum tube because conduction is by holes (positive charge) rather than by electrons (negative charge).

"The analogy was suggested by W. Shockley," he added. This was the climax of Bardeen's productive involvement with Shockley.

9

The Break from Bell

In the days following their invention, Bardeen and Brattain must have had an overwhelming urge to continue working, despite the holidays that had crept up on them. Their families wanted them home to help celebrate. Nature sided with their families. An unusually heavy snow began to fall on Friday, December 26. All staff members were sent home after more than a foot of snow had accumulated at Murray Hill.

Across the Hudson, New York City had been "utterly unprepared" for the storm that the *New York Times* reported as having "swept in from the Atlantic at a point where there are no weather observers." By Saturday morning, after a night in which the storm dropped 25.8 inches of snow, transportation had ground to a halt. "The city's towers wore tremendous tufts and beards of snow." Worse than the blizzard of 1888, which had set New York City's previous snowfall record of 20.9 inches, the blizzard of 1947 produced "the greatest snowfall in the city's recorded history."

From Summit, Jane wrote to her mother that they were having "one hell of a storm." The snow had turned to freezing rain, complicating New Year's Eve celebrations. Most residents were without power for nearly a week. The Bardeens had to borrow a neighbor's gas stove to prepare meals. Jane described it as "great sport (?) stumbling through the snow from their back door to ours with a tray of hot kettles." She said that the members of her hungry family were "all ready, with forks in hand, around the kitchen table when I came in. I told them I felt like the zoo keeper who throws fish to the seals!"

Shockley worked at home all through the storm and even after the sun finally came out on Sunday, December 27. The Christmas holidays of 1947 could not have been joyous for Jean Shockley or for Bill's and Jean's two young children, Billy and Alison. Shockley stewed in the knowledge that Brattain and Bardeen had made an important, perhaps revolutionary, discovery and that he had not been part of it. Because the work had been motivated by *his* designs, Shockley, as Bardeen later reported, "thought he would have the basic patent on the field effect principle." Shockley resolved to explain to Bell's attorneys why the patent should be written in his name.

While the children of New York and New Jersey dug tunnels in the snow, Shockley struggled alone to develop his idea for a new kind of bipolar transistor that would be a significant improvement over Bardeen and Brattain's design. Unlike the cumbersome point-contact device, his he expected would be commercially viable. It would use junctions of p- and n-type silicon or germanium in place of the electrically noisy point contacts. This idea had been working its way through Shockley's mind at odd moments over the last month while the point-contact transistor was born. He had kept his work hidden from the semiconductor group.

On Monday, December 28, Shockley ran his designs over to Murray Hill and got an experimentalist, Richard Haynes, to witness the documents. Then he took the train to Manhattan and boarded the Twentieth Century Limited to Chicago. There Shockley attended a meeting of the American Physical Society on December 30 and 31. Afterwards, he returned to his room at the Bismarck Hotel and worked for almost a week on his new design: a three-layer "sandwich" in which the "bread" was of p-type semiconductor and the "meat" was of n-type. One of the outside layers would supply the holes; the other would collect them. The n-type layer acted as the grid, the "faucet" that controlled the current by creating an electrical barrier whose height he later realized could be raised or lowered by changing the voltage. The idea would evolve into the junction transistor.

Back at Murray Hill, Kelly clamped down on communicating with the outside about the Bardeen and Brattain discovery. Imitating the secrecy procedures of the Manhattan Project, he declared their invention "Bell Labs confidential." He code named the work the "Surface States Project." Jane had already written to her parents about "John's invention." Now she hurriedly wrote back to ask that they keep the news confidential "until the patent arrangements have been made. The lawyers are working on that now." She added proudly that "all the top 'brass' of Bell has been out to see it, so it must be good."

Kelly also insisted that Bardeen and Brattain drop everything else and focus on drafting patent applications. In effect, by tying up their time, he was offering Shockley an open field to work on his new junction device. Bardeen and Brattain worked closely with the patent attorney, Harry Hart, who came to think of the two as one combined individual. Years later he wrote a joint letter to them, because he always "thought of you as a pair." He added, "Your personalities are as different as those of Jack Sprat and his wife. Perhaps that is the reason you were able, between you, to lick the platter so clean."

When Shockley returned to Murray Hill from Chicago, he took a further step that ruined any possibility of future teamwork with Bardeen or Brattain. He called each of them separately into his office and told them, as Brattain recalled, "that he could write a patent—starting with the field effect—on the whole damn thing." Bardeen and Brattain were stunned. Shockley added the final insult when he said, "Sometimes the people who do the work don't get the credit for it."

But Shockley's plan failed. The attorneys discovered that the field effect idea was not original after all, but had "turned up in a number of forms in the patent literature." In 1930 the idea had been patented by the Polish-American inventor, Julius E. Lilienfeld, then a professor of physics at Leipzig. Lilienfeld's 1926 patent application had described the field effect concept so well that "there was no way that Shockley would get a patent on the field effect principle," Bardeen later explained. It did not matter whether Lilienfeld's device actually worked. Bell Labs could not risk the rejection of a claim based on the already patented field effect idea. Bell therefore based its application on Bardeen and Brattain's bipolar design. That idea *was* new because holes—rather than electric fields—were functioning as the grid.

Shockley must have felt that he was losing on all sides. He continued to work in solitude on his junction device, while Bardeen and Brattain worked on the patents for their bipolar device. Then one day Shockley suddenly revealed his secret work.

On February 18, experimentalist John Shive demonstrated a design for a transistor to his Bell Labs colleagues in which the points were on opposites of a sliver of germanium. The fact that Shive's device worked indicated that holes were indeed traveling through the sample.

Shockley knew that Bardeen would instantly recognize the implications of Shive's demonstration, if indeed he had not already done so. Shockley could no longer afford to keep his work quiet, for Bardeen could easily scoop him. So at the demonstration Shockley abruptly rose to present his design for a new type of bipolar amplifier based on p-n junctions. The design resembled his New Year's Eve conception, except that it was an n-p-n rather than a p-n-p sandwich, in which electrons, rather than holes, carried most of the current. In addition, electrical contacts were attached to all three layers, allowing better control of the current.

Brattain was appalled that Shockley "went off by himself and worked at home" and especially that he had "ceased being a member of the research team." Bardeen also fumed, but he, characteristically, said little about it. Jane and a few others, including Brattain, knew how much Shockley's behavior disturbed Bardeen when he would mutter through his teeth that Shockley had "jumped in with both feet."

Bardeen and Brattain had other worries, too. With Kelly's secrecy order in place, they could not publish or even speak about their invention during the first six months of 1948. They also were anxious about being scooped, particularly by the physicists working under Karl Lark-Horowitz.

Ralph Bray, a member of the Lark-Horowitz group, had reported a "spreading resistance" in germanium. As a graduate student during the Second World War, Bray had observed that when small pulses of positive voltage are applied to points in contact with germanium, the electrical current spreads out under the points, lowering the resistance. He did not yet realize that injected holes were causing the phenomenon, but Brattain and Bardeen felt that sooner or later Bray or another member of his group would figure this out.

In January 1948, at an American Physical Society meeting in New York, Brattain discussed Bray's experiments in the corridor with Seymour Benzer, another member of the Lark-Horowitz group. Benzer said, "I think if somebody put another point contact down on the surface, close to this point, and measured the distribution of potential about the point, then we might be able to understand what this is all about." Brattain responded uneasily before walking quickly away, "Yes, I think maybe that would be a very good experiment."

Another group was indeed working on a similar device in Paris. Heinrich Welker and Herbert Mataré were creating a germanium device similar to the Bardeen-Brattain point-contact transistor. They filed a patent for it on August 13, 1948. Historian of science Kai Handel, who has written about Welker and Mataré's work, asserts that "it is probable that in any laboratory concerned with high purity Germanium and with experience in experiments with point contacts, somebody would perform experiments with two point contacts close to each other." Like the Purdue group, the Paris team did not have the benefit of a theorist like Bardeen with a thorough understanding of the phenomenon. Handel adds, "It is certain however that they didn't—unlike Bardeen and Brattain— succeed in understanding how their device really worked."

Military classification was another concern. Bell Labs felt obliged to get a military release because of the amplifier's possible military applications. While they waited for that release, Bardeen and Brattain sent a letter about their invention to the *Physical Review*, asking the editor to hold back publication until Bell Labs had officially heard that the invention would not be classified. Again, Brattain had difficulty containing his impatience. He felt "very strongly that most restrictions done in the name of national security turn out to be foolish." Not until June did they hear, with considerable relief, that their device would not be classified.

In the meantime it was crucial to find a good name for the invention. Suggestions included semiconductor triode, surface-state amplifier, crystal triode, and iotatron (to emphasize its small-ness). Many names were dropped because Brattain and Bardeen didn't like words ending in 'tron.' In May 1948, a Bell Labs electri-cal engineer found the best name.

John Pierce happened to stop by Brattain's office. "John, you're just the man I want to see," said Brattain. "Sit down." Pierce had a way with words. He would later write science fiction stories under the pen name of J. J. Coupling. As the two men discussed the naming issue, Pierce pointed out that the vacuum tube worked because of its transconductance, while the new device worked because of its transresistance; it was the electrical dual of the vacuum tube. Brattain said that he thought the name should fit with the other names used for solid-state devices, such as "varistor" and "thermistor." Pierce suggested "transistor." "Pierce, that is it!" Brattain exclaimed.

All the suggested names were circulated in a Bell Labs memorandum. Those on the distribution list were asked to vote on the best name for the new device. "Transistor" was the overwhelming favorite.

The invention was finally released to the public on June 30, 1948. May Shockley, Bill's mother, flew in from California and joined Jean, Keren, Jane, and their husbands for an elegant lunch in the executive dining room on the top floor of the West Street building. The women watched while the men were photographed. A few pictures were taken of them as well. Then all were called into the auditorium.

Ralph Bown, then the Bell Labs' director of research, was preparing to speak. He was smartly dressed in a tailored gray suit with a colorful bow tie. "What we have to show you today represents a fine example of teamwork," he began. He stressed "the value of basic research in an industrial framework." Then he went on to explain what "this cylindrical object which I am holding up" actually did:

> We have called it the Transistor, T-R-A-N-S-I-S-T-O-R, because it is a resistor, or semiconductor device which can amplify electrical signals as they are transferred through it from input to output terminals. It is, if you will, the electrical equivalent of a vacuum tube amplifier. But there the similarity ceases. It has no vacuum, no filament, no glass tube. It is composed entirely of cold solid substances.

Bown demonstrated the device by amplifying his own voice. He spoke through a telephone handset to members of the audience who were holding receivers into which he inserted, and then removed, one of the new transistorized amplifiers. He emphasized that the new transistors, unlike vacuum tubes, needed no warm-up. After the demonstration, Shockley fielded questions.

Bardeen did not initially notice the shadow that Bown's demonstration cast on his and Brattain's role in the development of the transistor. "Everything went well. I thought it was a very successful demonstration," Bardeen said. The Bardeens and the Brattains had "received VIP treatment and were very thrilled to be there," Jane gushed to her family. She mentioned that "the final perfect touch was being chauffeured home through the blazing heat in a seven-passenger Packard." To the public, however, Bown's presentation and Shockley's articulate handling of the audience questions gave the impression that the transistor's development had been orchestrated from above. The invention was presented as the result of teamwork in a group directed by one of Bell Labs' most brilliant physicists.

Their spirits dampened, recalled Bardeen, when "we read the *New York Times* the next day" and found "they had just a few words in the radio column about it, and that's all." The *Times* reported that "a device called a transistor, which has several applications in radio where a vacuum tube ordinarily is employed, was demonstrated for the first time yesterday at Bell Telephone Laboratories." The account, buried on page 46, near the end of the paper's regular News of Radio column, was given lower priority than the notice that on Monday evenings during the summer CBS would air the popular radio comedy *Our Miss Brooks* in place of its regular Monday evening feature, *Radio Theater*. The *New York Herald Tribune* was more upbeat, saying that "engineers believe it will cause a revolution in the electronics industry!"

Radio buffs responded enthusiastically to the news. Instead of waiting for transistors to hit the market, they "would just buy a germanium dial and make themselves a transistor," reported Brattain. "A radio set without vacuum tubes," Jane marveled in a letter home, "just John's gadget and a speaker. Won't that be something!"

Engineers were perhaps the most excited of all. In September 1948 *Electronics* magazine featured a carefully posed image of Bardeen, Brattain, and Shockley on its cover. They had been photographed in the laboratory wearing dress shirts and ties. But neither of the two inventors was at the center of the photo. Instead Shockley was posed at Brattain's workbench, while Brattain and Bardeen looked on from behind. The portrait was staged to convey a message consistent with Bown's press conference. "Walter sure hates this picture," Bardeen told a friend.

All three men now hit the lecture circuit. Bardeen spoke in Chicago at an American Physical Society meeting, at Purdue in

Indiana, and at the Oak Ridge Laboratory in Tennessee on the way home. In January 1949, Bell Labs, now welcoming publicity, sent *Life* magazine photographers to the Bardeens' home for a planned article on the transistor. The day before, John went downtown to "get a haircut so he'll look pretty."

Jane worried that John was exhausting himself. In late October she wrote home that he "has been under considerable pressure lately writing a paper for *Physical Review* on the transistor and giving lectures. He spoke twice last week at the labs and on Friday at Columbia." He was planning to be gone most of the month of November giving talks in San Francisco, Los Angeles, and Chicago and would "not be home again until the weekend after Thanksgiving."

Bell Labs prepared illustrations and demonstrations for use in the talks. In 1949, to highlight the transistor's portability, an engineer at West Street built three music boxes. Each contained an oscillator-amplifier circuit and two of the early point-contact transistors. Bardeen and Brattain were each given a box; the other was for Bell Labs. Bardeen's box (now preserved at the Spurlock Museum of the University of Illinois) was the only one that did not disappear over the years. "Dad was very proud of this music box," said his son Jim. John kept his box in his office safe and it still worked more than fifty years later.

Bardeen would use the box in lecture demonstrations to illustrate the transistor's use as both an oscillator and an amplifier. Referring to a small piece of paper taped on top, he would key in the tune "How Dry I Am," which could be heard over a loudspeaker. Then he would amplify the sound by switching the second point-contact transistor into the circuit. The Prohibition-era drinking song would usually produce laughter in the audience, easing John's way into his lecture.

Jane was given one of the first transistorized hearing aids. The earlier-generation vacuum tube hearing aids were large devices, usually worn around the waist or sometimes held in a pocket. Consumers were willing to pay the high cost of improving them. They were limited by the expense and weight of the battery needed to power them. Transistorized hearing aids would eventually become more affordable and less cumbersome, but the early transistors cost about $200 each and were about the size of a pencil eraser. In recognition of Alexander Graham Bell's commitment to helping the deaf

and hearing impaired, AT&T offered hearing aid manufacturers roy-
alty-free licenses.

Throughout 1949 and 1950 Bardeen grappled unhappily with the
stress of his continuing lectures and travel, and especially from the
deterioration of his relationship with Shockley. Other staff at Bell
Labs could see that Shockley "wanted to have everything under his
control," limiting the freedom of those who worked under him.
Shockley was breaking with the enlightened philosophy of research
that had drawn Bardeen to Bell Labs in the first place, but Kelly and
Bown did not intervene.

Conyers Herring was among those who noticed that Shockley
had "a great measure of confidence from the higher management,"
who were impressed "with his mental brilliance." That confidence
from above meant a lot to Shockley, who, Herring sensed, "felt all
alone." Shockley wanted recognition and ultimately to "become a
big name in the world of business." He thought he "was brighter
than anyone else, and in many ways he was." Contributing to his
growing isolation, Shockley would let others know that he did not
consider most of them intelligent enough to make good company
for him.

Bardeen and Brattain were distressed when Shockley banned
them from the work on the junction transistor, a project they saw
as the continuation of their work on the bipolar transistor. In the
fall of 1948, Shockley went so far as to assign Bardeen and Brattain
to offices on a different floor from his in the newly completed Building 2. Shockley directed them to work on the surface effects of
point-contact transistors, an area he "probably knew was a blind
alley."

When Bardeen complained, Shockley proposed that Bardeen
could go and work in development with Jack Morton or in radio
engineering with Russell Ohl, "but these solutions naturally did
not appeal to me," Bardeen later wrote. Like Shockley, Bardeen
had come to Bell Labs because of the promise to work on his own
basic research. Shockley's policy "meant that I could not play an
active role in the semiconductor research program." Bardeen was
directed to work only on problems "in which Shockley was not
interested himself."

Bardeen went to speak with Ralph Bown, but Bown was reluctant to interfere. Shockley, Bown said, was "in a highly emotional state" and working seventy or more hours a week. Bown assured Bardeen that "the difficulties would be resolved in time. Bardeen deduced that "it was the desire of the administration to give Shockley a free hand."

In April 1948 Bardeen had seriously considered accepting a permanent position at Oak Ridge, Tennessee, directing solid-state work in support of the reactor program of the Atomic Energy Laboratory. He thought about going there for six weeks, or possibly a year, to try the job out. Ultimately he decided not to accept the offer from Oak Ridge because he still hoped his position at Bell Labs would improve. "I am still interested in spending some time at Oak Ridge, but so far I haven't been able to get out from under the work on the transistor," Bardeen wrote to the director of the Oak Ridge laboratory, Alvin Weinberg. He held the door open. "If you want to give me a rain check, I will let you know when things let up here."

A year later, when Weinberg renewed his offer, the situation had not improved. Bardeen wrote back, "You make a very strong case for spending a period at Oak Ridge on the basis of a leave of absence from the Bell Laboratories. Although I am not certain of it, I believe that if the situation warranted it such a leave could be arranged."

With his Oak Ridge offer in hand, Bardeen wrote for advice to Jim Fisk, who was then at Harvard in the Department of Applied Physics. Bardeen explained that Weinberg "has been in contact with me, first in regard to a permanent position at Oak Ridge which I turned down, and now in regard to spending a period of a year or so there on leave from the Laboratories. He makes a strong case for the latter." Bardeen understood that his position at Oak Ridge would be to "act as a senior advisor for the solid-state group." His contributions "would be connected with fundamental investigations into the properties of matter subjected to nuclear bombardment." Bardeen asked Fisk, "What do you think of the proposition?" Fisk, then slated to return to Bell Labs as director of physical research, advised Bardeen to wait the crisis out.

Jane was secretly relieved about John's decision to stay at Bell Labs. She enjoyed living in Summit. She also worried about how John would fare in a government position. "The rising anti-intellectual sentiment makes any government service by a professor or scientist a risky business. Let's hope we're not heading toward a fascist state." Committed to liberal principles, the Bardeens were opposed to the growing atmosphere of intolerance caused by the "red scare" and Senator Joseph McCarthy's infamous hearings. The House Un-American Activities Committee had already attacked several scientists. Jane recommended that her family read a *New Yorker* article about William Remington, "one of the victims, last summer, of the House Un-American Activities Committee. I remember telling Betty that the Kittels knew him and were sure that he was not a Communist, nor guilty of giving information to the Communists as that crackpot E. Bentley charged."

At Bell Labs "conditions if anything got worse," but Bardeen did find one project to have a little fun with. Collaborating with Herring, he wrote a paper criticizing some aspects of an earlier article by Fred Seitz on the Kirkendal effect in alloys. Bardeen had discussed the problem with Herring, telling him "he wanted to study diffusion from a rather fundamental point of view." Herring had been working on the problem of diffusion from a similar perspective. And as Bardeen "rather liked" Herring's approach to the problem, he suggested they publish a joint paper. In it Bardeen dealt with the Kirkendal effect and Herring treated what became known as the correlation effect in diffusion.

Herring had by this time become the Bell Labs' authority on the latest research in solid-state physics. He was the one to whom people would usually come with questions about the physics literature. He took his role of guru seriously, but the field was growing so rapidly that he felt he could not reliably retain all the information in his memory. So he began to carry around a made-to-order roller-skate box filled with 3×5 index cards containing notes and references to the latest work in the field. The box had begun as a notebook when Herring was at Princeton. After the war he replaced the notebook with a set of filing cards, initially kept in a file box, but the cards soon outgrew the box. One day, while attending a meeting of the American Physical Society in New York, he "stopped in at a luggage store there, and had them make me to order there a suitcase so big that I could put 3×5 cards in." The box served Herring until the personal computer took over its function in the 1970s. But from the 1940s through the 1960s, everyone who worked on solid-state physics at Murray Hill depended on Herring to tote the institution's solid-state database around in his roller-skate box.

Bardeen found some relief from his anxieties at work in family activities. John bought Jane a corsage for her birthday in April 1948. When the family went out to dinner at the William Pitt in Chatham, the men wore pansies in their lapels; Betsy wore a pansy corsage. In January 1949, John took nine-year-old Jimmy skating. "Dad and I went skating at Lake Surprise three times," Jimmy wrote to Jane's mother. "The first time I could just stand up. The second time I learned to skate." In May 1949, to celebrate Jim's tenth birthday, the family picnicked along the Musconetcong River. "Even had candles on the chocolate cake!" A few months later, Jane wrote to her mother, "John wants to take the kids to the zoo in New York City this weekend."

Betsy adored her father. Like him, she "could be very determined." At age four she refused to take off her new patent leather shoes until her daddy could see them. John, who never lost his love of sweets, abandoned his diet when almost-five-year-old Betsy offered him tidbits from a box of her own candy.

The larger community of solid-state physicists formed another kind of supporting family. Bardeen became a father figure there too, one to whom other physicists came with questions of all kinds. He was always willing to help a young scientist. Nobel laureate Walter Kohn recalled his first interaction with Bardeen. Kohn was then working under Van Vleck on a project that used a table published by Wigner and Seitz.

> One day V^2 [Van Vleck] remarked that he believed that through some misstep this table contained a *wrong* (perhaps preliminary) set of numbers and that I should write to John [Bardeen] who, V^2 thought, had the *right* numbers. Of course, John was already very famous thanks to the creation of the transistor, while I was an absolute zero. With considerable trepidation I followed V^2's advice. Very promptly the reply came from John in a brown manila envelope. Contained therein was the correct table of numbers (V^2 had been right!), accompanied by a long, handwritten letter which explained how the mix-up had occurred and described in detail a mathematical procedure for making the best use of these numbers (I have used that procedure ever since).

Kohn later wrote to Jane, explaining that "this—in some ways—small experience had a big effect on me." Bardeen's gesture "showed me a great scientist patiently giving a leg up to a raw beginner whom he had never even met."

That kind of interaction with Bardeen became legendary—so much so that physicists at Bell Labs developed a number to describe it, the "Bardeen number." Bell Labs physicist Ravin Bhatt (a student of Bardeen's student William McMillan) described it as "the ratio of substance to self-advertisement." Bhatt explained that, in general, if a scientist even made number one on that scale "you're doing good. Most people these days are way below that." But "Bardeen was infinity on that scale."

But while Bardeen's interactions with his family and physics community often improved his mood, they did so only briefly. He grew increasingly unhappy throughout 1948, 1949, and into 1950. "Bardeen was fed up with Bell Labs—with a particular person at Bell Labs," said Brattain.

One day early in 1950 Bardeen just stopped working on the problems that Shockley had relegated to him. With Fisk's approval, Bardeen turned back to the problem he had been studying at the time he was called to Washington, D.C. Explaining superconductivity was still the outstanding unsolved riddle of solid-state physics. Bardeen had not thought extensively about superconductivity since 1941, but when he pulled out his decade-old notes it felt like coming home.

In his prewar study of superconductivity, Bardeen had come to focus on the energy gap in the electronic structure of superconducting materials, a notion highlighted in the work of the London brothers. Bardeen had tried to explain the gap using a single-electron quantum-mechanical model. He had explored whether applying a small periodic distortion to the lattice could lower the energy and introduce band gaps near the Fermi surface. That approach had failed. Still he felt certain that the energy gap was crucial to the phenomenon. He could not explain why. Nor could he account for the gap's presence. Confronting this challenge distracted him from the aggravation of working under Shockley. It also led him to his most important scientific work.

Bardeen approached his return to superconductivity in the same way he approached every problem. He considered how to break the problem down into smaller parts and visited the library to see what others had done before him. He was pleased to find much new evidence supporting the Londons' notion of the energy gap. By now many experimenters were working on superconductivity. They included A. Brian Pippard, William Fairbank, Emanuel Maxwell, Paul Marcus, J. C. Daunt, Kurt Mendelssohn, B. Goodman, A. Brown, Mark Zemansky, Henry Boorse, W. Corak, Michael Tinkham, and Rolfe E. Glover.

Pippard, David Shoenberg's student, whom Bardeen had met in Cambridge in 1947, during his trip in Europe with Shockley, was now a professor at Cambridge. Pippard's studies of the surface impedance of normal and superconducting tin supported the London theory, as did related experimental work in the United States by Fairbank at Yale and by the MIT team of Maxwell, Marcus, and John Slater.

On May 15, 1950, Bardeen received a phone call that riveted his attention on superconductivity. Bernard Serin, an experimental physicist in the Department of Physics at Rutgers University in nearby New Brunswick, New Jersey, wanted to discuss the implications of certain recent findings of his group. In studying mercury isotopes Serin's group had found a clear and definite "isotope effect": the lighter the mass, the higher the temperature at which the material becomes superconducting. The stable and pure mercury isotopes they had used, having mass numbers between 198 and 202, had been prepared at Oak Ridge by bombarding gold with neutrons and separating the products using a mass spectrometer. They were available as a consequence of the wartime atomic bomb program. Emanuel Maxwell, then working at the National Bureau of Standards, had independently found the isotope effect studying isotopes from Los Alamos.

Serin's colleague, Henry Torrey, who during World War II had been nicknamed the "Crystal Crackin' Papa" of the MIT Rad Lab's crystal rectifier program, remembered "Serin coming to me when he got the results." He "was very excited." The fact that the mass was involved hinted that the lattice was too! As Torrey recalled, the observations "immediately suggested to Serin the relationship to the electron-phonon interaction." That was what Serin wanted to discuss with Bardeen; Serin had heard that Bardeen "knew a lot about" superconductivity.

The next day Bardeen scratched a note to himself about Serin's call: "These results indicate that electron–lattice interactions are important in determining superconductivity. . . . It is important to include their effect on the free energy of the electrons." Bardeen then spent the next several days trying to use the new clue to revive his earlier theory of superconductivity. In his research at Minnesota, he had failed to achieve a lowering of the energy using a periodic lattice distortion. Now he tried lattice fluctuations. That also failed, but he was sure he was on the right track. To secure priority, he dashed off a letter about the work to the *Physical Review*.

In the history of science, major discoveries often come in pairs or higher multiplets. The countless examples of simultaneous discovery of important scientific theories include the conservation of energy principle (Mayer, Joule, Kelvin, Helmholtz, and others), the entropy principle (Clausius, Kelvin), the calculus (Newton, Leibnitz), the theory of evolution (Darwin, Wallace), the mathematical formulation of quantum mechanics (Heisenberg, Schrödinger), and quantum electrodynamics (Feynman, Schwinger, Tomonaga).

What causes simultaneous discovery? Historians of science, including Thomas S. Kuhn, have found this question compelling. Is the coincidence the result of direct, or indirect, communication? Does it come from the "ripeness" of a field or of a body of research? Does it depend upon internal connections in the state of knowledge or thought—"something in the air"? Or is it but a matter of chance? Whatever the explanation, by the early 1950s several researchers, including Bardeen, had arrived at the hypothesis that electron–phonon interactions are crucial to superconductivity.

Even before Maxwell and Serin had completed their experiments, Herbert Fröhlich, a British theoretical physicist on leave from Liverpool and spending the 1950 spring semester at Purdue University, had written a theory of superconductivity that predicted the isotope effect. When Fröhlich learned about the confirming experimental results a day or two after they appeared in the *Physical Review*, he promptly sent a letter to the *Proceedings of the Royal Society* to claim priority. The competition between Fröhlich and Bardeen was on.

Fröhlich happened to visit Bell Labs a week after Bardeen submitted his letter. "It was very common for people to come to Bell Labs and pass through," Louise Herring recalled. On that particular visit, Fröhlich and Bardeen discussed their respective work on superconductivity, undoubtedly with a degree of tension. Without reference to any competition between them, Bardeen later wrote about their conversation, "Although there were mathematical difficulties in both his method and mine, we were convinced that at last we were on the road to an explanation of superconductivity."

Although Bardeen's and Fröhlich's theory both explained the isotope effect, neither could explain superconductivity. The problem was that both focused on *individual* electron energies. To explain superconductivity, it would be necessary to treat the energy that arises from the interaction of many electrons. The available mathematical formalism was still too limited.

The basic problem on which both Bardeen and Fröhlich became stuck was to find an interaction that made the *total energy* of the superconducting state lower than that of the normal state. They realized that if the electron–lattice interactions were to accomplish such a lowering of the total energy, there had to be an attractive force. Moreover, this interaction had to dominate the ordinary electrical repulsion between electrons. Bardeen and Fröhlich's theories in 1950 could not show this.

Bardeen continued to feel isolated at Bell Labs as he struggled with the problem of superconductivity. He later explained to Kelly: "There are very few people in the laboratories who are interested in the problem." Longing for colleagues with whom to discuss superconductivity, Bardeen asked Bown whether he could have his own theoretical group. Again the response was negative. When Bardeen brought up his problems with Shockley, Bown insisted that Shockley needed flexibility in directing his group.

On one bright moment in this otherwise gloomy period they learned that the patent for the point-contact transistor had finally been approved. Bardeen had been concerned that it might not be, for the discovery of the Lilienfeld patents had brought rejection to two of the four transistor patent applications (the ones concerning the Brattain-Gibney-Bardeen field effect work in November 1947). Betsy, then six, remembered her father coming home one evening and announcing happily, "We got the patent!" She admitted, "I had no idea what he was talking about, but I remember hugging him around the knees." The relief was temporary.

In October 1950 Bardeen attended a conference on "Crystal Imperfections and Grain Boundaries" held at a resort in the Pocono Mountains of western Pennsylvania. He and Shockley were both scheduled to present papers. Also at the meeting was Fred Seitz, Bardeen's old friend. Surrounded by the brilliant leaves of fall, Fred and John sat down for a heart-to-heart talk.

John told Fred what the last two years had been like for him at Bell Labs. He spoke about his frustrations with Shockley and of his exclusion from the junction transistor work. He also told Fred about his new work on superconductivity and of his desire for colleagues with whom to discuss the problem. "I am convinced I want to go back into academic work," he concluded. "What is the job situation like?"

Seitz was just the right person for Bardeen to speak with. He had known Shockley since 1932 and could easily imagine the situation at Bell. He could also identify with the frustration of working in a nonreceptive environment. Just a year earlier Seitz had left his position at the Carnegie Institute of Technology and moved to the University of Illinois. Among other frustrations at the Carnegie Institute, an aide to the president had "decided to change the physics curriculum drastically without either our input or approval."

Seitz told Bardeen that he didn't know of any academic jobs off-hand and would contact him if he heard of any. (He did not want to disappoint his friend.) But Fred, on his return to Illinois, went directly to William Everitt, the dean of engineering, and to F. Wheeler Loomis, the head of the Department of Physics. Everitt and Loomis were both enthusiastic about offering Bardeen an appointment.

Unfortunately, there was no money available in the regular budgets for hiring new professors. Student enrollment was down because of low birth rates during the Depression, and the GI Bill had run its course. Everitt and Loomis agreed, however, that the opportunity to acquire Bardeen should not be missed. Everitt assured Seitz that he would find a way to piece together a package for Bardeen.

When Seitz relayed the message from Everitt back to Bardeen, he told John that he should "keep cool, that the dean was working on it." Bardeen's response was, "Well, Illinois would be perfect. It's the kind of place I'd like to be at."

With the help of Coleman Griffith, the provost of the university, Everitt pieced together parts of budgets and arranged a joint position for Bardeen, half in physics, half in electrical engineering. The offer included support for an applied group in engineering. Seitz felt the dean's offer of $10,000 a year "was not quite enough for a person of John's stature," which had grown with the invention of the transistor. Fred told John, "Look, the dean has 10,000. If you hang on maybe you'll get more." John replied, "That's enough."

The negotiations continued over the next five months. One drawback was that "moving at this stage of my life is a difficult and costly business. . . . I will have to start building up a pension fund over again." Seitz wrote to Bardeen about his own recent experience at Illinois. "I have found more than enough outlets for all of my interests with the greatest possible freedom. I think there is no doubt that you would have exactly the same experience." He assured Bardeen that he would have colleagues to work with. "I need hardly add that the solids group would be more than thrilled to see you here." Bardeen leaned toward accepting when both physics and electrical engineering assured him that his work would be self-directed.

Late in March 1951, Bell Labs complicated Bardeen's decision by finally taking some steps to improve his situation there. Fisk, the new Bell Labs director of physical research and thus Shockley's new boss, recognized that Bell Labs might actually lose Bardeen. "One Friday," recalled Brattain, he and Bardeen had "walked into Fisk's office and told him that we did not wish to report to Shockley any longer." And on "Monday morning we weren't reporting to him." A memorandum from Fisk, dated Wednesday, March 28, shows the Bell Labs solid-state group divided into two separate groups: the Physics of Solids, led by Stanley Morgan, and Transistor Physics, led by Shockley. Bardeen and Brattain were in the Morgan group.

"I haven't reached a final decision on whether to leave the Laboratories or not," Bardeen wrote on April 6 to Gerald Almy, then the acting head of the Illinois physics department. "There was a reorganization here about a week ago which makes things much more favorable from my point of view, but I am still inclined toward Illinois." When Seitz ran into Fisk at Los Alamos, he warned him that Bell Labs was in danger of losing Bardeen. "Oh, don't you bother," said Fisk. "We've got that under control."

The decisive factor was Dean Everitt's letter to Bardeen on April 16 assuring him that he could lay out his own research program and "that in no case would you be asked to work on contracts which were not of your own choosing." Everitt specified that "the programs which we now have would not indicate in any way the type of work we would expect you to do here." As for teaching, "You would be pretty much your own boss," he wrote, "and would give such formal courses as you felt fitted into your program." The regular course load for physics faculty was two courses per semester. However, "If you felt that only one course a semester was the best way for you to work, that would be quite all right."

Ten days later Seitz wrote to Bardeen again with assurance that "everyone I have spoken to is more than enthusiastic and will do everything within his power to make you enjoy life here." Seitz described the collegial atmosphere and the opportunities for increasing one's academic income—for instance, working with the military controls project for part or all of the summer. Seitz enlisted his wife Betty to write to Jane about housing opportunities. Betty described their recent house-shopping experience. They had found a comfortable, inexpensive, two-story stucco in a tree-lined neighborhood on Iowa Street, a short walk from the campus.

Bardeen accepted the Illinois offer in a telegram sent on April 28. His initial salary of $10,000 would be evenly split between the physics and electrical engineering departments. On hearing from Everitt of Bardeen's acceptance, Seitz immediately sent off a warm letter of welcome.

Bardeen wanted Mervin Kelly to know why he was leaving Bell Labs, for when Kelly hired Bardeen, he had promised he would have full research freedom. Shockley had made a mockery of that promise. Now that Kelly was president of Bell Labs, Bardeen felt he deserved to know the full story.

Bardeen's carefully worded three-page letter, posted on May 24, began with a simple explanation of the Illinois offer. He emphasized the university's promise to be supportive of his research, no matter what direction it took. That was important, he explained, because he expected that his research would be on superconductivity "at least for the next year or two." He also noted that the "financial prospects appear to be as good as those at the Laboratories." Then he got to his main point: "I would not leave if I were not dissatisfied with conditions here. In fact I would not have received the offer if I had not let it be known that I was considering leaving the Laboratories."

"My difficulties stem from the invention of the transistor." He explained in detail how the work under Shockley had changed after his and Brattain's invention. "Before that there was an excellent research atmosphere here." Afterwards, Bardeen wrote, Shockley reserved all the interesting research for himself:

> Shockley at first refused to allow anyone else in the group (1170) to work on the problem (that is, aside from Brattain, Gibney and myself), and then did so only as he thought of problems of his own that he wanted investigated experimentally. In most cases these were problems in which he either already had done theoretical work or in which he wished to do the theoretical work himself in the future. In short, he used the group largely to exploit his own ideas.

Each time Bardeen tried to go over Shockley's head and gain entry into the research on the junction transistor, he had been rebuffed. Shockley "was well aware of the situation," wrote Bardeen; it "was a deliberate policy." Bardeen explained that he was in an "intolerable" position in which he "could not contribute to the experimental program unless he wanted to work in direct competition with my supervisor."

Alluding to his Oak Ridge offer in the spring of 1949 Bardeen told Kelly, "I seriously considered leaving the Laboratories about two years ago under much less favorable circumstances." Bardeen also told how his "intolerable" working conditions grew even worse after he decided to work on superconductivity, a new direction he had "discussed at the time with Fisk." Given the lack of support for his work on superconductivity, Bardeen explained that he "felt somewhat isolated." But even if there were experimental work on superconductivity going on at Bell, "I feel that I can work on superconductivity more effectively in a university. The problem is one of more scientific than practical importance and there is great interest in it in academic circles."

Bardeen made it clear that he intended to continue research in semiconductor physics. Even in this area he believed he would be in a more "favorable" position at Illinois, given Shockley's monopolistic attitude about the transistor.

> Before making the decision to leave, I again explored the possibility of working on the semiconductor program with Shockley. His attitude had not changed. He felt that he could supply all the ideas required, that he would want the people in his group to work on his ideas, and that I would not be happy in this situation. It has been suggested at various times that an independent semiconductor group be set up under my direction, but I did not feel that this was a satis-

factory solution from my point of view. In the discussion with
Shockley he indicated that he was unwilling to give up any significant part of the work.

In closing, Bardeen offered an orderly recapitulation of his main
points:

> To summarize, the invention of the transistor has led to the semi-
> conductor program being organized and directed in such a way that I
> could not take an effective part in it. I could work on superconduc-
> tivity, but I feel I could do this better in a university where it is of
> primary rather than secondary interest. I also feel that university
> work, in which one can set one's own pace, becomes relatively more
> desirable as one gets older. Therefore I have decided to leave.

Fisk and Kelly both tried to change Bardeen's mind. They
offered him a salary increase and his own theoretical group. But
Bardeen was no longer interested. "And when Bardeen makes up
his mind," said Brattain, "there is no use doing anything about it. It
is too late."

Bardeen's resignation may have contributed to the rise of the semiconductor industry in California. After his departure, Bell Labs
executives recognized the hazards of entrusting an important
research team to Shockley. By 1955 Shockley perceived that he had
reached a "glass ceiling" on the Bell Labs ladder. He was also annoyed
that his transistor work was not making him rich. He determined
to strike out on his own. That year he moved to California.

On the trip west, he stopped off in Urbana to see Seitz. With
him was his new partner, Emmy Lanning, a psychiatric nurse
Shockley had met a few years earlier in Washington, D.C. She had
helped him work out a theory of productivity and creativity based
on wartime theories of operations research and current theories of
psychology. Bill and Emmy developed a close emotional and intellectual relationship in a time when Shockley's marriage to Jean
(then suffering from uterine cancer) was deteriorating.

Bill invited Emmy to take part in his plans to create a pioneering research company in California devoted to semiconductors and
transistors. It would be staffed by scientists and engineers having a
high "mental temperature." She agreed. They eventually married.

Seitz remembered the day "Bill called up and asked if they
could stop here for a few days. And I said yes." Shockley had told

Seitz that he now "was on his way West." He was returning home to Palo Alto. He had "already prepared the ground for receiving proposals." More than twenty years had passed since Bill and Fred had traveled across the country together in Shockley's 1929 DeSoto convertible. In 1932, as students, the two had been heading east, from Palo Alto to their East Coast schools. Seitz recalled the adventure of their earlier trip—and the uneasiness with which he had viewed the loaded pistol Shockley had stowed in the glove compartment. On this visit in 1955, Seitz again felt considerable ambivalence in his interplay with Shockley, but it was of a different sort.

Betty Seitz "distrusted" Shockley. She had agreed to the visit, but she wasn't pleased about it. "She did it for me," said Fred. "We couldn't do other than treat them as guests." The visit also caused some tension between the Seitzes and the Bardeens, but "not in any serious way," said Seitz.

Shockley had started out on the trip by flying to Columbus to pick up Emmy. They then drove in her car to Urbana and settled into the Seitz's stucco on Iowa Street. Betty was dismayed when Bill tied up the phone for several days. "He just got on the phone and began talking to people up and down the coast," recalled Fred. The Seitz's teenaged son enjoyed the visit, for Bill "would amuse Jack with his sleight of hand tricks."

Seitz recalled that during one of Shockley's countless telephone calls, he completed a deal with the wealthy chemist Arnold Beckman, who agreed to fund the new company that became Shockley Semiconductor Laboratory. It was the first semiconductor company to be established in the region later known as Silicon Valley.

In California Shockley proved to be an even worse research director than he had been at Bell Labs. He assembled a brilliant research group, and insisted that all job applicants take creativity tests administered by psychologists in New York or San Francisco. Once they had been hired, he treated them like assistants. Most left after a short time. Two of the first eight who departed (Shockley called them "the traitorous eight"), Robert Noyce and Gordon Moore, went on to create Fairchild Semiconductor and later Intel Corporation. After Shockley's company failed, he joined the faculty at Stanford University. In the late 1960s, he focused his work on eugenics and achieved notoriety with his theories claiming a genetic correlation between race and intelligence.

After Bardeen's decision to leave Bell Labs, the family got on with the business of transforming their lives. A possibly apocryphal, but nonetheless revealing story features John showing the house on Primrose Place to a prospective buyer. When the buyer made a good offer, Bardeen fell into one of his thoughtful trances. A period of time passed during which Bardeen said nothing. The buyer thought he had offered too little and raised the offer. Still Bardeen remained silent. The offer was raised again. Eventually the house sold for much more than the Bardeens had expected.

The family bought a new car, a green DeSoto, for their move to the Midwest. Driving first to Madison, they spent some time renewing friendships and reinforcing family ties. At his alma mater Bardeen took part in a research seminar on low-temperature physics. He also discussed his evolving work on superconductivity with colleagues. John particularly enjoyed seeing Wigner, who visited Madison that summer.

At the University of Wisconsin, Bardeen taught his first formal course on the electrical properties of solids. About 20 students enrolled in the summer course and several more audited. Shockley's new book, *Electrons and Holes,* served as the text. Just inside the front cover of Bardeen's copy Shockley had inscribed: "To John Bardeen, Who made a book like this a need. Bill Shockley, Dec. 1950."

10

Homecoming

The grand sweep of the prairie came as an unexpected thrill to Jane when the family drove into Champaign-Urbana in June 1951. She "gasped because it was so glorious." She had had misgivings about the move, especially after John's visit to Urbana the previous spring when he had brought home dire warnings from faculty wives who hated the endless fields upon fields of corn and soybeans that grew there on some of the nation's richest farmland. Many found the Midwest a cultural void.

Jane, however, loved the sense of space and freedom evoked by the central plains and the region's wide skies, at times dramatically animated by electrical storms. She appreciated the white farmhouses that decorated the verdant landscape and the wildlife of the region. Here and there a section of wooded river valley had been set aside as a park or greenway. Along roadsides and railroad rights of way, tall grasses provided cover for a variety of wild animals. In the cool air of dawn and dusk, deer, raccoons, foxes, and pheasants quietly explored the fields. Falcons rested alertly on telephone lines, waiting for a meal, while redwinged blackbirds whistled shrilly from fence posts.

John experienced moving to Illinois like coming home. He was immediately comfortable in the small twin cities of Urbana and Champaign. In some ways the atmosphere resembled the Madison of his youth.

The two Illinois cities dated back to when the Illinois Central Railroad came through the region in 1854. The tracks ran two miles west of the Urbana border. A small community sprang up around the railroad depot, and its residents resisted when Urbana tried to annex the area in 1855. Five years later the new depot town was incorporated as Champaign. With the railroad providing transportation to Chicago, Champaign became the county's commercial center, much to the chagrin of Urbana residents. Both towns grew rapidly with subdivisions replacing soybeans at a breakneck pace, especially in Champaign. Soon the towns bumped up against one another, until all that divided them was their separate systems of governance. The university, straddling Wright Street, the border between the two towns, became a center of activity and growth.

The Bardeens had purchased a large, pie-shaped lot at 55 Greencroft, in a new subdivision on the western edge of Champaign. The modest ranch-style house, already under construction there, would meet their needs for the next forty years. It was near the university, but not close enough to be disturbed by campus bustle. As the neighborhood aged, the vast fields of corn and soybeans that had once stretched from Greencroft into the distance were gradually replaced by new developments, but Champaign held on to its small-town atmosphere. That suited the Bardeens just fine. "I don't like big-city living," John said bluntly.

The new house would offer something for every member of the family. Jane and John's large bedroom overlooked the backyard. The boys shared a room with "Hollywood" beds, and Betsy had a room of her own. The basement made a perfect recreation room. A ping-pong table soon became its central feature. It was a room that Bardeen's students and postdocs would come to know well. The kitchen was small but convenient. Eventually the family added a spacious sunroom to the back of the house, accessible from both the kitchen and the living room. Nearly floor-to-ceiling windows and an expansive view of trees and flowers gave the room a breezy natural feeling.

Jane couldn't wait to sink her hands into the rich black soil. That first spring she planted every kind of vegetable she could think of in her garden plot. But when the family took a two-week vacation, she returned to a tangle of overgrown plants. She learned to plant more discriminatingly and soon had a thriving garden full of flowers and vegetables.

John's and Jane's different ways of appreciating the outdoors expressed their separate interests and needs. John would increasingly turn to golf, an activity that served as an outlet and training ground for his intense focus, competitiveness, and drive toward mastery. Jane preferred gardening—working with living, growing plants and nurturing them into their full potential. Although the two shared many things over the course of their marriage, Jane and John kept a respectful distance, allowing one another separate lives in their different spheres.

The children quickly found friends in the neighborhood. Their yard, large for Champaign, was a good place for Jim and Bill to play catch with friends. John occasionally helped the boys polish their Little League baseball skills by playing catcher to their pitcher, or chasing and catching fly balls as they practiced their batting techniques. Bill found the neighborhood a splendid place to get into scrapes, from which he sometimes had to be rescued or even carried to the doctor. Once, when they were living in Summit, John had walked into the house, looked at Jane's harried face, and asked, "What has he done now?" John had found it his unpleasant duty to occasionally administer a spanking to the accompaniment of Bill's exaggerated screams.

Once in a while John brought home books that he thought might interest the children. Many were about science. In the evenings he liked to play records on the family phonograph. He always claimed that "his instrument was the record player." Mozart was a favorite, and Betsy often drifted off to sleep to the "blissful" sounds of chamber music.

John liked to play with the children at their own level. "He would wrestle with us," Betsy recalled many years later when she had children of her own. He would "toss us through the air, and let us climb all over him." It was a favorite entertainment after dinner. "I had always thought that all parents treated their children this way," she confessed, "until I reduced a child and her mother to hysteria when I flipped her over in one of our milder 'tricks.'"

The house on Greencroft was within walking distance of the Champaign Country Club, which at the time Bardeen joined had nearly 500 members and a new air-conditioned building. With its eighteen-hole golf course, the club provided John with countless hours of exercise and companionship. In later years the Bardeens often entertained their visitors by taking them to dinner there.

When John was off on the greens, Jane usually worked in her garden—"my golf," she called it.

Marriage and family life formed the center of Jane's world. She took pleasure in knowing that the home she attended to was the supporting stage for John's creative work. At the same time she continued to feel peripheral to John's engagement with his physics. Their quietly companionable relationship resembled the one that John's father, Charles, had shared with Althea. Jane would have preferred to have more conversation, but John was "a man of almost fewer words at home than when he was with his fellow scientists," she once said wistfully.

They sometimes squabbled. During the summer of 1956 Jane came down with a mysterious ailment during a stay in Les Houches, near Chamonix in the French Alps, where John was teaching at a summer school. John "interpreted my lack of interest in activities and my lack of enthusiasm for Les Houches as evidence of bad temper and a mean disposition and was very blunt in telling me so. Consequently I stubbornly resolved to keep my troubles to myself and to just stick it out until this 'flu,' 'virus,' or whatever had cured itself." When she was finally diagnosed with typhoid, she insisted that he and the children go on with their European tour while she recovered in Switzerland. Her contrite husband sent letters and care packages from the road and offered to come stay with her. "Thanks a million for offering to come, John," she wrote back, "but really there isn't a thing you could do here."

When not needed by her family, and when she was not putting together dinner parties or cookouts for John's students and colleagues, Jane turned her energies and organizational skills outward into the community. She joined a number of local clubs, including the Score Club, a small women's society originally organized around the musical interests of its members. By the time Jane became a member, the club had evolved into more of a reading group. She worked on problems of racial discrimination through the League of Women Voters and was active in the University Women, a university-wide organization that was "helpful to newcomers, because it gives them a door into their new community." Even after she was no longer active in the club, Jane participated in its affairs now and then, for example, modeling the gown she wore for the 1956 Nobel ceremony and to a 1962 White House dinner.

The Bardeens were not religious, but Jane believed that the

church provided opportunities to meet people and a venue for teaching the children history, culture, and ethical standards. Soon after their move to Champaign, she "took the kids down to the First Presbyterian Church, and the next thing she knew she was teaching Sunday school."

Religious tradition, which had been an issue at the time John and Jane were married, arose again with the question of religious training for the children. Back in Summit, Jane had taken the children to a Presbyterian church. John surprised her with the comment that she "might have considered the Unitarians." "Why?" she asked. "You're not a Unitarian." "No," he replied, "but my father was." It was the first time she had heard John mention a religious preference. Several years later, when Jim asked Jane whether he could quit Sunday school, because everyone there just wanted to "horse around," Jane said yes. If he was not learning anything, he needn't go. Bill followed suit. Betsy, however, agreed that she would go to Sunday school as long as Jane went to church. Jane became a church elder, and Betsy stayed in Sunday school.

John's mother, Althea, had been reared in the Quaker tradition, and his stepmother, Ruth, was Catholic, but John was resolutely secular throughout his life. He was once "taken by surprise" when an interviewer asked him a question about religion. "I am not a religious person," he said, "and so do not think about it very much." He went on in a rare elaboration of his personal beliefs.

> I feel that science cannot provide an answer to the ultimate questions about the meaning and purpose of life. With religion, one can get answers on faith. Most scientists leave them open and perhaps unanswerable, but do abide by a code of moral values. For civilized society to succeed, there must be a common consensus on moral values and moral behavior, with due regard to the welfare of our fellow man. There are likely many sets of moral values compatible with successful civilized society. It is when they conflict that difficulties arise.

The warm and collegial physics department welcomed Bardeen. He was free to work there on whatever subject he wanted—alone or, as he typically preferred, in collaboration with others. He could teach as much or as little as he wished and offer whatever courses he wanted to teach.

The department had been built up recently into a world-class institution by the popular and energetic F. Wheeler Loomis, who served as its head from 1929 to 1940 and again from 1947 to 1957. Loomis had been in Zürich, on leave as a Guggenheim fellow from New York University's department of physics, when the University of Illinois first invited him in 1929 to come as a full professor and head of the Department of Physics. Loomis was not impressed. Under Albert P. Carman, the Illinois physics department had stagnated from 1897 until 1929. It had been completely bypassed by the quantum mechanics revolution. Loomis also was not thrilled with the idea of moving from New York City to the small prairie town of Urbana.

He was, however, challenged by the opportunity to transform a backward department into a world-class research institution. He learned that Roger Adams had, under similar circumstances, taken the Illinois chemistry department from rags to riches in a short time. The dean of the College of Engineering, Milo Ketchum, assured Loomis that he would have a free hand and generous funding to do so for physics. He took the job. (Later he learned that his name had been last on the list of possible candidates; all the others had declined.)

The job proved harder than Loomis had anticipated, for shortly after his arrival the stock market crashed and the university could not deliver the promised resources. He also discovered that it was extremely difficult to attract physics "stars" to Urbana. Isadore I. Rabi had humorously rebuffed his invitation with the comment, "I love subways and I hate cows."

Loomis managed to hire a few new faculty using a successful hiring strategy similar to the one Charles Bardeen had used in shaping the Wisconsin medical school. He focused on promising postdoctoral-level scientists: "young, competent but relatively unproven people with fresh ideas, rather than less-risky, established—and more expensive—scientists." In 1930 Loomis brought in Gerald M. Almy, a young experimental nuclear spectroscopist, later one of Loomis's successors as head of the department. Around the same time he hired Gerald Kruger, another spectroscopist, and the nuclear physicist Donald Kerst, who would later invent the world's first betatron, an electron accelerator for studying elementary particles. Loomis hired twelve young physicists between 1937 and 1941. The median age of the department dropped from fifty-five in 1929 to thirty-two in 1940.

Then World War II scattered most of the physicists that Loomis had brought to Illinois. He himself, along with several of his hires, landed at the MIT Radiation Laboratory, where Loomis became associate director. Almost two-thirds of the faculty that Loomis had added in the 1930s went elsewhere during the war, forcing him to undertake yet another massive rebuilding effort after his return to Illinois in 1947.

This time Loomis drew on a network of influential colleagues that included his friend Louis Ridenour from the MIT Rad Laboratory, who had received his doctorate at Caltech in the up-and-coming field of high-voltage physics (which would evolve into high-energy physics). Loomis encouraged Ridenour to accept an offer to become dean of Illinois's graduate college. After he did, Ridenour convinced William L. Everitt, who had been an electrical engineering professor at Illinois since 1944, to become dean of the College of Engineering in 1949. Already known for numerous inventions in telephony, radio, and antenna systems, Everitt now proceeded to build one of the foremost engineering programs in the United States.

For a few fertile years between 1949 and 1951, Loomis, Ridenour, and Everitt worked as a team building the Illinois School of Engineering into a world-class institution. They decided that the time was ripe to establish a strong group in solid-state physics. Ridenour was able to offer generous seed funding—including salaries for two senior and three junior positions and resources to set up a new laboratory. The administrative decisions were made easily and rapidly within the "old-boy network," with the justifying paperwork typically appearing after the fact.

When bargaining for resources, Loomis drew freely on his personal connections with the important physicists of his day. In an era when discrimination against Jews and other minorities was common, his integrity gained him the respect of many colleagues. He was angered by the occasional "warning" about the Jewish background of a job applicant. Over the years his refusal to make hiring decisions on the basis of religion or ethnicity paid off, for he was able to attract gifted candidates that other universities would not consider. Similarly, Loomis refused to be intimidated by the scare tactics of the anticommunist fanatics of the early 1950s. Edwin Goldwasser, hired by Loomis in 1951, said that he believed "Loomis made the decision [to hire me] on the grounds that I was refusing to sign a loyalty oath." Another physics professor had a student who belonged to the Communist Party and "the department swung behind [that professor] pretty strongly."

Ridenour's first major solid-state hire was Fred Seitz, whom he had known in the late 1930s at the University of Pennsylvania, during the war at the MIT Rad Lab, and in the late 1940s at the Carnegie Institute of Technology where Seitz was the head of the Department of Physics. Ridenour persuaded Seitz to move to Illinois, with an offer that included "a lot of money by the standards of the day" and the opportunity to make several departmental appointments in solid-state physics.

Despite Betty Seitz's reservations about living in the Midwest, she "fell in love" with the area. She and Fred decided that the landscape had "many of the qualities of the sea":

> ... the same kind of rich interplay between sky and rolling land as is seen between sky and undulating water. Brilliant sunsets and magnificent storms hover over the vast spread of the land. And, like the sea, the land presents markedly different aspects at different times of day and at different seasons.

Seitz found the intellectual atmosphere congenial. "There were no rivalries of any significance." He became chair of the new solid-state group. Robert Maurer and James Koehler followed Seitz from Carnegie Tech to join the solid-state group as associate professors. Four young instructors also arrived in 1949 as part of the Seitz package: David Lazarus from the University of Chicago, Dillon Mapother from Carnegie Tech, and James Schneider and Charles Slichter from Harvard. Lazarus, who was appointed at Seitz's suggestion, arrived even before Seitz, who had contacted Loomis on Lazarus's behalf after the young physicist told Seitz he was looking for a job. By the time Bardeen arrived, the Illinois physics department was, with Cornell, one of the top two academic departments in America in the area of solid-state physics.

The department was also growing in other areas, including nuclear physics. In 1950 its new 300 MeV betatron went online. In 1951, through Ridenour's energetic leadership, the University of Illinois established its commanding position in the computer revolution by designing and building two large digital computers. The first, ORDVAC, went to the Army's Aberdeen Proving Grounds in Maryland; the second, ILLIAC (Illinois Automatic Computer), stayed in Urbana and became "one of the busiest machines on campus."

Morale was unusually high under Loomis. He kept each member of the department involved in its operation. He invited faculty to help determine his "Loomis list," which ranked faculty according to the quality of their work and their overall contributions to the department. His "wonderful parties" fostered a sense of family. He encouraged a tradition (still in place today) of faculty stopping in the halls or meeting in one another's offices or labs to chat informally about physics. Around 10:00 A.M. most physicists would wander over to the departmental lounge to drink coffee while catching up on the work of colleagues or discussing recent journal articles. At lunch they would meet regularly in small groups for conversation.

Abundant monetary resources during the Korean War supported the department's high morale. In the 1950s much of the funding for the solid-state group came from the Office of Naval Research (ONR) or from the Atomic Energy Commission (AEC). The flow of money within the government–university circuit was eased by the personal relationships that developed among scientists and government officials during World War II. For example, Seitz was one of the many friends of Emmanuel Piore, chief scientist at the ONR, who with others (such as Mina Rees and Randal Robertson) made day-to-day funding decisions. "Everyone was Manny's friend." Scientists attached to well-regarded programs usually needed to write no more than a page describing a particular project to receive funding. Most modest requests (e.g., to finance a researcher and one or two graduate assistants) were funded as a matter of course.

Bardeen's new physics community extended to the entire world. After his month-long trip to Japan in 1953 to attend a meeting of the International Union of Pure and Applied Physics, he went abroad nearly every year. For that first memorable Japanese meeting, he and Seitz hired a military plane for the thirty-seven-hour trip. According to Seitz, Bardeen was "given very special attention, as Japan was now greatly interested in the transistor." "I've never seen so many flashbulbs in my life," John wrote Jane. He told Japanese scientists, "It was like 'carrying coal to Newcastle' to speak in a field in which you are so outstanding."

On the trip Bardeen met many Japanese scientists who would remain colleagues and friends throughout his life. Among them was Sadao Nakajima, who presented a short paper on interactions between electrons and phonons in superconductivity. There was no time to discuss the work at the conference, so Bardeen asked Nakajima to meet him at the train station in Nagoya, where he would be stopping on his way back to Tokyo from Kyoto. "So I went to the station," Nakajima later said, and "talked with him through the train window for a very short time while the train stopped." Nakajima also presented Bardeen a written version of his work—in Japanese. (Bardeen had it translated back in Illinois.) Later, Nakajima accepted Bardeen's invitation to visit Urbana.

Bardeen also met Michio (George) Hatoyama, whose uncle was Japan's premier, and Makoto Kikuchi, with whom he would have lifelong friendships. They both worked at the Japanese Ministry of International Trade and Industry's (MITI) Electrotechnical Laboratory (Denki Shikenjo) in Tokyo. Hatoyama entertained his American colleague at his home during the long 1953 trip. Later, Bardeen wrote to Hatoyama that the visit to Japan had been "one of the great events of my life."

In 1960 Hatoyama left Denki Shikenjo to direct Sony's then new research laboratory. One of his best friends, Kazuo Iwama, had been the head of the company's transistor group since its early years. The company had been founded as Tokyo Tsushin Kogyo Company (Totsuko) in 1952, when Masaro Ibuka and his partner Akio Morita invested in patent rights for the junction transistor. They made a fortune by producing transistorized consumer products and in 1958 renamed the company Sony. Kikuchi took over as director of the Sony Research Center in 1974. Bardeen was a frequent and honored guest at Sony, and the company occasionally sent students or scientists to the University of Illinois. Bardeen often sent his Japanese colleagues letters of introduction for friends visiting Japan. He always tried to accommodate Japanese visitors to the States.

Three years after Bardeen's memorable first trip to Japan, he would write to Hatoyama and Kikuchi to introduce his student, Nick Holonyak, who had recently been drafted into the U.S. Army Signal Corps. Holonyak was to be stationed in Japan. Years later he would become the first to hold the prestigious Sony Professorship at the University of Illinois, created in honor of John Bardeen.

Holonyak was a second-year graduate student in electrical engineering at the time he met Bardeen in the fall of 1951. Taking

Bardeen's undergraduate atomic physics course (Physics 381) that fall of 1951, he discovered that he "learned very effectively" from Bardeen. He rushed to sign up for Bardeen's semiconductor course to be offered for the first time in the spring of 1952. It was among the first such courses at any university.

Bardeen had begun to prepare for the course during the summer of 1951. In August he wrote to Jim Fisk to ask whether Bell Labs could supply materials: several kinds of transistors, including point-contact and junction transistors, germanium blocks and rods, and other germanium and silicon samples with which he could perform resistivity and Hall measurements. He wanted these soon, "so that we can get the experiments organized." Fisk obliged.

During the fall of 1951 Holonyak also heard Bardeen give his first Urbana seminar on the transistor. The lecture became vividly imprinted in Holonyak's memory. Bardeen arrived with a transparent box, "about eight or nine or ten inches long and about six or seven or eight inches high and three or four inches thick." It was the music box that Bell Labs had made for him in 1949. The box had "a little loudspeaker on the front, and it had two of these point contact transistors sitting in some sockets and a battery." But what it did was infinitely more exciting than how it looked.

> As soon as he flipped the switch on, the thing started to play, and I almost fell out of my seat because I was already a grad student, and I knew how to build things out of vacuum tubes. You turn it on and you wait for the filaments to warm up, and you wait for a considerable time before the thing does anything. This thing, whatever it was doing, the magic, turned on immediately as soon as the power was turned on. I knew right away, "Oh-oh, this man has got something different sitting here in our face."

Holonyak remembered that Bardeen taught the new course on semiconductors and transistors using a loose-leaf notebook cradled in his left forearm, fingers curled around the top. He again used Bill Shockley's text, *Electrons and Holes*. During the Second World War, John told the class, scientists had been studying metal–semiconductor junctions for use in radar. To illustrate, he drew a diagram on the chalkboard, with the "barrier facing the metal on the left and the semiconductor on the right." Holonyak said, "I remember him pointing with the chalk at the holes and smiling a little bit, just a little hint of a smile, saying that if Schottky had looked there to see what the holes were doing, the transistor would have been invented."

Bardeen's mention of the German physicist Walter Schottky was an understated reference to his own major contribution to transistor history, to the recognition that in his work with Walter Brattain the holes entering the germanium were the key to the transistor's action. Schottky and the British physicist Nevill Mott had been among the first to develop a viable theory of the rectification due to semiconductor-to-metal interfaces. Working at the Siemens Laboratory in 1938, Schottky had attributed the rectification to an observed potential barrier at the interface, increased by an applied voltage in the back direction of the current and decreased by voltage in the forward direction. In 1945, when Shockley sketched the field effect design that started Bardeen on his transistor research, the Mott-Schottky theory was at the cutting edge of rectification physics. But the theory did not take into account the surface states that Bardeen postulated in 1946 or the holes that would feature prominently in Bardeen and Brattain's transistor.

When Holonyak learned that Bardeen was going to set up a semiconductor laboratory, he recognized it as an "opportunity to get into something fresh and original." He asked his advisor—the electrical, cybernetics, and computer engineer Heinz von Foerster—if it would be all right for him to become Bardeen's student. Von Foerster agreed.

Bardeen and his first graduate student became instant and unlikely friends. While Bardeen came from a middle-class academic family, Holonyak was the son of a southern Illinois coal miner. Unlike the mature, soft-spoken, and meditative Bardeen, Holonyak was young, garrulous, and exuberant. Years later the *Champaign-Urbana News Gazette* described Holonyak "as animated as Bardeen is dry; as effusive as Bardeen is self-contained." He was another of the "voluble and extroverted and enthusiastic" people for whom Betsy said her father had an affinity—like John's brother Bill, Walter Brattain, or Walter Osterhoudt.

Holonyak intuitively understood both sides of the strong bond between Bardeen and Brattain. He recognized that Brattain's frequent visits to Urbana in the early 1950s were because "he missed his partner." On one particular visit, Holonyak happened to observe Brattain and Bardeen working together in Bardeen's semiconductor laboratory. Brattain was scribbling on a blackboard while talking and gesturing excitedly. He was having trouble with a calculation. Finally, Bardeen inserted himself into Brattain's problem, mumbling the explanation as he scratched numbers and symbols on the board. Brattain eyed Bardeen over his half-glasses as both stepped back to admire the solution. "Goddammit, John," Brattain burst out, "how in the hell did you do that?" To Holonyak it appeared that Bardeen was the "head" in the relationship, while Walter was the "hands," as well as the one who did most of the talking.

In the electrical engineering department, Bardeen continued to work on semiconductor physics. One goal was to establish an experimental semiconductor group in the Electrical Engineering Research Laboratory (EERL). Holonyak and Richard Sirrine were the first two graduate students in the semiconductor lab; the first two postdoctoral researchers were Harry Letaw, Jr., a physical chemist, and S. Roy Morrison, a solid-state physicist. Bardeen directed them in an informal paternal fashion.

The semiconductor laboratory, when it began functioning in the fall of 1952, was initially housed in a large, empty room that had previously contained the university's historic ILLIAC computer. The ten-foot-long, two-foot-wide computer, which was just becoming operable at the time Bardeen's group began setting up their laboratory, left behind ample work space when it was moved to another building. Bardeen's work replaced the ILLIAC in another way, too. Based on vacuum tubes, the ILLIAC would soon become obsolete in the wake of transistor technology. The behemoth computer had required so many (2,800) vacuum tubes to drive its forty-bit parallel memory that a parade of students had to regularly come through with bushel baskets full of tubes to replace the defective ones.

Bardeen wrote grant proposals to support his students and equip the semiconductor lab. Holonyak worked with the others to construct items that were not yet available commercially. Although the lab was well funded by government agencies, such as the Office of Naval Research and the Air Force Office of Scientific Research and Development, it was not fancy. Bardeen never lost the "sense of economy" that Althea had noticed when he was a boy. Some members of the group grumbled that his funding requests were too modest. They felt that a scientist as eminent as Bardeen should have been given an exceptionally well-equipped lab with cutting-edge materials and equipment. Holonyak soon figured out that his mentor "didn't need anything lavish." What he cared about was that the problems were good ones. "He didn't need anything to pump up his ego," because "John Bardeen knew that he was John Bardeen."

Bardeen believed that both theorists and experimenters needed a solid grounding in experiment, as he had had. He sent some of his early theory students, including Robert Schrieffer, over to work in the semiconductor lab for a while and get their hands dirty. "I thought that even theorists should have some lab experience." Later, Bardeen recognized that studying both theory and experiment was too much to expect if a student was to finish a thesis expeditiously.

An apocryphal story features John Bardeen offering his two early graduate students—Holonyak and Schrieffer—a choice between semiconductor research and the superconductivity problem. "The semiconductor topic," he is reputed to have said, "is guaranteed to generate results if you work hard enough. The superconductivity topic is different. It can lead to an unbelievably great work. But it also has a risk of generating no results at all." The story is dramatic but fanciful. Holonyak worked in Bardeen's semiconductor laboratory from his earliest association with his mentor and never considered the more theoretical topic of superconductivity. Moreover, by the time Schrieffer got to the point of thinking about a thesis project, Holonyak had already finished his thesis (1954) and moved on to Bell Labs.

James Bray, a much later theory student of Bardeen's, felt that he absorbed Bardeen's predilection to root his work in experiment even without having laboratory experience. "For whatever reason, I had by then developed, and still have, a strong desire to be around experimentalists," Bray acknowledged. "It just happened naturally. There are theorists who like to work with theorists and be around theorists and be theoretical all the time." That had never been Bardeen's style. "John would occasionally call those people 'theorist's theorists'—and he wasn't using the term in a complimentary fashion as some other people do." Bray explained that Bardeen "thought it was very important for theorists to immerse themselves in experimental data and be guided by that." He did not appreciate a theorist "who just sat down and would plot down a bunch of equations . . . whether or not it bore any relation to reality ultimately."

In the first years of the semiconductor lab, Bardeen came by almost every day to find out what the students had been doing and whether they needed help. He always asked about the materials being used. Although he never "picked up a pair of pliers," he encouraged his students to find out for themselves how to design an experiment and interpret results. Bardeen greatly appreciated Holonyak's hands-on ability in the laboratory, as he had, similarly, appreciated Brattain's. Holonyak soon learned that his mentor was not the best person to answer questions about the details of an experiment. Before he got to know Bardeen well, Holonyak "would go up to him when he would come in, and I would say, 'what do I use to do this, a hydrogen atmosphere or a vacuum?'" Bardeen "would just look at me. He wouldn't say anything." After a while, "I began to understand that he didn't do experimental work. I was asking him things that he wasn't familiar with. And he didn't say anything just because there wasn't anything to say, and he wasn't going to talk just for no good reason at all."

When Holonyak left for Bell Labs in 1954, Paul Handler came on board as a postdoc to work in the semiconductor lab. Bardeen emphasized the work's importance for the government when he wrote a letter in support of extending Handler's military deferment. "It is known," Bardeen wrote, "that Russia is training scientists at a much faster rate than we are, so that it is urgent that we make the best possible use of our trained people." Handler recalled that Bardeen "knew what were important problems in semiconductor physics." Moreover, he was "very aware of other experimental work." For example, Handler "didn't know about the ultra-high vacuum work that had been done at Brown University," but Bardeen did, and he "sent me out there to find out how it was done."

Bardeen gradually relinquished control of the lab to Handler. At first, "he'd come by about twice a week, maybe three times a week, stop in and see how things were going." After some time, "he came in once a week." By 1962, Bardeen was spending little time in the semiconductor lab. Nor did he need to worry about writing grants or obtaining funding for graduate students. Eventually Handler "wrote them with just my name alone. So essentially there was a transition."

Holonyak had not been at Bell Labs a year when he was called to serve in the army. Bardeen was concerned about Holonyak's career. He told Holonyak to try to get assigned to a government laboratory where he could continue to do research. Bardeen gently prodded him, "If you can possibly do it, I hope you can revise your thesis for publication before you leave. . . . You did a very good piece of work; it should not lie buried in government reports." He also made the first of a series of offers to Holonyak to come "home" to Urbana. "If you ever have a desire to return to Illinois please let me know. I feel sure that we could work something out that would combine teaching and research."

Five weeks later Bardeen wrote again: "I know the indirect pressure to produce can be very great. We would certainly welcome you back if you would like to join the staff here." In October 1955 Holonyak married Katherine Jerger, who was then attending graduate school at New York University. She went home to Chicago and took a teaching position while he was away in the army for two years.

When Holonyak learned that he would be stationed in Japan, Bardeen encouraged him to call on Hatoyama and Kikuchi. Hatoyama, Bardeen wrote Holonyak, "is a very nice fellow, and I am sure he would enjoy meeting you." For the next eight months Holonyak visited both Japanese scientists "every other weekend," and they became good friends. Hatoyama wrote to Bardeen that Holonyak was "a nice fellow" and that "he was always talking of his wife."

In the fall of 1957, Holonyak returned to industrial research, working at General Electric for the next six years. Periodically Bardeen would urge him to consider returning to Illinois. "I think there are some real advantages to being in a university community," he wrote Holonyak in 1959. Four years later when Holonyak finally returned to Urbana, Bardeen resumed his habit of regularly visiting his young friend's semiconductor lab.

Teaching large groups of students remained a chore for Bardeen and for some of the students. Many could not interpret his muffled speech nor his odd silences. For some students, his frequent pauses to think or scratch an equation on the board had a soporific effect. Sometimes he pressed the chalk to the board so lightly that it was barely possible to make out the characters. Sitting close to the front of the room was therefore important. Andy Anderson recalled that a class he took with Bardeen was on the third floor, "and when our previous class was over, we all ran like hell up the stairs to try to get a front seat."

In the mid-1950s physics graduate students were called upon

to spoof their professors in skits at the annual department party. Bob Schrieffer, who had been working with Bardeen since the fall of 1953, "found a dark blue suit, a white shirt, and a dark blue tie." He climbed to the stage and "began to lecture before a group of students on the stage, saying in a soft voice, 'You are going to sleep, *you are going to sleep.*'" He was a little concerned about how his mentor would react to the parody, especially when he saw Bardeen flush bright red at the burst of laughter that greeted Schrieffer's gentle mockery. At the next class meeting Bardeen announced with a mischievous grin that he had reserved the auditorium for his lecture, "so that they could sleep more comfortably." Schrieffer suspected that Jane had "had a hand in this genius move."

Some students found it unnerving when Bardeen would explain something in exactly the same words every time he was asked. "He would simply repeat, almost like a tape recorder," a teaching method not helpful to a student who had not understood the explanation to begin with. Many students ended up dropping out of his courses.

Other students found Bardeen's teaching methods exceptionally stimulating. His habit of stopping to think offered them the chance to do the same. "He was fascinating to listen to," recalled Holonyak, "because you were invited into what he was doing. If he paused and he was thinking, you realized that he was thinking, and *you* thought." Bardeen viewed student participation as one of the most important aspects of the learning experience, for undergraduates as well as graduate students. He helped them to develop confidence by encouraging them to find answers for themselves. "You can figure it out," he would prod them.

Bardeen and his students—and their students—illustrate Harriet Zuckerman's theory of "mutual influences," in which scientists pass their values on to succeeding generations in a kind of socialization process. Bardeen's students learned from him how to identify important problems, how to attack them, and what constitutes a solution. According to Zuckerman, master scientists reinforce in their students "not only cognitive substance and skills but the values, norms, self-images, and expectations that they take to be appropriate for this stratum in science."

Bardeen's style of teaching was reminiscent of his father's. Dean Bardeen had been well known for his muffled speech and hands-off approach to teaching. Charles hoped that his students would "take on the spirit of the investigator, the delight in first hand information about natural phenomena, the willingness to work hard to get at the truth." But he pointed out, "This spirit will not be aroused in the student by one who is not himself filled with it." The teaching method of both father and son required students to have confidence and persistence. Bardeen "would not impart motivation," said Bray, "if you didn't have it." The students who came away with an understanding of Bardeen's lessons were the ones who participated in the learning experience. He encouraged them to stretch their imaginations and bury themselves in the literature to which he directed them.

Even the best students had to work hard to navigate Bardeen's leaps of thought. "Often," Schrieffer said, "it would sound like he was giving a logical explanation when it was nothing but reading out steps 3, 17 and 42." Many students and colleagues had difficulty filling in the missing steps. Schrieffer himself adopted a "standard technique" to get more detail from Bardeen. By saying "things that were slightly wrong," he exasperated Bardeen just enough so that "he'd talk more."

To teach students to solve problems, Bray said that Bardeen "would sort of outline the problem and then indicate the directions you ought to go work on—what sort of things you ought to try to compute." One time, "he just walked in one morning, wrote a suggestion on the black board and walked out." Although "he did this in five minutes," it was just what they needed to help them with their calculation. "From what he wrote there we had it done by the end of the day."

Schrieffer had at first been unnerved by his mentor's communication style. "Often he would sit for 8 or 10 minutes. He'd say nothing, and I found this terribly disquieting." Students would ask themselves, "Have I said something stupid?" Bardeen's students learned that "he was just thinking. Sometimes you'd walk to the door and the door would be half way closed and he would start speaking to the empty chair." For many people conversation is a way of organizing thoughts that are incomplete out loud. Schrieffer came to "realize that usual conversations about serious research move much too rapidly for people to think deeply." Bardeen presented a different model. "You should rather get your thoughts in place and eliminate a lot of silly nonsense that you talk out." After Bardeen's long stretches of silence, what he finally did articulate was "meaningful."

Bardeen's public presentations were often even more concise. He uttered "one sentence," recalled his colleague Chih-Tang Sah, "one phrase or one sentence, that defined the problem—the whole thing. The details you can work out or he'll give you a simple example on the blackboard."

Student experiences with Bardeen varied. Some hesitated to ask him questions, worrying that he would think them incompetent or lazy. For others, despite his policy of keeping his office door open, "he always *appeared* too busy to encourage idle questions." Bardeen politely hid any irritation he may have felt for less able or less confident students who approached him with questions he considered trivial. But he did not waste time explaining things that had already found their best exposition in the literature. Dan Mattis, who worked with Bardeen in the 1950s, tried not to approach him with a problem until he already had some reasonable approximation of an answer. "The best strategy appeared to be to try to understand just what he had had in mind, then return, as soon as possible, with the final results."

Other students always felt welcome in Bardeen's office. "Bardeen was extremely accessible," said David Allender, who finished his doctoral dissertation in 1975. "If I wanted to talk to him he always made time." Henry Pao, who completed both his undergraduate and graduate degrees at Illinois, recalled that when he had questions he "always would go to Bardeen and he wasn't afraid to ask." Bardeen was not given to lengthy explanations, but he "would give you leads to things that you can research and then understand." Pao suggested that engineers are "more practical" and therefore not as inclined as physicists to put a man like Bardeen on so high a pedestal that they were afraid to approach him with practical questions.

Colleagues would consult with Bardeen when they reached a barrier in their research. "He was the one I would go to when I was stuck," recalled Lazarus. Paul Coleman would sometimes walk over from engineering to ask a question. Typically they would chat for a moment about their mutual passion for golf, then Coleman would pose his question. He appreciated the fact that Bardeen never embarrassed him by saying, "You mean to tell me you don't know that!" Dan Alpert thought that "one of the reasons that people respected him so highly is that if you went in and asked him questions that reflected whatever level of knowledge you had, you never had to prove yourself to him."

Asked once what Bardeen was like as a teacher, Alpert said it "depends on who you think his students were." To him, Bardeen "was one of the great teachers on this campus and his students were the faculty. He had some graduate students of course, but the whole physics faculty were his students." Bardeen treated every colleague as an equal. "When I had a specific technical question to ask, he got to the heart of the issue. He didn't mess around with all kinds of talk, and the last thing on his agenda was to impress you with how much he knew."

Bardeen took a humane approach in working with students. When he sat on George Russell's thesis committee, he made no comment when Russell told the committee that he could not find any literature about one aspect of his proposed study. But "after it was over, [Bardeen] came up and gave me a reference to some obscure journal." To Russell it appeared that Bardeen "knew everything that was going on in his field."

Bardeen occasionally helped the students of other professors. John Wheatley suggested that his student, Andy Anderson, go see Bardeen about the interpretation of some of his experimental results. They differed by a factor of two from those of the eminent Russian theorists, I. L. Bekarevich and I. M. Khalatnikov, given in a calculation of the Kapitza thermal boundary resistance at the interface between liquid ^{3}He and a solid. Bardeen watched silently while Anderson scribbled his figures on the blackboard. After the younger man stopped speaking, Bardeen sat gazing into space.

Anderson backed toward the door uneasily and reached for the knob. Suddenly Bardeen started talking. He told Anderson that he knew the Russians' work and that it was unlikely they had made such a mistake, but he was unwilling to make any definite pronouncements on a subject that he had not thoroughly thought out himself. "I then escaped feeling very embarrassed at having bothered him with my stupid ideas," recalled Anderson. However, some months later Bardeen assigned the problem of the factor of two to a new French postdoc, J. Gavoret, who found the source of the error and published it in the *Physical Review*. "Only then did I realize that Bardeen had taken me seriously."

Bardeen never took on more than a few graduate students at a time (see Figure 10-1). One of his star students in the early 1960s was William McMillan. A true individualist, McMillan had a pronounced stutter, a magnificent beard, and a penchant for applying

FIGURE 10-1 Bardeen's graduate students, 1952-1974

Name of Student	Dissertation Title	Year
Paul J. Leurgans	Interaction of Excitons and Impurity Atoms with Phonons in Germanium	1952
Nick Holonyak, Jr. (electrical engineering)	Effect of Surface Conditions on Characteristics of Rectifier Junctions	1954
Thomas Nolan Morgan	Photoconductivity in Germanium at Liquid Helium Temperatures	1955
Newton Bernades	Theory of the Specific Heat of Superconductors Based on an Energy Gap Model	1957
Daniel Charles Mattis	Conductivity Problems in Quantum Theory	1957
J. Robert Schrieffer	The Theory of Superconductivity	1957
R. C. Sirrine (electrical engineering)	An Investigation of the Surface States on N-Type Germanium	1957
Michael Francis Millea	Effect of Heavy Doping on the Diffusion of Impurities in Silicon	1958
Milton William Valenta	Effect of Heavy Doping on the Diffusion of Impurities in Germanium	1958
Ling Yun Wei (electrical engineering)	Diffusion of Silver, Copper, Cobalt, and Iron in Germanium	1958
Stephen Reynolds Arnold	Effect of Ion Bombardment on Semiconductor Surfaces	1959
William Manos Portnoy	The Conductance of a Cleaned Germanium Surface	1959
Richard Elmo Coovert	Surface Magnetoconductivity Experiments on Silicon	1960
Piotr B. Miller	Frequency Dependent Hall Effect in Normal and Superconducting Metals	1960

continued on next page

FIGURE 10-1 *Continued*

Name of Student	Dissertation Title	Year
Rolland A. Missman, Jr.	Galvanomagnetic Properties of Carriers Associated with the Cleaned Germanium Surface	1960
Kendal True Rogers	Superconductivity in Small Systems	1960
Toshihito Tsuneto (thesis submitted to Kyoto University, Japan)	Part One: Transverse Collective Excitations in Superconductors and Electromagnetic Absorbtion. Part Two: Ultrasonic Attenuation in Superconductors	1960
Jerome Luther Hartke	Drift Mobilities of Electrons and Holes and Space-Charge-Limited Currents in Amorphous Selenium Films	1961
Kenneth Rose (electrical engineering)	A Microwave Technique for the Study of Deviations from Ohm's Law in High Resistivity Superconductors	1961
Peter Vance Gray	Tunnelling from Metal to Superconductors	1962
Daniel Warren Hone	Spin Diffusion and Other Transport Properties of Liquid He^3	1962
William L. McMillan	Ground State of Liquid He^4	1963
John Warren Wilkins	Part I: Tunnelling in Superconductors: Second-Order Processes and Lifetime Effects Part II: A Contribution to the Theory of the Electron-Phonon Interaction	1963
Wayne Earl Tefft	A Theory of the Electrical Resistivity of Liquid Metals	1964
John Richard Clem	Effects of Anisotropy of the Superconducting Energy Gap	1965
Wesley N. Mathews, Jr.	Bogoliubov Equations and Their Application to a Normal Superconducting Boundary	1966

FIGURE 10-1 *Continued*

Name of Student	Dissertation Title	Year
Sang Boo Nam	Theory of Electromagnetic Properties of Strong Coupling and Impure Superconductors	1966
Reiner Kümmel (thesis submitted to Goethe-Universität, Frankfurt, Germany)	A: Schichtdicken-abhängiger Quantisierungseffekt in Tunnelkontackten B: Untersuchungen zum Zwischenzustand und gemischten Zustand von Supraleitern I. und II. Art.	1968
David William Allender	Model for an Exciton Mechanism of Superconductivity in Planar Geometry	1974
James William Bray	Fluctuation Conductivity from Charge Density Waves in Pseudo-One-Dimensional Systems	1974
Jared Logan Johnson	Current Flow in Inhomogenous Superconductors	1974

ingenious methods to challenging problems. For his thesis he convinced Bardeen to let him apply Monte Carlo statistical sampling methods to calculate properties of the ground state of superfluid helium. In this state, analogous to the state of superconductivity except that the particles have no charge, the fluid flows without friction. Bardeen knew little about Monte Carlo methods but felt he could rely on McMillan to judge whether they would apply in this case. McMillan completed his doctorate in 1964 and took the department's annual award for best Ph.D. thesis.

Following a stint at Bell Labs, where he worked with John Rowell deducing lattice vibration spectra in superconductors from tunneling data, McMillan returned to Illinois in 1972 as a professor. Once back in Urbana he continued to work on liquid crystals, light scattering, heat capacity measurements, and charge density waves, as well as electrically driven structural phase transitions and their interactions with superconductivity. Bardeen had the highest regard for McMillan, whom he once described as "very bright, original—needs self-discipline." He later retracted the last bit and said that McMillan "was probably putting in his time to a better advantage than studying the things required for courses." He told McMillan in 1978 that his "work on deriving phonon spectra directly from tunneling data" has had a great "impact and is one of the main reasons why superconductivity has been called the best understood of solid-state phenomena."

The strength of the bonds that formed between Bardeen and his many students and postdocs brought many of them back to the University of Illinois, at least for a time. The Bardeens often invited students and colleagues as well as friends to their house, especially in their early years in the community when Champaign-Urbana offered few restaurants or entertainments. Jane would do the cooking and planning, usually keeping it simple. "It was very informal and very spontaneous." Everyone appreciated her hospitality and warmth. In her presence people felt at ease and attended to. On social occasions, "I always immediately ran to Jane, because she was easy to talk to," said Karl Hess, a postdoc from Austria, recalling his early days in Champaign-Urbana. To Hess and his wife, their first Christmas in the United States looked dismal until the Bardeens invited them over to celebrate Christmas Eve. Hess remembers John showing the two toddlers his medals and a strange golf ball that had a transistor radio inside. Another year, when postdoc Ludwig Tewordt and his wife found it impossible to visit family in Europe at Christmas, John and Jane stopped by their apartment on Christmas Day bearing gifts for their three children.

In recalling social events at the Bardeen home, students typically used the words "relaxed" and "informal." The house was furnished for comfort rather than elegance. The kitchen was the focal point. Photos cluttered the piano top. The walls held art from the Harmer side of John's family. Shelves overflowed with books and mementos from trips. The entire effect was modest. Neither John nor Jane ever thought of themselves as wealthy or socially upper crust. Hess recalls that he and his wife arrived dressed up for their first party at the Bardeens' and were shocked to see everyone else in shorts. Even more surprising was to find Bardeen, the great Nobel Prize winner, in the kitchen serving drinks.

Bardeen continued to be extremely popular with young children, with whom his reserve melted. He often viewed the world through childlike eyes, diving under the bedcovers to play "loon" with his grandson, or playing jacks on the porch floor with the children of friends. Tom Bardeen, John's nephew, basked in the feeling that he had "Uncle John's" undivided attention whenever they interacted. Ned Goldwasser remembered one "absolutely superb" experience at a department picnic. When one of his children screwed up his courage and asked Bardeen a question about relativity, the Nobel laureate did his best to explain the theory in terms the ten-year-old could understand.

When the Bardeens entertained they often issued special invitations to the children of their guests. One of the authors will never forget the afternoon in 1985 when Bardeen personally telephoned to invite her young son, Michael Baym, and infant daughter, Carol Baym, to a party. Some years later when Bardeen passed away, nine-year-old Michael felt that he had lost a friend and insisted on attending the memorial held in Bardeen's memory.

For his own children, John was the model of a scientist engrossed and happy in his work. As the children grew older he often had answers for their questions about nature. His approach, as Jim interpreted it, was "opening doors, but not trying to force us through them." The method worked well enough to interest all three Bardeen children in the physical sciences. Jim and Bill became physicists. Betsy studied anthropology, then computer programming. In 1966 she married a physicist, Tom Greytak.

11

Cracking the Riddle of Superconductivity

The network of support Bardeen found at the University of Illinois formed the backdrop for his continuing work on superconductivity. A year after Serin's call in May 1950 with the insight about the lattice vibrations, Bardeen was still struggling to make further progress. In July he complained to Rudolf Peierls that all the methods he had tried for treating superconductivity had failed. Still he sensed the work was moving in the right direction. "I believe that the explanation of the superconducting properties is to be found along the lines suggested by Fritz London."

The hint that bolstered Bardeen's confidence was small but significant: "The wave functions for the electrons are not altered very much by a magnetic field." From this "rigidity" of the superconducting wave function Bardeen derived his sense of certainty about the nature of the solution. The rigidity followed from the energy gap. Somehow, he believed, it had to do with the long-range ordering. That was, he guessed, why superconductors expel magnetic fields (the Meissner effect). But these were only intuitions. He could not yet express them in any language, not even that of mathematics.

Bardeen's notes and letters about superconductivity let others follow the painstaking route he now took in approaching the riddle. In 1951 his handwritten notes include the following list of smaller problems he thought would be useful to solve along the way:

(1) Derivation of London equations for multiply connected body

(2) Proof of current and effective mass theories for one-electron wave functions

(3) Analysis of diamagnetic properties of gas of electrons with small effective mass

(4) Extension of 3 to include high frequencies and scattering of electrons (effect of electric field)

(5) Boundary energy for thin films

(6) Better calculation of interaction energies

(7) Specific heat and other thermal properties.

All these "sub-problems" involved situations that were accessible through experimental or theoretical research.

The first one deals with experiments in which currents were observed to flow indefinitely around a ring (a multiply connected body). The current and effective mass theories in the second item refer to Bardeen's wish to fully understand the theory of single-electron wave functions. Diamagnetic properties were critical because of the observation that superconductors expel magnetic fields. Examining the case of high frequencies enabled drawing on experiments offering information about the dynamics of superconductors. Boundary and interaction energies were important to research, especially for thin-film experiments in which one can study the region near the surface where magnetic fields are not completely excluded, as in the thick sample case. Finally, the specific heat and other thermal properties were measurable properties that could test one's understanding of the superconducting state.

These problems could not be treated using a free-electron model. They resulted, Bardeen knew, from the countless interactions between electrons and between electrons and the lattice. Bardeen worried that in using the standard Hartree-Fock approximation method, which assumed that electrons move independently in a self-consistent field, he might be eliminating a crucial—perhaps the most crucial—aspect of the system. As he stated in a talk given later that year, he hoped to draw on the new field theory tools developed shortly after World War II by Richard Feynman, Julian Schwinger, and Sin-itero Tomonaga. "It was becoming clear that field theory might be useful in solving the many-body problems of a Fermi gas with attractive interactions between the particles."

Routinely taught to theory graduate students by 1950, field theory had not been part of Bardeen's training. The fastest way for him to draw on the new formalism would be to take on a collabora-tor. He was particularly interested in exploring the many-body for-malism that David Bohm, one of Robert Oppenheimer's doctoral students, had developed in Berkeley.

Bohm's original theory concerned the electron–electron inter-actions in the ionized gases known as electron plasmas, a problem of interest to the Manhattan Project, for plasmas were relevant in separating uranium isotopes electromagnetically. Bohm had con-tinued to work on the problem after he joined the faculty at Princeton in 1946. With his graduate students, Eugene Gross and David Pines, Bohm extended the techniques he had developed to modeling the electron interactions in the electron gas.

Learning of Bohm's work on a visit to Princeton during the spring of 1950, Bardeen became interested in how Bohm's theory separated the long-range Coulomb interactions from the short-range single-particle excitations. He thought the techniques might also work in treating the electron–electron interactions in super-conductors. He asked Bohm in 1951 whether any of his students might be interested in a postdoctoral position at Illinois. Bohm suggested Pines, who arrived in Urbana in July 1952.

Bardeen asked Pines to start by looking at a simpler problem that appeared to involve some of the same physics as super-conductors. The problem of the polaron, which Herbert Fröhlich had studied earlier, dealt with how an electron distorts a crystal lattice as it moves along in it—in analogy with the way that a child distorts the mattress of a soft bed as he or she walks on it. In the polaron problem it is the ions in the crystal that cause distortion of the lattice around the electron that moves along. In employing the technique of "canonical transformations" and writing down a de-scription of the energy in terms of operators that can remove or add particles to the system (the method of "second quantization"), Fröhlich had been among the first to introduce field theory con-cepts into solid-state physics. Bardeen wanted to extend this work to how the electrons couple with the movement of the ions to form waves ("lattice vibrations").

In discussing the problem with Tsung-Dao Lee, a young theorist (and future Nobel laureate) who was spending the summer of 1952 in Urbana as Bardeen's postdoc, Pines realized that the "intermedi-ate coupling method," which Lee had recently used in his own field theory studies, could be adapted for the polaron problem. Francis

Low, an assistant professor in the Illinois physics department, also joined the discussion. In their subsequent paper, Lee, Low, and Pines formulated the polaron problem in such a way that it could later be applied directly to the ground-state wave function specifying the state of lowest energy in a superconductor.

Bardeen also worked with Pines to extend the Bohm-Pines formalism to treat the coupling of electrons to the lattice vibrations. One calculation compared the two kinds of electron–electron interactions: the attractive interaction induced by the lattice vibrations (or "phonons") and the ordinary repulsive interaction deriving from the fact that like charges repel.

Gradually Pines and Bardeen came to understand that the attractive phonon-induced interaction arises in situations like that of the polaron, in which an electron moves along in the crystal pulling the positive lattice in toward it. When the lattice moves in, it pulls along other electrons as well, so that the electrons are *in effect* attracting one another. They also repel one other in the ordinary way because they are like charges. But in their calculations Bardeen and Pines found that in cases where the energy transfer is small, the attractive interaction is actually *stronger* than the repulsive one. This exciting result implied that for pairs of electrons near the Fermi surface the net interaction is attractive! Could this be the mechanism behind superconductivity? It would take Bardeen five more years to fully answer this question, in the affirmative.

In 1953 Bardeen undertook an extensive literature study of superconductivity for an article that he agreed to author for the *Handbuch der Physik*, a major review encyclopedia. Each year the editor, then Sigfried Flügge, selected a research topic that he judged ripe for review and commissioned authors to write on the topic. For the 1956 *Handbuch*, Flügge selected low-temperature physics as the topic. He asked Bernard Serin to review the experimental situation and Bardeen the theoretical outlook on superconductivity.

Bardeen concentrated on the review during 1954, writing almost 100 pages. Trying to identify the missing pieces in the current understanding of superconductivity and the concepts that would be needed to develop a complete theory, he argued for the London notion of an "ordered phase in which quantum effects extend over large distances in space." He ventured to say that superconductors are "probably characterized by some sort of order

parameter which goes to zero at the transition," admitting, however, that "we do not have any understanding at all of what the
order parameter represents in physical terms."

Bardeen's article also explored the nature of the phase transition between the normal and superconducting states. Following
Fritz London, he tried to unravel the meaning of the energy gap in
the electronic excitation spectrum resulting from the "rigidity" of
the wave function. At this stage Bardeen could not yet explain such
a gap, but by assuming its existence he could show how to develop
both the electrodynamic properties of superconductors and a gen-
eralization of the London equations, resembling the empirically
based nonlocal formulation of superconductor electrodynamics
that Pippard had put forth several years earlier.

Another focus of Bardeen's review was the theoretical machinery for computing the interactions between electrons or between
electrons and phonons (lattice waves) in the superconductor.
Appealing to recently developed field theory techniques, including
Tomonaga's strong-coupling approach and the Bohm-Pines theory,
he underscored the importance of considering the electrons as electrically screened by the cloud of positive charge surrounding them
when they travel through the system. Bardeen's review concluded:
"A framework for an adequate theory of superconductivity exists,
but the problem is an exceedingly difficult one. Some radically new
ideas are required."

In the period when Bardeen was working on his *Handbuch*
article, J. Robert Schrieffer—the "S" of the BCS (Bardeen-Cooper-
Schrieffer) theory of superconductivity—came to Urbana from MIT.
He arrived in the fall of 1953, having written an undergraduate
thesis under John Slater on the energy-level spacing in the multiplet
structure of transition metal atoms. His exposure to the work of
Slater's group had whetted Schrieffer's appetite for theoretical physics.
He developed "a model in my own mind of how creativity takes
place and how this structured process of going from the known to
the unknown in finite steps is very important." He also came to
realize that "one can't make the intuitive leap immediately, and
there is a heck of a lot of hard tough work that goes in between."

Schrieffer was not completely happy with Slater's approach.
He felt that the people in Slater's group seemed to be "doing
numerical calculations on and on." Schrieffer decided that he
"wanted to have a graduate education which perhaps allowed more

flexibility for individual creativity." Applying to several graduate programs, he was "shocked" to receive a letter from the University of Illinois inviting him to work with Bardeen. He had "heard outstanding things about Professor Bardeen."

On the day Schrieffer came to meet his new advisor, Bardeen's office appeared to be empty. But while Schrieffer wandered around the building, "I passed a gentleman on the steps three times in a row, and it happened to be Professor Bardeen." Schrieffer spent the next year and a half taking courses, working out physics problems, and experimenting in Bardeen's semiconductor laboratory. Schrieffer met with Bardeen "at least weekly and often more frequently." He never felt "any problem of coming and chatting with him. He was sympathetic, always suggesting how to get around difficulties."

In the spring of 1955, after working with Bardeen for a year and a half, Schrieffer approached Bardeen about selecting a thesis topic. Well prepared, Bardeen reached into his bottom drawer and produced a list of about ten problems that he considered suitable. When Schrieffer glanced at the list, he noticed that one question involved explaining why a metal with a magnetic impurity shows a resistance minimum in the electrical resistivity versus temperature curve. Normally the resistivity rises with temperature in a smooth curve. Bardeen and other many-body theorists recognized that explaining this resistance minimum was an important problem years before Jun Kondo solved it in the late 1960s.

Schrieffer's attention fixed on the last item on the list, superconductivity. "Why don't you think about it?" said Bardeen. Schrieffer understood that selecting superconductivity for his thesis would be risky. The problem "had been worked on for a long time and workers had met a lot of failures." He also knew something about the problem already, having helped to proofread Bardeen's *Handbuch* article. "The question I had in my mind was: was there something that I might do and that I might contribute? The other question was whether Professor Bardeen himself would be concentrating on this area, so that we would have in some sense a useful interaction."

Schrieffer sought the advice of Francis Low. "How old are you?" Low asked him. "Twenty-four." "Well, you can waste a year of your life and see how it goes." Schrieffer took this answer "as reasonable evidence that he felt there might be some chance of doing something."

"OK, fine," said Bardeen when Schrieffer told him he would work on superconductivity. Bardeen mentioned that Leon Cooper, a young theorist trained at Columbia University would be joining them. He explained that Cooper "had a field theory background and that this might be useful." Cooper had studied nuclear theory under James Rainwater, who, in 1975, would share a Nobel Prize with Aage Bohr and Ben Mottelson for developing a theory of nuclear structure based on connecting ideas of collective motion and particle motion in the nucleus.

The addition of Cooper—the "C" of BCS—had been another consequence of Bardeen's creative use of teamwork. Pines was scheduled to leave Urbana at the end of the 1954–1955 academic year to take a teaching post at Princeton. Still concerned about his own limited training in field theory and Feynman's and Fröhlich's considerable advantage in this area, Bardeen had looked for someone else "versed in field theory who might be willing to work on superconductivity." In the spring of 1955, Bardeen telephoned the theorist Chen Ning Yang, then at the Princeton Institute for Advanced Study, and asked for a recommendation.

Bardeen recalled that "one of the active many-body problems at that time was the structure of the nucleus—many neutrons and protons making up a nucleus. I thought that someone with a background in that field could provide useful information to the problem of the interaction of the electrons and the vibrations of the crystal lattice of the metal and what made it superconducting." Yang recommended Cooper, who was spending a postdoctoral year at the institute. Cooper was up-to-date in "the latest and most fashionable theoretical techniques" (at that time, Feynman diagrams, renormalization methods, and functional integrals).

When Bardeen stopped by the institute to meet Cooper, the younger physicist began by saying he didn't know anything about superconductivity. Bardeen told him "that didn't matter, that he'd teach me everything and that he was looking for someone who was familiar with current field theoretic techniques." Cooper also told Bardeen that he doubted field theory would be of any use in explaining superconductivity. But he added that he was ready to leave Princeton. By then, "progress in field theory seemed rather discouraging." Although most of the theorists at the institute were still working on the consequences of the great breakthroughs in field theory, the "first flush had already passed." Reflecting on the work of Schwinger, Feynman, and Tomonaga, they "used to sit around saying, 'These guys solved all the easy problems and left the hard ones for us.'" Einstein's death that spring also cast a shadow over the institute.

Cooper pondered Bardeen's offer for a few months. Then early in the summer, while in Sicily, he decided that superconductivity would be "my problem I was going to solve." He returned to Princeton and drove to Urbana.

When Cooper arrived, he found he "really didn't like the geography" of this "corn-field place." But he soon discovered that "the department was wonderful. The people were wonderful. It was a fantastic environment." He vividly recalled the "parties and comradery. And we were always together."

Bardeen asked Cooper to offer a series of informal seminars on field theory as it was used in electrodynamics and to speculate on how it might possibly inform problems of many-body theory. Cooper talked about Feynman diagrams that corresponded to virtual excitations of the Fermi gas. He included a discussion of Feynman's notion that a positron (antielectron) or a hole are equivalent to an electron going backwards in time. Cooper said that the diagrammatic methods were "by and large perturbative techniques rather than techniques which would lead to phase transitions or a qualitative change of the nature of the matter in question." Schrieffer found Cooper's talks "very clear," but he was bothered by his pessimism about whether the techniques could help with superconductivity.

Cooper did think that Schwinger's work on coupled Green's function methods might be useful, as they were not necessarily perturbative methods. But "we didn't want to use totally unfamiliar mathematics," Schrieffer recalled, so that line was dropped. It was later taken up successfully by Lev Gor'kov in the Soviet Union.

To help Cooper educate himself about superconductivity, Bardeen suggested he read two recent texts, one by Fritz London, the other by David Shoenberg. London's book, published in 1950, offered a reformulation of his earlier ideas about electrons ordered in a system over large distances in a macroscopic quantum state. London called superconductivity a "quantum structure on a macroscopic scale." The state resulted from "a kind of solidification or condensation" of the average momentum distribution of the electrons. Cooper, who also took his turn proofreading Bardeen's *Handbuch* article, agreed that "if we got something like London's long-range order then we'd get superconductivity." The mathematical problem was to fill in the intermediate steps.

The notion of broken symmetry proved a fertile framework for conceptualizing superconductivity. It describes, for example, the case of a spherical magnet in which the magnetic field is free to point in any direction. But any given spherical magnet always has a north and south pole: the fact that it chooses a certain direction for its axis breaks the system's symmetry.

A more familiar example of broken symmetry is found at the bottom of an ordinary wine bottle, one in which the center of the bottom edge curves upwards. If the bottle were symmetrical, the residue particles would all sit just at the center, on top of the bump. But for any real bottle, things are not symmetrical, so most or all the particles roll down and come to rest at one of the infinite number of places around the edge, which have lower elevation. By choosing any one place to roll down to, the particle breaks the symmetry.

Cooper focused on the energy gap. For the case of a normal metal, "you sort of knew what that was like." But for the superconductor, it "was qualitatively different. The question was how does that come about?" He argued that if there was "such a radical difference as a single particle energy gap, chances are everything else is going to come out." It seemed to Cooper "that this was really a very simple problem," one that should be soluble by "just elementary quantum mechanics. And why am I throwing all of this apparatus at it?"

In later years, Cooper described his own approach to problem solving as first examining the simplest possible version of the problem that retained the essential features. "I think one of the ways people delude themselves is to try to solve the more complicated problem before trying to solve the simpler problem that's along the way." The approach was a version of the one Bardeen claimed to have learned from Wagner.

In the last months of 1955 Cooper turned to "what happens if you have a highly degenerate system with an attractive interaction." He examined the simple case of two electrons just outside the Fermi surface, spinning in opposite directions. He tried "to see how, under what assumptions could you get a Meissner effect." Making certain assumptions, he showed that if the net force between the two electrons is attractive, they will form a bound state lying below the normal continuum of states and separated from the continuum by an energy gap. By late February or early March of 1956, "it seemed clear that if somehow the entire ground state could be composed of such pairs, one would have a ground state with qualitatively different properties from the normal state." And this ground state—the state of superconductivity—would be separated from the excited states by an energy gap.

Cooper "was very excited," said Schrieffer, who remembered the day Cooper discovered the bound state. It "was there regardless of how weak was the coupling." Cooper recalled, "I was reasonably excited by these results. But I became aware, painfully, in the months that followed how long the road still was." For at this stage, Cooper's communication with Bardeen temporarily broke down. "We went through a period where he thought I was out of my mind."

It was not that Bardeen did not recognize the importance of the Cooper pairs, but rather that he did not want to let the enthusiasm of this step obscure the need, and difficulty, of developing a detailed solution that would consolidate all the known facts. He recognized the vast amount of work that still needed to be done before the problem could be considered solved. As Cooper saw it, "Bardeen couldn't figure out what I was doing."

Despite the tension that now ensued, Cooper still felt that all three formed a remarkable team in which "each contributed parts that were so essential." Years later he said, "I can't imagine any more cooperative feeling. The advance of one was the advance of another." The team's interplay recalled the collaborative atmosphere of the Bell Labs semiconductor team before the invention of the transistor.

Bardeen structured the BCS team on his favorite social model, the family. He was the patriarch and the children were expected to pitch in. In making assignments, motivating members, and planting theoretical seeds, he tried to use the unique talents of each member. Like his own father, Bardeen treated his students as fellow explorers, giving them latitude and allowing them to suffer consequences of their actions. He shared with them his deep understanding of the quantum theory of solids, which went "all the way back to the very beginning, back to the 1930s," said John Miller.

When discussing a problem, Bardeen often impressed his students and younger colleagues with his ability to cite specific steps from particular sections of a work he had read.

Breaking down the problem of superconductivity into smaller parts, he asked Schrieffer to look into the "t-matrix methods" that Keith Brueckner had recently developed in studying nuclei. He asked Cooper to examine the Bohm-Pines theory and the Bardeen-Pines work in 1954 on the electron–electron interaction. Meanwhile he sought other leads. Schrieffer emphasized "the enormous fun it was for the family to work together."

Jane Bardeen, "one of the most affable, lovable, outgoing people," played the role of the nurturing mother. "From the very beginning I was invited as part of their family," said Schrieffer. He recalled telling Betsy stories "about snakes and alligators in Florida."

Typically Bardeen and Cooper would work independently in their shared office, Room 307 in the Physics Lab on Green Street. Schrieffer worked in a larger room upstairs, between the third and fourth floors, shared with other graduate students. A sign on its door read "Institute for Retarded Study." For a graduate student in that period, "if somehow you were able to move to the Institute for Retarded Study, you had made it." Whenever a desk opened up, "everyone would sort of scramble around to see who could get in there." Schrieffer found that he learned as much as he did in his courses from talking with other students in that room.

When Schrieffer came to Bardeen's office, to speak with either Cooper or Bardeen, both would "wheel around their chairs" and join the discussion. "It was a sort of round robin, where I think John and Leon probably didn't talk too much more than that," said Schrieffer. Schrieffer also recalled the collaboration as "a very happy relationship, which largely came about because there weren't enough offices for everyone."

"He was very stubborn," Cooper said, recalling the work with Bardeen, but "I was very stubborn too!" Cooper was constantly amazed at how quickly the older physicist could calculate and learn. "For example, if I would suddenly understand something he didn't understand, then I'd show him, you know, one morning—BAM!—he had it instantly. He was using it in the afternoon. He was very fast."

Schrieffer claims that he and Cooper both gained an appreciation for Bardeen's style of doing physics—his taste in choosing problems, his experiment-based methodology, the way he broke problems down, and his approach of using as little theoretical machinery as possible, "the smallest weapon available in your arsenal to kill a monster."

Schrieffer also found useful the advice that Bardeen passed down from Wigner: "to divide up the problem into small pieces, and attack each one." In the version that Schrieffer later recalled, there was an added maxim "to focus your shots so that each one counts, and then reassemble." Bardeen "was always very leery of someone who started out with a grand formalism and deduced proof." He felt that it was more effective to work on the "little problems along the way." By synthesizing the smaller components into the larger picture, "the big problem will get solved by nature if you just keep at it." Of utmost importance was "carefully deciding which is the essential piece." It seemed to Schrieffer that this was "how [Bardeen] went about everything—thinking about it carefully and isolating those pieces which were relevant. And then coming back."

Bardeen took very seriously the suggestions laid out in the last chapter of London's book on how to proceed toward a microscopic theory. He felt that "those suggestions were exactly the direction one should go," although how to "transcribe that into mathematical reality and detailed theory was far from clear." Schrieffer remembered that Bardeen "kept saying in effect, 'This system has to be rigid. There has to be a gap.'" In agreement with London, Bardeen stressed that the condensation of electrons had to occur in momentum space, not ordinary space.

Bardeen adopted other guiding ideas of London as well. One was that "in an isolated simply connected superconductor and for a given applied field, there is only one stable current distribution" (as a consequence of the Meissner effect). Another posited that in thermal equilibrium there is no "permanent current in an isolated superconductor (in agreement with Bloch's theorem) except in the presence of an applied magnetic field." Another was that "there is no conservation of these currents; they differ for every variation of the strength or direction of the applied field."

Bardeen pressed his team to clarify the notion of long-range order using a "phase coherence" parameter that determines the size of a Cooper pair over whatever distance their motions are correlated. Pippard had proposed that this distance was of the order of a micron. Schieffer remarked that Bardeen "had a feeling" the condensation involved only the electrons close to the Fermi surface. Out of roughly 10^{23} electrons in every cubic centimeter, only 10^{19} will be close to the Fermi surface, but this is still an enormous number of condensed electrons!

In the 1920s Niels Bohr had helped younger physicists of his generation seek the as-yet-unknown quantum mechanics by formulating a theoretical bridge known as the "correspondence principle." The principle created a link between wave mechanics and Newtonian mechanics as one approaches macroscopic systems, the limit of high quantum numbers. Similarly Bardeen helped his team strike out into the unknown by offering a principle that would bridge the poorly understood theory of interacting electrons in the normal state with the properties of states of noninteracting electrons. Anticipating Lev Landau's theory of Fermi liquids, Bardeen realized that, although in the normal state electrons are not free, one could assume a one-to-one correspondence between interacting states and free electron states. One can arrive at the interacting gas by turning on the interaction slowly and deforming the free-electron gas continuously into the interacting system. He expected a one-to-one correspondence between the states of real superconductors with simplified states in which one included only the weak interactions responsible for superconductivity.

What they needed to do, Bardeen said, was "to think of the normal state as like a free electron gas, with the excitations being in one-to-one correspondence with those of a free electron gas—the interactions between the actual electrons are enormous, but we have to think in terms of effective excitations which include these strong interactions insofar as they enter the normal state." In other words, Bardeen assumed that the bulk of the interaction effects would shape the normal state and that small residual interactions which occur in the normal state, but are not fully taken into account theoretically, would lead to superconductivity. He felt that one should start by simply studying the effects of these residual interactions on free electrons.

Bardeen was, Schrieffer explained, introducing a basic tenet of the Fermi liquid theory, to be developed by Landau, the idea of "quasi-particles," "effective electrons" existing in a one-to-one correspondence with the electrons of a free electron gas. Bardeen introduced this formulation independently of Landau, at a time when the notion of quasi particles was not yet firmly grasped in the Western world.

Cooper continued to feel that Bardeen was ignoring the breakthrough he had made several months after his arrival in Urbana. He spent almost a "year in the wilderness" trying to convince Bardeen that pairing held the key to the solution. He recognized the psychology of his position: "Here I am, totally unknown, this wild-eyed kid. . . . And I say, well that's the solution."

Bardeen insisted that Cooper's solution was not complete. They did not yet know how to go from a single "Cooper pair" to a full-blown many-electron theory. "We tried many techniques," Schrieffer recalled but "things just didn't jell." A major hurdle was coping with many pairs at the same time, most of which overlapped. Even conceptualizing the situation was very difficult.

More than a year later Schrieffer found a way to portray the problem using the analogy of a crowded dance floor on which many couples are doing the Frug. In this popular late-1950s dance, the members of couples dance separately but remain bound to one another, even when they are far apart and other dancers come between them. The problem was how to represent this situation mathematically. "It was very perplexing to us." Schrieffer remembers Bardeen saying, "Well, it's in momentum space. You shouldn't think about the coordinate space so much. It may not be confusing if you view it in the right language."

In seeking wave functions, they worried about the validity of their approximations. If they worked only with the part of the system responsible for pairing, might they be ignoring something else that was even more important? Schrieffer recalled, "We were feeling a little bit downtrodden because things weren't breaking so quickly after Leon's contribution."

On November 1, 1956 Schrieffer ran into Bardeen on the street. Bardeen smiled, "Oh, I just wanted to mention—I won the Nobel Prize. " He was still adjusting to the news, which had taken him by surprise at 7 a.m. while he was in the kitchen frying eggs for the family. He had assumed this duty temporarily because Jane was still under doctor's orders to rest after her bout with typhoid fever the previous summer. Suddenly Betsy and Bill rushed into the kitchen shouting that he won the Nobel Prize. It had been announced on the CBS *World News Roundup*. John dropped the frying pan.

"Well, I guess I better go shave," John drawled during breakfast, when the whole family heard the announcement repeated. Jane scribbled down some notes about that day. "The children were jubilant, John a bit pale with a dreamy abstracted look." After she sent them all off to school, "receiving congratulations was my pleasant chore all day." Brattain was among the many who phoned.

Wheeler Loomis had asked Jane to keep John home. That evening she served a steak dinner in an attempt at "a normal Thursday evening." When the doorbell rang, Bardeen opened the door. He saw Loomis followed by a parade of "about 60 physicists and wives marching down the road carrying flashlight 'torches' and singing 'For He's a Jolly Good Fellow.'" They brought "champagne, cups and cakes" for an "instant and gala party." "It was really thrilling," Jane recalled, to see "all those people in the dark, coming down the hill, singing and waving their lights." Paul Handler reported, "Everybody was elated, everybody was on a high."

Reporters clustered outside Bardeen's home and office in the days following the announcement. The *Washington Observer*, the local paper in Washington, Pennsylvania, announced, "Husband of Former Local Girl Nobel Prize Winner." Congratulatory notes and telegrams poured in from colleagues and friends. Wigner wrote, "I can't tell you how proud I am of our past association and how much pleasure I derive from this honor that came your way." The prize was "a matter of great personal pleasure and satisfaction to me," wrote John Van Vleck. "I like to recall that both you and Brattain studied Quantum Mechanics with me, although I would not for a moment claim that this had anything to do with your careers." He added, "Also, it is fine to have a recipient who is a Madisonian, a Wisconsin graduate, and a member of the Harvard Society of Fellows. Furthermore, one should not overlook the Minnesota angle, since you and I both taught there, Abigail [Van Vleck's wife] graduated there, and this is where Brattain took his Ph.D. All told, the award is most satisfying." Dutch Osterhoudt, Bardeen's college friend and Gulf Labs colleague, wrote, "Gretchen and I got a thrill when we heard the first announcement over the radio."

Bardeen was deeply gratified by the prestigious award, but a part of him could not fully engage in the celebrations and congratulations. Deep within, "he felt he didn't deserve a Nobel for the transistor," reflected David Lazarus. In late November Bardeen wrote to E. J. W. Vewey at Philips Laboratory, "I suspect that the worry of many of those on the committee was whether the science itself was worthy of the award. I have my doubts of that myself."

The celebrations were a huge distraction from his work on superconductivity. Schrieffer and Cooper may have felt stuck, but Bardeen sensed that they were in fact nearing breakthrough. He was keenly aware of the competition—especially from Feynman, who in September had spoken on superfluidity and superconductivity at the International Congress on Theoretical Physics in Seattle. Feynman ended the talk with a statement about the problem of superconductivity: "When one works on it—I warn you before you start—one comes up finally to a terrible shock: one discovers that he is too stupid to solve the problem."

It appeared to Jane that John worked on his research all the time in the days before going to Stockholm. She wrote in her Nobel diary that he "really worked, day and night, until we left Champaign November 29." When one reporter who had discovered Bardeen's passion for golf, remarked, "Oh! Just like President Eisenhower!" Bardeen grumped, "Yes, only I don't have as much time for it as he does."

Schrieffer, now a fourth-year student, also had "mixed feelings" about Bardeen's prize. He was enormously pleased for Bardeen and the excitement "sort of boosted us up." But he had a personal matter to resolve. He wanted to accept an attractive National Science Foundation fellowship he had won for study in Europe during the following year. But a condition was that he be done with his thesis. That seemed to be going nowhere. He went to see Bardeen shortly before the latter's trip to Stockholm. Cautiously, Schrieffer asked whether it might make sense for him to switch to another problem—perhaps ferromagnetism, which he had started "to quietly work on." Bardeen mumbled, "Give it another month, or a month and a half. Wait 'til I get back and keep working. Maybe something'll happen."

The preparations for Stockholm absorbed much energy. A "three-way phone call with Walter and Bill settled the division of labor on the Nobel speeches, which had to be written." Appropriate clothes had to be found for all the events. For the formal ceremony, Jane wanted something "significant." Seitz recalled, "The girls all piled in to help her." Then John started to worry that Jane would "spend all the money before we get it." On one rainy day, after much fruitless searching for a dress in town, Jane took a Greyhound bus to Indianapolis and found a long royal blue silk faille gown and several other formal dresses and accessories.

The celebrations began even before the Bardeens reached Stockholm. On November 30, William Baker, then vice-president in charge of research at Bell Labs, hosted a pre-Stockholm dinner in honor of the Nobel laureates in New York City. Jim took a break from his studies at Harvard for his father's occasion. Many friends and colleagues from Bell attended.

A series of talks were part of the celebration. Ralph Bown spoke on "the origin of the transistor discovery," Conyers Herring on "the contribution of the transistor discovery to basic scientific knowledge," Jack Morton on "measures of the transistor's impact," and Karl K. Darrow on "history of the Nobel Prizes, particularly in physics." John and Jane stayed in New Jersey, at the home of Philip and Joyce Anderson. The Andersons could sense Bardeen's impatience with all the fanfare. At one moment, when John thought he was alone, Joyce heard him swear ferociously at a shoe.

Bill and Betsy, then 15 and 12, stayed home in Champaign with friends. "We didn't expect that it would be proper" to bring them along, said Jane, who also felt that, as the two younger children had been in Europe the previous summer, they "had had their share of knocking around over there." Jane later regretted not bringing them when the King of Sweden, Gustav VI Adolf, asked, "Mrs. Bardeen, do you have children?" "Yes, we do. We have three." "And why didn't you bring them?"

John and Jane coordinated their travels with Walter and Keren Brattain, who brought along their young son, Billy. The five prefaced the celebrations in Stockholm with vacation time in New York City and Copenhagen. In New York, after a leisurely breakfast, John went off on an "expedition to buy a vest for his full dress outfit."

On the plane to Copenhagen, after their airplane dinner and a little nap, they all awoke thirsty. "Did the steward have champagne?" Brattain wrote: "He not only did but it was a vintage year." They learned it was only $4 a bottle,

. . . just the thing. After one bottle, John pointed out that we were losing $4 for every bottle we did not drink, as it would certainly cost $8 in the U.S.A., so we had another. After this John was still worried about saving another $4 but the rest of us said he would have to do his saving alone, so we quit and went to sleep, waking up at about 12 midnight New York time for breakfast and then our landing in Copenhagen.

In Copenhagen, Keren and Jane enjoyed "a shopper's paradise," while Walter and John visited Niels Bohr at his institute. On a side trip to the town of Göteborg in Sweden, Brattain delivered a lecture at Chalmers Technical University. They met with Professor Wallman, who had been a graduate student at Princeton with Bardeen, Seitz, and Robert Brattain. Worried that Walter was getting more publicity than John, Jane wrote home, "I hope John gets a square deal when we get to Stockholm."

Their December 6 arrival in Stockholm, where they traveled next by train, left them a few more days to sightsee and acclimate as they settled into Stockholm's Grand Hotel. Brattain noted that Sweden was a "low hilly country with many lakes—evergreen forests of pine and fir—not a dense forest as one would find in the Pacific Northwest."

The big ceremony was to be held, according to tradition, on December 10, the anniversary of Alfred Nobel's death. The Swedish Foreign Office assigned each Nobelist a guide to help them through their busy days filled with press conferences, photo sessions, ceremonies, luncheons, and dinner celebrations. "Flash bulbs were popping at random as we moved out of the station and waited for our cars," Brattain wrote.

When Bardeen and Brattain visited Hafo, the Institute for Semiconductor Research, they were each presented with a "stickpin with a single crystal of silicon cut on definite planes as the jewel." Shockley missed this party because of a flight delay. He and Emmy had flown to Paris on Air France, but as the Paris airport was fogged in, they had to spend the night in Bordeaux, in a cheap hotel with no hot water. When they arrived in Stockholm on Saturday, they barely had enough time to wash up and begin celebrating.

More like brothers than collaborators, Bardeen and Brattain were enjoying more time together than they had in years. On Monday, the day of the ceremony, they both suffered queasy stomachs. Brattain "hit on the idea of a bottle of quinine water to settle my stomach and Bardeen had some too, then a light lunch and up to our rooms to climb into our monkey suits." The phone rang in the Brattains' suite while Keren, Billy, and Walter were getting dressed. It was "a last minute call from John for a spare tie because of some accident to his." Brattain was used to such crises, for "Bardeen had already borrowed one of his vests." The one John had just bought in New York "had laundered green."

The award ceremony took place in the concert hall. The stage had been decorated lavishly with yellow chrysanthemums. All the laureates had been instructed in the proper procedure for bowing to King Gustav VI Adolf, who presented the awards, and to the Queen. Brattain fretted over getting it all right. It was the only occasion at which "the King stands to receive, so the honor is very great," Jane wrote. Brattain watched as Shockley "went first and made all his bows properly." Then he cringed for Bardeen as the King accidentally stepped in between Bardeen and the Queen, making John's bow to the Queen awkward. Brattain followed Bardeen uneventfully, and the most nerve-wracking part of the ceremony was over.

It had been rumored that the King's dinner that evening, usually an affair of a thousand people at the town hall, would be cancelled because of the current political upheavals in Hungary, whose abortive revolution had been brutally suppressed by the Soviet Union. The dinner was not cancelled, but it was reduced to 175 guests and relocated to the Stock Exchange Building.

After the formal dinner the party moved to the dining room of the Grand Hotel. The Brattains sent Billy to bed and joined the Bardeens as well as other laureates and their attendants. They celebrated into the wee hours, plying one another with champagne. "It was a grand time; we were certainly in a hilarious frame of mind," Brattain later reported. He woke up "very bubbly" the next morning.

The next evening the King entertained the Nobelists in his private apartments. Jane wrote home that "conversation was not difficult," as the royal family consists of "very genuine people." She admired the King's art collection and his orchids and noted that "their living room is much like ours, in a sense, books piled on all the tables."

The group wound down a bit over the next few days. John and Jane took some time to write postcards to a few friends and family, alluding to the "fabulous life" they had been leading during the celebrations. "We have only time enough to sleep, eat, and go to parties." Stockholm was "wonderful but strenuous," John wrote in a postcard to Nick Holonyak. The Nobel whirl, he wrote, had "been like living in a different world." He also said that they "had a dinner in the King's apartment in the Palace shown in the picture. Returning soon to a more normal existance [*sic*]."

Holonyak was then a member of a small unit of the Army Signal Corps in Yokohama. A stodgy lieutenant he was working for in a special operation "would go out just before noon and get our mail, and then he would come back with the mail and throw it at us." During the winter of 1956–1957, the lieutenant came back one day and shouted, "Holonyak! Who in the hell do you know by the name of John in Sweden?" In later years Holonyak would smile fondly when he recalled Bardeen's misspelling of the word "existence" on the postcard. He told his friends that John said his inability to spell was a legacy of his having been skipped in elementary school.

When John and Jane returned to Champaign-Urbana, it took Bardeen a few days to adjust to his normal routine. He brought his Nobel medal into the office to show his colleagues. As he told them about the heady experience they had had in Stockholm, Charles Slichter thought Bardeen was "sharing in the nicest possible way."

Bardeen finally bought a television as a sort of consolation gift to the children who had missed the festivities in Sweden. For some years John had resisted the family's entreaties for a television set, saying he was waiting for the cost of transistorized color TVs to drop to a reasonable price. Afterwards the family noted with amusement that John watched TV more than anyone else. Although he rarely cooked, he especially enjoyed watching Julia Childs's cooking show. He spent many happy hours watching sports, but never home games, for those he attended in person, always rooting with enthusiasm.

Not long after that, Bardeen achieved one of his lifelong goals, a hole in one. It happened on the university golf course near the Champaign airport. According to Bob Schrieffer, "He thought that was almost as good as the Nobel." Years later Bardeen was asked which he considered the greater accomplishment, a Nobel Prize or a hole in one. He replied, "Well, perhaps *two* Nobels are worth more than one hole-in-one."

By Christmas Bardeen was again deeply immersed in superconductivity. The children could tell the problem "was desperately important to him." Jim, home from Harvard for the holidays, could see a big difference from the way his father had worked on the transistor nine years earlier. "I remember him being much more uptight" and trying "to push harder to get things done." Betsy remembered that her father was off in another world that Christmas. But the problem did not yield during the holidays.

Then in the last days of January the turn came. Schrieffer and Cooper were on the East Coast attending physics conferences—one on the many-body problem was held in Hoboken, New Jersey, on January 28 and 29 and hosted by the Stevens Institute of Technology; the other was the annual American Physical Society meeting in New York City, from January 30 through February 2. Schrieffer often took public transportation while commuting between the two meetings and from them to Summit where he was staying with a friend. One day, while riding on the Hudson Tubes, Schrieffer wrote down the wave function for the superconducting ground state.

He recalled the process of developing the wave function as a sort of intellectual tinkering. Talks he had just listened to on the nuclear interaction between pi-mesons, protons, and neutrons, as well as other ideas tumbled around in his head. Among them were the Cooper pairs and the variational approach that Tomonaga had used in the pion-nucleon problem. He was thinking constantly about superconductivity.

Calculating on a pad while on the subway, "I realized that the algebra was very simple." He called on Bardeen's bridging principle and formed "the wave function as a coherent super-position of normal state-like configurations." Then, following Tomonaga, he tuned the expression up. Trying a product, he noticed that his construction did not conserve the number of electrons. He fixed this problem by adding in a "chemical potential term," as in the grand canonical ensemble of statistical mechanics.

Schrieffer worked more on the expression that night at his friend's house. In the morning he did a variational calculation to determine the gap equation. "I solved the gap equation for the cutoff potential. It was just a few hours work." Expanding the expression, he found he had written down a product of mathematical operators on the vacuum that expressed adding electrons to the vacuum. In his sum of a series of terms, each one corresponded to a different total number of pairs. He could hardly believe it. The expression "was really ordered in momentum space" and the ground-state energy "was exponentially lower in energy," as required for the state to be stable.

By chance Schrieffer and Cooper flew into Champaign at the same time. Schrieffer had to show his wave function to Cooper right there in the terminal. Cooper was enthusiastic. "I knew immediately we could calculate that," he recalled. "It was this marvelous, elegant, subtle solution that you could calculate. And essentially what it said is don't drive yourself crazy, just take the pairs, put them together so they satisfy the Pauli principle." Cooper said, "Let's go and talk to John in the morning."

The next morning Bardeen looked at the wave function and said "he thought that there was something really there. So we chatted around about that for a few hours." Schrieffer remembered that Bardeen "was quite convinced that there would be an energy gap in the excitation spectrum." Then using the wave function to compute the gap, "John showed that the gap was exactly the same parameter, delta—we called it epsilon zero at that time—that I'd found entering in the ground state energy!"

Not long after that Cooper walked into the office he shared with Bardeen. "How would you like to write a paper together on superconductivity?" asked Bardeen. Cooper replied, "I would like that just fine. And then we began to calculate and we calculated for just about six months, day and night."

The most exciting moment in the work occurred several days later when Bardeen calculated the condensation energy in terms of both the energy gap and the critical field, obtaining a relationship between these two experimental quantities. "We had the experimental number from the Tinkham group for the gap. We knew the critical field, and the whole problem was converting units." At first Bardeen "was very upset that he couldn't get the numbers to work out." When they finally did, "something like 9 compared to 11 in the appropriate units, we were really overjoyed, and sort of hit the roof. Things looked like pay dirt." All the pieces of the puzzle were fitting together.

Once Bardeen "felt that this was the right direction, then it was clear that the mode of operation changed." Bardeen knew just what to do, recalled Schrieffer, having had "almost all the pieces" assembled for some time in his mind. He had an idea of "how the theory must ultimately work out," and he knew "what you should work out, what theoretical predictions of which experiments you should go after, and what experimental predictions are such that they will be critical of different parts of the theory." And, said Cooper, "because he had done all these calculations for normal metals, he knew everything that happened there." Their job was to "take that normal calculation" and rework it using the super-conducting wave functions. "The point is he just knew everything that had to be done and he was very good and very fast. I mean I was the young hotshot and let me tell you he could calculate paths faster than I could. I was amazed. And sometimes he used very old methods and still got the answer first."

It was only a week after Schrieffer first wrote down the wave function, but they still felt pressed. They knew Feynman was working on the problem "using all sorts of complicated field theory." Worrying that Feynman "might break the problem from another point of view" and scoop them, the team made an effort to bulldoze through their calculations, working intensively over the next six months.

Bardeen laid out a program of tasks. To save time he divided them among the members of the team, assigning Schrieffer thermodynamic properties, asking Cooper to work on the Meissner effect and other electrodynamic properties, while he worked out the transport and nonequilibrium properties. Bardeen's colleagues knew something was up when they asked him a question and were told, apologetically, that he was too busy to think about anything else just then. Schrieffer's wave function initiated "a period of the most concentrated, intense and incredibly fruitful work" that Cooper had ever experienced.

Two weeks after Schrieffer's breakthrough, the team was still a long way from completing the work. They had not yet succeeded in deriving the second-order phase transition. Bardeen decided to publish an announcement of their breakthrough in the form of a letter to the *Physical Review*. On February 15, he sent off the letter. He wrote in a cover note to Samuel Goudsmit, the editor of the *Physical Review*, "I know that you object to letters, but we feel that this work represents a major breakthrough in the theory of super-conductivity and this warrants special handling."

The letter, received by the journal on February 18, explained

Plate 1

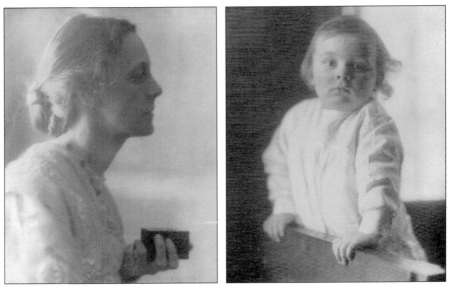

John Bardeen's parents were his earliest and most important mentors. Above left: Althea Bardeen taught him to break problems down, an approach she had emphasized to her students at the progressive Dewey School in Chicago, where she had taught before marrying Charles Bardeen. Above right: John Bardeen as a toddler in 1909. Below: The family of Charles R. Bardeen, Dean of Medicine at the University of Wisconsin, around 1914. From left to right, John's sister Helen; his paternal grandfather, educator Charles W. Bardeen; Charles R. Bardeen with his youngest son, Tom, seated in front; William; Althea; and John.

Note: All photographs © the Bardeen Family Collection except as otherwise noted in parentheses following captions.

Plate 2

Left: John Van Vleck introduced John Bardeen to quantum physics when John was an engineering student at the University of Wisconsin. Van Vleck later supported Bardeen's pursuit of a physics career at Princeton, Harvard, Minnesota, and elsewhere. *(AIP Emilio Segrè Visual Archives, Francis Simon Collection)*

Right: John Bardeen at 17 as a University of Wisconsin student of engineering.
Below left: Bardeen's swimming ability earned him letters at Wisconsin, even though he was younger than most of his team-mates. *(University of Wisconsin–Madison Archives)*

Below, Bardeen's thesis advisor, Eugene Wigner, whose first three graduate students at Princeton (1932–1936)—Frederick Seitz, John Bardeen, and Conyers Herring—were in the first generation of theoretical physicists to refer to themselves as solid-state physicists. *(AIP Emilio Segrè Visual Archives, Physics Today Collection)*

Plate 2

Left: John Van Vleck introduced John Bardeen to quantum physics when John was an engineering student at the University of Wisconsin. Van Vleck later supported Bardeen's pursuit of a physics career at Princeton, Harvard, Minnesota, and elsewhere. *(AIP Emilio Segrè Visual Archives, Francis Simon Collection)*

Right: John Bardeen at 17 as a University of Wisconsin student of engineering.
Below left: Bardeen's swimming ability earned him letters at Wisconsin, even though he was younger than most of his team-mates. *(University of Wisconsin–Madison Archives)*

Below, Bardeen's thesis advisor, Eugene Wigner, whose first three graduate students at Princeton (1932–1936)—Frederick Seitz, John Bardeen, and Conyers Herring—were in the first generation of theoretical physi-cists to refer to themselves as solid-state physicists. *(AIP Emilio Segrè Visual Archives, Physics Today Collection)*

Plate 1

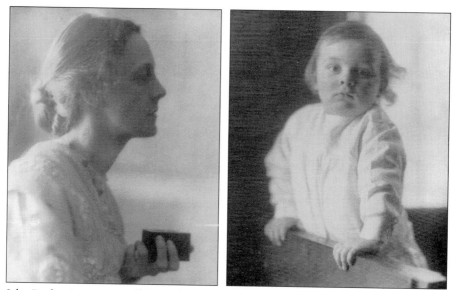

John Bardeen's parents were his earliest and most important mentors. Above left: Althea Bardeen taught him to break problems down, an approach she had emphasized to her students at the progressive Dewey School in Chicago, where she had taught before marrying Charles Bardeen. Above right: John Bardeen as a toddler in 1909. Below: The family of Charles R. Bardeen, Dean of Medicine at the University of Wisconsin, around 1914. From left to right, John's sister Helen; his paternal grandfather, educator Charles W. Bardeen; Charles R. Bardeen with his youngest son, Tom, seated in front; William; Althea; and John.

Note: All photographs © the Bardeen Family Collection except as otherwise noted in parentheses following captions.

Plate 3

Left: John Bardeen and his bride Jane Maxwell pose for a snapshot at their wedding in July 1938, at the end of Bardeen's time as a Harvard Junior Fellow (1935-38). His physics career at Minnesota was cut short by wartime work in Washington, D.C., a tense time in which he often read, above right, the "train book" to Billy, who had memorized every word.

Above: Late in 1945 Bardeen joined Bell Telephone Laboratories, where he worked in a small semiconductor subgroup headed by William Shockley. In the summer of 1947 Shockley and Bardeen visited European physics laboratories to catch up on the latest European research. Shockley took this photograph of Bardeen during a hike in the Alps.

Plate 4

Below: The first transistor, a point-contact device invented by Walter Brattain and Bardeen in December 1947.

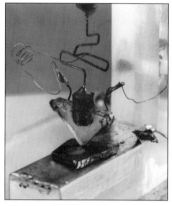

Above right: Bell Labs often used this photo, featured on the cover of *Electronics*, in its efforts to promote the transistor. Shockley sits at the laboratory bench, while Bardeen and Brattain look on. Brattain hated this photo. He would have found the photograph below—where Brattain is handling the apparatus, Bardeen is entering data, and Shockley looks on—a more accurate depiction of the conditions under which the transistor was invented. *(All photos this page courtesy of Lucent Technologies' Bell Labs)*

Plate 5

Left: Bardeen accepts the 1956 Nobel Prize in physics for the invention, with Shockley and Brattain, of the transistor; colaureate Brattain waits behind him.

Above right: Posing in front of the student union building on the University of Illinois campus are (left to right) Bardeen, Mikoto Kikuchi of Sony Corporation, and Nick Holonyak, Bardeen's first graduate student. Sony endowed a $3 million chair at the University in Bardeen's name in 1990. *(Nick Holonyak)*

Stockholm. Kungl. Slottet *Dec 13, 1956*

> Dear Nick:
> We have had a wonderful but strenuous time here for the past week. It has been like living in a different world. Had a dinner in the King's apartment in the Palace shown in the picture. Returning soon to a more normal existance.
> John.

N. Holonyak. Jr.

US 55 510 784

HQ Co. Sig. Supply Center
8084. A.U. A.P.O. 503
San Francisco Calif.
U.S.A.

Above: Bardeen sent this postcard from Stockholm to his former student Nick Holonyak. Note the misspelling of "existance," rather than "existence." Bardeen claimed that his spelling suffered from his having skipped several grades in primary school. *(Nick Holonyak)*

Plate 6

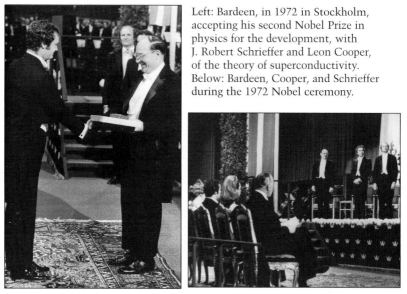

Left: Bardeen, in 1972 in Stockholm, accepting his second Nobel Prize in physics for the development, with J. Robert Schrieffer and Leon Cooper, of the theory of superconductivity. Below: Bardeen, Cooper, and Schrieffer during the 1972 Nobel ceremony.

Below: John and Jane enjoy sharing the event with their grandchildren, Karen and Chuck Bardeen. Jane always regretted not taking her children the first time they went to Stockholm in 1956.

Plate 7

Left: Bardeen was an avid golfer and a good one. Whenever possible he sought out golfing opportunities during research or consulting trips. Below: Paul Handler (left) and Schrieffer present Bardeen with a "Superconductor" collage, featuring Bardeen as the "superconductor" of his "grand canonical ensemble" of students. Schrieffer assembled the collage in 1968 for Bardeen's sixtieth birthday party.

Below: Theorists in the University of Illinois Physics Department pose with Hans Bethe. Seated in front are Bardeen (left) and Bethe; standing behind (left to right) are Christopher Pethick, Shau-Jin Chang, and Gordon Baym. *(Gordon Baym)*

Plate 8

Right: Bardeen demonstrates his historic music box containing some of the original point-contact transistors.

Above: Bardeen's office at the University of Illinois, where he worked nearly every day, even after his official retirement in 1975. Left: John Bardeen, an American genius.

how superconductivity arises from the coupling between electrons and phonons, an interaction in whose presence the system forms a coherent superconducting ground state in which individual particle states are occupied in pairs, "such that if one of the pair is occupied, the other is also."

The letter summarized the advantages of the theory:

(1) It leads to an energy-gap model of the sort that may be expected to account for the electromagnetic properties.

(2) It gives the isotope effect.

(3) An order parameter, which might be taken as the fraction of electrons above the Fermi surface in virtual pair states, comes in a natural way.

(4) An exponential factor in the energy may account for the fact that kT_c is very much smaller than $\hbar\omega$.

(5) The theory is simple enough so that it should be possible to make calculations of thermal, transport, and electromagnetic properties of the superconducting state.

By the time the letter was received, they understood the second-order phase transition. As Schrieffer later recalled, shortly afterwards, the Bardeens were entertaining a distinguished Swedish scientist at their home.

John was somehow off on Cloud 7 that night, and there were long gaps in the conversation where John was staring into space and the conversation was going on but in a very strange sort of way. And it was clear that John was thinking hard about something. And what he was thinking about was how to get the second order phase transition, and exactly how to write the wave function down.

The next morning Bardeen phoned Schrieffer. "He was really excited." Schrieffer remembered that the phone call "woke me up early in the morning." Bardeen had had the Eureka moment. "I think at that point he had felt everything was correct."

Bumping into Slichter in the hall, Bardeen announced the breakthrough in his characteristic way. As usual, he struggled for words. Then he drawled: "Well, I think we've figured out superconductivity." Slichter and his student Charles Hebel were among the first to confirm the new theory experimentally. Measuring the rate at which nuclear spins relax in aluminum, they found that as they lowered the temperature below the point at which aluminum becomes superconducting, instead of falling, the nuclear magnetic resonance rate *increased:* it rose to more than twice its value in the normal state. But as the temperature was further reduced and passed below the transition temperature, the rate again began to fall.

The Bardeen-Cooper-Schrieffer theory, which became known as BCS, explained the effect easily. When one tries to make nuclear spins line up, they try to relax by banging into the electrons. In the process they transfer their spins to the electrons. When this happens in a normal metal, the nucleons exchange spins until half point up and half point down. In the BCS state, however, the number that can be flipped is greatly reduced because the electrons are locked together. In consequence, the relaxation takes much longer. Cooper said that Slichter became so "heavily involved" in the computations that "by April or May he was calculating along with us and pointing out errors."

Within a few months many experimenters at other institutions were validating the BCS theory. Among the first major confirmations outside Champaign-Urbana were those of R. W. Morse and H. V. Bohm at Brown University and those by Rolfe Glover and Michael Tinkham at Berkeley. Morse and Bohm studied the acoustic attenuation as a function of temperature. In accordance with the BCS theory, they found a rapid decrease of the attenuation in pure superconductors as the temperature fell below the transition value. Experiments by Glover and Tinkham, studying far-infrared transmission through thin superconducting films, provided direct evidence for an energy gap corresponding to just twice the frequency at which the absorption sets in. These experiments seemed to be saying, "Come on now. The wave function looks right."

The team announced the BCS theory in March 1957 at the annual meeting of the American Physical Society devoted to solid-state physics. It was held that year in Philadelphia from March 21 to 23. Seitz phoned Eli Burstein, the secretary of the society's Division of Solid-State Physics to tell him about the breakthrough. The abstracts were sent to Karl K. Darrow, the secretary of the society. Two post deadline papers were arranged, one to be given by Schrieffer, the other by Cooper. Schrieffer, who had gone to New Hampshire to write up his thesis, received word too late to attend the meeting. So Cooper had to deliver both papers.

Bardeen chose not to attend the March meeting. "He wanted to make sure that the young people got the credit." Schrieffer considered it remarkable that "having finally come to the pinnacle of achievement in his professional life," Bardeen now "steps aside for two young people." Schrieffer saw Bardeen's act of saying, "OK, you go out and tell the world, and I will stay here in Urbana" as a "striking example" of giving credit, "even to an extent beyond that which is due and also pushing young people as fast as they can to become professionals, and treating them as professionals, and making them rely on themselves."

During March the team worked on understanding the coherence factors, the size of the Cooper pairs. They were concerned that the superconducting wave functions did not appear to conserve the number of particles and thus were not obeying gauge invariance, a condition referring to the symmetry of Maxwell's electrodynamics. They wondered whether their formalism was completely valid because it demanded superposing states having different numbers of electrons. Eventually it became clear that the superposing of states was but a trick that simplified the mathematics. It did not really imply that electrons were not being conserved.

The intense work that spring continued through all the ups and downs of personal events. Bardeen was saddened when he learned in April that Keren Brattain had died of cancer. After enjoying the trip to Stockholm in December, she suddenly fell ill two months later and "in the space of a few months she was gone." John dropped everything to visit Walter. During the visit young Billy became fascinated with John's solar battery-powered transistor radio. John spontaneously presented it to the boy. Billy never forgot the gesture.

Cooper's first daughter was born around 4:30 A.M. on May 6, 1957. After making sure that mother and baby were fine, he decided to go in to work. He had always felt awkward when he arrived at the office about 10 A.M. and found that Bardeen had been there since 8:00. Bardeen had assured Cooper, "I could work any way I wanted to—as long as I worked." Still Cooper felt embarrassed by his pattern.

That morning Cooper rushed home to change. He grinned as he showered and shaved. "This time I would be first." But when he arrived at the office around 7:30, he found "the door was already open; John was sitting at his desk working. He had chosen this morning to come in early." (Cooper had not realized that Bardeen was now coming in earlier every day.)

The feverish pace of the work continued into the summer. By July they were ready to send their full-length article to the *Physical*

Review. In this masterpiece of modern physics, they showed in more detail how their theory explained (1) the infinite conductivity that Kamerlingh Onnes had discovered in 1911; (2) the diamagnetic (magnetic-field expelling) effect found by Meissner and Ochsenfeld in 1933; (3) the second-order phase transition at the critical temperature; (4) the isotope effect first observed by Serin and Maxwell; and (5) the energy gap in the excitation spectrum, first postulated by the London brothers. Their paper also showed how the BCS theory offers quantitative agreement for many experimentally determined quantities, including the specific heat and the penetration depth of the magnetic field near the surface.

Most of the experimentalists who had worked on superconductivity were enthusiastic about the BCS theory. But many of the theorists, especially those who had invested much time on the problem, met the new theory with skepticism. Pippard (both a theorist and an experimentalist) wrote a series of pointed letters to Bardeen. He apologized for the detail with which he probed BCS. "In case what follows gives you the impression that I am bent solely on destructive criticism, let me say at once that I am sure you have made an important step forward towards a theory of superconductivity; I am not wholly convinced that you have its final answer, but I think it will be found along the lines you suggest."

John wrote Jane from a September conference in Madison that the skeptics were "mostly those who have tried and failed themselves and do not yet really understand what we have done." He was confident about dealing with the resistance. "I expect they will come around eventually. Experimental checks have been remarkably good." He also wrote to Cooper about the Madison meeting, which he described as "quite interesting and well attended." Commenting on the reception of the BCS theory: "There was considerable interest in the superconductivity theory. While very few had had a chance to study the theory carefully, the reaction was generally favorable. Experimentalists were particularly enthusiastic about the theory. Objections were raised mainly by those with preconceived notions of what the theory should look like."

Bardeen, Cooper, and Schrieffer spent more than a year "defending the fort" against the "enormous number of people who had vested interests in themselves solving that problem, and in our having not yet solved it." The team faced an impossible problem, for their opponents "wanted to be really convinced that this was a correct solution," and there just was no simple and rigorous way to prove BCS. "It was an intuitive leap." They could only appeal to a long series of experimental "tiepoints" and "ultimately you hope there is a theoretical deductive way of getting there, but it was certainly far from there." As Bardeen wrote to Pippard in late September 1957: "In formulating our theory, we have attempted to construct the very simplest model which we believe has the essential features of superconductivity as it exists in actual metals. With this approach we would hope to find a qualitative and perhaps even semiquantitative agreement with experiment. The remarkably close agreement we have found for most properties surprised us."

One major objection was the theory's apparent lack of gauge invariance, the issue concerning conservation of the number of particles, which the team had worried about. Attempts to address the issue eventually bore fruit in several major papers by Philip Anderson, Pines and Schrieffer, and others, addressing the gauge invariance. They showed that deep down in the formalism the gauge invariance is indeed present. Bardeen wrote to Anderson in October that his manuscript "certainly gives the answer to gauge invariance."

An important by-product of the BCS theory and its reception concerned "broken symmetry," which Yoichiro Nambu, one of the original objectors to BCS, subsequently introduced into the theoretical framework of particle physics. Viktor Weisskopf would lament in 1960: "Particle physicists are so desperate these days that they have to borrow from the new things coming up in many-body physics."

When Schrieffer "came through Copenhagen" between May and August 1958, after spending the previous fall in Birmingham working with Rudolf Peierls and the winter in Italy, "where the weather was a little bit better," he spoke about the theory with Niels Bohr, whose long-standing interest in superconductivity dated back to the 1920s. "It just can't be true. I don't believe it," Bohr said to Schrieffer. "It's an interesting idea, but Nature isn't that simple." Schrieffer then "wrote about two hours of notes trying to recall what Bohr had said."

Based on these notes, Schrieffer composed a long letter to Bardeen explaining, "Unfortunately, it's wrong. Bohr has told me it's not right." Schrieffer got back a prompt reply from Bardeen that said, in essence, "Bohr doesn't know what he's talking about." The problem was that Bohr's remark hit a sore spot. Schrieffer admitted that he himself "thought it was too simple and this can't be the answer."

In the years that followed, all three members of the BCS team continued to clarify their understanding of the meaning and implications of their theory. After returning from his National Science Foundation fellowship year in Europe, Schrieffer went to the University of Chicago for a year. After that he returned to the University of Illinois for three years as an assistant professor, before moving on in 1963 to the University of Pennsylvania. Cooper taught at Ohio State for a year after finishing his postdoctoral research at Illinois. He then moved to Brown University, where he eventually changed fields and became prominent in neural networks research.

BCS became recognized as among the "big ideas" of physics, a theory applicable to problems in many fields of physics. In 1958 the superfluidity (superconductivity without charges) of nuclear matter was proposed by Aage Bohr, Ben Mottelson, and David Pines and developed further in 1959 by Arkady B. Migdal as well as Spartak T. Belyaev. Then Migdal and later Vitaly L.Ginzburg and D. A. Kirzhnits applied the idea of superfluid pairing to the matter inside neutron stars, a theory fleshed out subsequently by Baym, Pethick, and Pines in Bardeen's many-body theory group at Illinois. It seemed natural that the system of ^{3}He, which is neutral, would also become superfluid at low temperatures. The phenomenon was discovered experimentally by Robert Richardson, David Lee, and Douglas Osheroff (below one millikelvin), and the nature of that helium state was explained theoretically by Anthony Leggett.

It became clear that all these problems are tied together by the fundamental property of electrons that had allowed Wolfgang Pauli to open the development of the quantum theory of solids in the first place, the fact that the particles involved obey Fermi-Dirac statistics: one cannot put two of them in the same quantum state. If there are attractive interactions in a system, these "fermions," whether charged or not, will form superconductors or superfluids by forming a system of pairs with one another. BCS became recognized, as Lazarus pointed out, as "a tremendous universal."

12

Two Nobels Are Better Than One Hole in One

B ardeen gradually settled into the role of physics guru, the authority to consult on almost any problem. In both the physics and electrical engineering departments at the University of Illinois, "John will know" became a familiar refrain. Bernard Serin, who spent the 1958–1959 academic year at Illinois as a visiting research associate professor, observed the flow of colleagues and students who passed through Bardeen's office seeking advice. Serin later told a colleague it was like "sharing an office with Buddha."

Bardeen's scientific court extended beyond the university and even beyond the physics community. His many invitations included a White House dinner in April 1962, at which forty-nine Nobel laureates dined with a distinguished group of additional guests, including the astronaut John Glenn, the poet Robert Frost, and the author John Dos Passos. President Kennedy entertained the group gathered in the State Dining Room and neighboring Blue Room with his famous extemporaneous remarks. "I think that this is the most extraordinary collection of talent, of human knowledge, that has ever been gathered together at the White House— with the possible exception of when Thomas Jefferson dined alone." Jane wrote to her family, "We felt greatly honored to be there in the company of so many distinguished people."

In 1963 Bardeen's office moved into a larger space that he would occupy for the next twenty-seven years, in a new building constructed for the physics department several blocks east along Green

Street. The building would later be named the Loomis Laboratory of Physics in honor of Wheeler Loomis.

The old physics lab, built in 1909, had provided ample work space in its first two decades. But by the 1940s the department was making use of all manner of additional space to accommodate its needs. Its cyclotron had to be housed in a garage. When, by the 1950s, "every nook from basement to attic" became fully occupied, a new building was planned. As the department initially could secure enough funds to construct only half of the new building, the eastern portion that was built first had exposed concrete blocks forming its western wall. These blocks became the walls of interior rooms when it was possible to build the western half of the building between 1961 and 1963.

In September 1963 the many-body group, including Bardeen, moved into a multi-office suite on the third floor of the new part of the building. Three of the suite's four offices were lined up next to one another along the south wall. The fourth office faced west. Bardeen chose the first of the three south-wall offices. He lined several desks up along the east wall and organized his library on shelves above. Blackboards decorated the other wall. Two file cabinets stood near the window. Bardeen usually worked beside the window, through which he could see the afternoon sun brighten the trees near a residence hall. He usually left his office door open while working.

Bardeen's move created an unexpected problem for some of his colleagues. In the crowded old building, Bardeen had worked in a shared office. When a colleague came to ask him a question, it was not necessary for them to wait for Bardeen's answer, which could take up to half an hour or more to arrive. Bardeen's office mate could receive the answer and later deliver it to the questioner. In the new office this system didn't work.

The new arrangement did, however, promote collaboration within the many-body group. David Pines, in the office next to Bardeen's, enjoyed frequent discussions with Bardeen over the years on a wide range of many-body problems, including quantum plasmas, electrons in metals, collective excitations in solids, superconductivity, nuclear structure, compact helium liquids, and high-temperature superconductivity. Pines had returned to Urbana in 1959 after four years at Princeton (as an assistant professor and then as a member of the Institute for Advanced Study).

Two recently hired young theorists, Leo Kadanoff and Gordon Baym, moved into the suite's remaining two offices. After completing graduate study at Harvard, both had worked in Copenhagen during 1960 as postdoctoral associates at the Universitetets Institut for Teoretisk Fysik (now the Niels Bohr Institute). They had collaborated on problems of transport in many-body systems using Green's functions, including "gauge invariance" and other symmetries. They also coauthored a research monograph on quantum statistical mechanics. By the time Kadanoff and Baym accepted permanent positions at Illinois, in 1961 and 1962, respectively, they were already recognized as leaders in many-body theory. The department had become one of the three most important centers for many-body physics, along with Bell Labs and the Landau Institute in Moscow.

Bardeen, Pines, Baym, and Kadanoff worked in a loose collaboration, often wandering into one another's offices to talk problems through. One memorable collaboration involved John Wheatley, an experimentalist who was studying the Fermi liquid properties of ^{3}He at low temperatures. When Bardeen learned that David Edwards of Ohio State had demonstrated that ^{3}He can dissolve into superfluid ^{4}He, he immediately recognized the opportunity to study a new Fermi liquid, one that he thought could become a superfluid.

The three theorists, Bardeen, Baym and Pines, formed a collaboration with the experimentalists, Wheatley and Edwards, one that also included Wheatley's former graduate student, Andy Anderson, who was now a postdoc. Wheatley's lab was just downstairs in the physics building, and the three theorists regularly looked in on his experiments. Bardeen, Baym, and Pines developed a theory for the intriguing ^{3}He–^{4}He mixtures—intriguing because the particles of ^{3}He are fermions (obeying Fermi-Dirac statistics), while those of ^{4}He are bosons (obeying Bose-Einstein statistics). Although the ^{3}He particles form a normal fluid, the calculations, drawing on a suggestion of Bardeen's to use an "effective interaction" for describing the ^{3}He, indicated that the ^{3}He would become superfluid at very low temperatures (several millikelvin). The system would thus become a composite of two independent superfluids, each with different properties. (This phenomenon has not yet been observed experimentally in dilute solutions.) Once again, Bardeen served as the leader of the team, the one who knew what questions to ask.

The Illinois many-body group continued to grow around Bardeen. Christopher Pethick joined as a postdoc in 1966–1967. In early 1966 Michael Wortis, who specialized in magnetic systems and eventually phase transitions, came as a research assistant professor. In 1972 three new permanent members joined as assistant or full professors: Vijay Pandaripande, an expert in the nuclear many-body problem. Frederick Lamb, a student of astrophysics, and Bardeen's former student Bill McMillan, who returned from Bell Labs, where he had gone after taking his degree in 1963. In 1982 Anthony Leggett, who had been a postdoc with the group in 1964–1965 and again in 1967, joined permanently.

Bardeen spent part of most days answering questions posed by his colleagues and students. Because of his detailed technical knowledge of physics, he could often point directly to the heart of a problem—and to its solution. But there were times when Bardeen's colleagues could not grasp the master's meaning. A few pressed for elucidation; many simply accepted his judgment—after all, he had been right countless times before. Colleagues and students were not the only ones to trust Bardeen's intuition. He did so himself. Occasionally this habit led to embarrassment.

In the realm of classical physics, particles cannot pass through concrete walls. But they can do so in the quantum-mechanical world if they try often enough. In fact, in this strange world, an electron or other elementary particle has a finite (if typically extremely small) probability of "tunneling" through a physical barrier. For the case of a supercurrent, however, it would appear that traversing a barrier between two superconducting materials would be too improbable to occur in nature, because a *pair* of electrons would have to pass through at the same time.

In 1962 Brian Josephson, a 23-year-old graduate student of Brian Pippard, looked into this matter carefully. His theory predicted that electron pairs in superconductors *should* be able to tunnel through a thin barrier separating two superconductors. The phenomenon became known as the "Josephson effect." Its existence became the subject of a famous debate between Bardeen and Josephson. The physicist Donald G. McDonald has studied this debate, staged in September 1962 at a major physics conference. McDonald interviewed many of those involved and reviewed documents from the period. His work has informed our account.

Bardeen became involved in the question of superconducting

tunneling during the summer of 1960, while on a consulting visit to the General Electric Research Laboratory in Schenectady, New York. There he conferred with one of Seitz's recently graduated Ph.D. students, Walter Harrison. They discussed experiments that Ivar Giaever had done while he was a graduate student at Rennselaer Polytechnic Institute. They revealed quantum-mechanical tunneling of electrons from a normal material into a superconducting one.

When Harrison pointed out that for certain systems Giaever's assumptions would need modification, Bardeen said he was interested in taking a closer look at the whole question of tunneling in superconductors. He checked with Harrison to make sure it was not "cutting in" on him. Harrison assured him it was not. That fall Bardeen developed a theory of superconducting tunneling based on the assumption that in the region of the barrier the electrons "are not paired and the wave function is essentially the same as in the normal state."

Bardeen's paper on superconducting tunneling raised criticism from a theorist in the physics department at the University of Chicago, James C. Phillips, who felt that Bardeen's argument was unclear. Phillips joined his colleagues Leo Falicov and Morrel Cohen in a discussion of the tunneling problem that resulted in February 1962 in a joint paper. Their work treated the case of tunneling through a barrier separating a superconductor and a normal metal.

Five months later Bardeen responded to Phillips et al. in another paper on the same subject. This one included a footnote referring to a recent publication by Josephson. Following up on his earlier assumption that in the region of the barrier the electrons are not paired, Bardeen intuitively judged Josephson's suggestion "of superfluid flow across the tunneling region [between two superconductors] in which no quasi-particles are created" as impossible. Bardeen stated that "pairing does not extend into the barrier, so that there can be no such superfluid flow."

Josephson was then at the end of his second year at the Royal Society's Mond Laboratory in Cambridge. He had taken a course on many-body theory from Philip Anderson, who was also in Cambridge that year (1961–1962), on a sabbatical leave from Bell Labs. Anderson found teaching this course "a disconcerting experience, I can assure you." For "everything had to be right or he [Josephson] would come up and explain it to me after class." As part of the course Anderson had introduced the concept of broken symmetry in superconductors. Josephson "was fascinated by the idea of broken symmetry, and wondered whether there could be any way of observing it experimentally."

After learning of Giaever's tunneling experiments and having also studied recent experimental work by Hans Meissner and theoretical work by Robert Parmenter, Josephson went to Pippard with his idea that supercurrent flow might indeed be possible. Pippard had himself thought it "perfectly possible for this superconductor to infect that normal metal with superconducting pairs so that supercurrent could pass from one side to the other." But he had argued that the probability of two electrons tunneling simultaneously through an insulating barrier of the kind Giaever had used would be so small as to be unobservable.

Josephson was not so sure, especially when Anderson showed him a preprint of the Cohen, Falicov, and Phillips paper on the superconductor-barrier-normal metal system. He tried to extend the calculation to the situation where both sides of the barrier were superconductors and was surprised to find a term dependent on the phase. The current, an even function of the voltage, did not vanish when the voltage was zero. There did indeed appear to be a supercurrent, despite Pippard's suspicion that pair tunneling would have too small a probability to be observable. Josephson discussed his puzzling finding with both Pippard and Anderson. "It was some days before I was able to convince myself that I had not made an error in the calculation," Josephson said.

"We were all—Josephson, Pippard and myself, as well as various other people who also habitually sat at the Mond tea and participated in the discussions of the next few weeks—very much puzzled by the meaning of the fact that the current depends on the phase," Anderson wrote. "By this time I knew Josephson well enough that I would have accepted anything else he said on faith. However, he himself seemed dubious, so I spent an evening checking one of the terms that make up the current." Together Anderson and Josephson "discussed how broken symmetry made this peculiar behavior of the current possible." Josephson's result stood fast. He was ready to publish.

"The embarrassing feature" of the work was "that the effects predicted were too large," Josephson later wrote in his Nobel address. Because of some uneasiness with the results, he submitted his paper on June 8, 1962, not to *Physical Review Letters* but to the new journal *Physics Letters*. He wrote that in a tunneling experiment when both metals are superconducting, "new effects are predicted, due to the possibility that electron pairs may tunnel through the barrier leaving the quasi-particle distribution unchanged." This paper was received by *Physics Letters* on June 8, 1962. Two weeks later Bardeen submitted his paper to *Physical Review Letters*, with the footnote challenging Josephson's theory.

Bardeen's challenge came to the attention of the organizing committee of the Eighth International Low Temperature Physics Conference (LT-8), held September 16–22, 1962, at Queen Mary's College of London University. The organizers had "wanted macroscopic quantum phenomena to be a major theme of the conference." The Harvard physicist Paul Martin was chair of a session that included, among other topics, tunneling. He invited Josephson and Bardeen to appear face to face in a public debate. At that point the effect predicted by Josephson and questioned by Bardeen had not yet been confirmed.

The debate, although not mentioned anywhere in the conference proceedings, was well attended. Bardeen, who had opened the conference with the Fritz London Memorial Lecture, met briefly with Josephson shortly before the debate. "I introduced Josephson to Bardeen in London, when people were milling around in a big hall," Giaever recalled. When Josephson tried to explain his theory to Bardeen, "Bardeen shook his head slightly and said 'I don't think so,' because he had carefully thought about the problem. I stood there during the short conversation. Then Bardeen left, and Josephson was quite upset."

They proceeded into the conference room, which "was crowded late in the afternoon, in anticipation of the debate," wrote McDonald. Bardeen sat near the back. Josephson began with a brief outline of Anderson's notion of broken symmetry in superconductors. Drawing on Lev Gor'kov's formulation of the BCS theory, he explained how his calculations "predicted that pair tunneling would be a large effect." Then "Bardeen rose to describe his theory of single-particle tunneling including his previously published comment that pairing does not extend into the barrier." Josephson interrupted with questions. Bardeen came back with answers, and with questions of his own for Josephson. "The exchanges went back and forth several times, with Josephson answering each criticism of his theory. The scene was quite civil because both men were soft spoken." But as Morrel Cohen recalled, the two "seemed to speak different languages."

"Bardeen was outspokenly skeptical," according to Wolfgang Klose, a professor of thermodynamics at the University of Kassel. But "Josephson did not give in." When Bardeen said that he believed Josephson's idea "was quite impossible, Josephson repeatedly asked Bardeen, 'Did you calculate it? No? I did.'" Josephson won the debate. Afterwards, Klose noticed, "Bardeen put his arm around Josephson, like a father to his son. In this attitude they left the lecture-room."

Josephson later explained that "Bardeen's basic error was to ignore the non-locality inherent in the Gor'kov theory, and assume a local connection between the potential and the pairing." The issues would be fully spelled out in 1963 in a paper by Anderson and John Rowell, a Bell Labs experimentalist, who observed the tunneling supercurrent predicted by Josephson. Anderson and others explained the Josephson effect within the framework of Gor'kov's formulation of the BCS theory.

Two months after Anderson and Rowell submitted their paper to *Physical Review Letters* Bardeen wrote to Anderson. "Your evidence, particularly the effect of a magnetic field on the supercurrent, is quite impressive." He suggested that to reconcile his view with Josephson's it was necessary to take "into account the superconductive energy gain from the matrix element for pair transitions across the barrier. . . . It is this energy that I did not take into consideration in my previous estimates of the possibility of a tunneling supercurrent."

Josephson's effect was now certain. In July 1963 Sidney Shapiro published a paper reporting "several startling accompaniments of the Josephson effect." Later that year Bardeen publicly withdrew his opposition at a conference held in August 1963. Two years later Josephson accepted Bardeen's invitation to come to Illinois for the academic year 1965–1966 as a postdoc in the many-body group. While there, Josephson wrote a now-famous paper on the second-order phase transition to the superfluid state.

The Josephson effect has had important practical applications. Two of the most prominent are SQUID (Superconducting Quantum Interference Device) magnetometers that can measure very weak, almost static, magnetic fields and switches based on Josephson junctions.

Bardeen's colleagues sometimes went out of their way to show their appreciation for him. In 1968 the physics department threw a gala sixtieth birthday party for him attended by many friends and colleagues from around the world.

Bob Schrieffer created a work of art as a birthday present for his mentor. He "sent off for little photographs" of all of Bardeen's students. Then he "cut out small bodies from brown felt and made them hold violins." He arranged them on a large poster board. It included "John in a set of tails with a red sash as the superconductor in his grand canonical ensemble." Bardeen hung the collage in his office. Another unusual gift came from John Van Vleck, who distributed to all the dinner guests a microfiche card containing Bardeen's collected life's work. "Makes one feel humble," Bardeen wrote.

George Hatoyama of Sony Corporation, whom Bardeen had befriended on his first trip to Japan in 1953, brought a remarkable golf ball created specially for Bardeen. The gift was designed to recognize Bardeen's work on the transistor and to commemorate the many happy times Bardeen had played golf with his Sony friends. The size and shape of a normal golf ball, this one had a transistor radio inside with a tiny speaker and a rechargeable battery.

In later years Bardeen would sometimes use the golf ball radio in lecture demonstrations or to play tricks with it on the golf course. In one trick, he would replace one of his opponent's balls with the radio, so when the person arrived to make his putt, the "magic" ball would be playing some radio station. Bardeen's hesitant manner did not lend itself to joke telling, but he would often express his impish sense of humor in such practical jokes.

In time the radio developed a loose connection or short that made it necessary to squeeze the ball or hit it in just the right way to make it work. "I will have to take it apart and repair it," he wrote to Hatoyama. In the fall of 1985 he accepted Hatoyama's offer to have the radio repaired at Sony's service company. By Christmas he was using it again to entertain friends or to illustrate transistor history.

In 1971 Bardeen directed James Bray and David Allender, his

last students, to a controversial problem involving a new mechanism for superconductivity. In this picture the attractive electron interaction responsible for the state comes not from the electron–phonon interaction but from the virtual excitation of electron-hole pairs. Bray confessed that he had no concept of the complexity of this excitonic superconductivity at the time he began work on it. But he knew that when Bardeen offered a problem to a student, it was "not just to throw you into the wolves" but "because he had an idea or two." In fact, Bardeen had "an outline in his mind of how to pursue the excitonic superconductivity problem" and worked on the problem right along with them. When it came time to publish, he insisted that his name appear last on their joint paper.

One day later that year, Myron Salamon, one of Bardeen's Illinois colleagues, recalled, "I was sitting in coffee hour up here in 208 MRL [Materials Research Laboratory] and John came and said, 'Have you seen this? We really ought to get involved.'" Alan Heeger at the University of Pennsylvania (who would in 2000 win a Nobel Prize in chemistry) had just announced his discovery of the conducting organic, one-dimensional material, TTF-TCNQ. He believed it was superconducting. Salamon and others measured the ratio of the thermal to electrical conductivity (the Wiedemann-Franz ratio). They "showed these results to John," who happened to be nearby looking at his mail. "I said, 'We're finding that the Wiedemann-Franz law holds' and he looked up and he says, 'Oh, I guess the entropy is a constant of the motion.' And he walked away." Salamon asked Bray, "What does this mean, what's going on here, how can this be? What is he doing?' It took us a week, and we figured out what John meant."

Bardeen had an intuitive idea for explaining charge transport in the exotic material, which was later shown not to be superconducting after all. He proposed treating the transport in the one-dimensional organic system like current-carrying sliding charge density waves. The idea harkened back to his earlier work on superconductivity in 1950. Heeger's discovery opened a new field of research on "lower dimensional materials," which would be extensively studied for their nonlinear properties. These materials, in which charge is transported along long molecules, are among those that exhibit charge density and spin density waves. Because Bardeen recognized the opportunity, "we jumped in," said Bray. Subsequently Bray studied the "excess conductivity in these one-dimensional organic systems." Bray and Allender finished their thesis work in 1974, shortly before Bardeen retired.

In the early 1970s Baym and Pethick were working on the theory of superfluid ^{4}He. They were puzzled by the question of the momentum carried by a "roton," a short wavelength excitation. "So we went to John," who they assumed had thought deeply about this. "And indeed he had." Bardeen answered their question in the form of a riddle. "Well, it's just the same as the difference between **B** and **H**," he said, the two forms in which the magnetic field is expressed in Maxwell's equations. "We had no idea what he meant," Baym confessed, "and it was very difficult to find out." They made several attempts to decipher Bardeen's cryptic remark, but eventually they put the question aside.

About a decade later Pethick and Baym were trying to understand the long-range magnetic interactions between electrons in metals. "So we went back to John for advice." Again he replied, "Well, that's just the difference between **B** and **H**." Again they could not understand his meaning. "But the second time was different because he gave us a reference, the review article that he and Bob Schrieffer wrote in *Progress in Low Temperature Physics*." The article, written in the late 1950s, included a discussion about the difference between **B** and **H**, but the meaning was still unclear to them.

"Fortunately, I was going to see Bob Schrieffer in two days in Santa Barbara, so I took along a copy of the article," Baym said. He was excited as he walked into Schrieffer's office. He expected to learn at last what the difference between **B** and **H** was all about and how it could inform problems like the roton or the magnetic interaction between electrons. "I showed it to Bob and said, 'Okay, explain to me what this means.'" Schrieffer laughed. "Well, John kept saying that all the time, it was the difference between **B** and **H**." He turned to Baym and said, "I never understood what he meant, but he insisted we put it in the review article." For some time after that, whenever members of the group encountered a problem whose explanation was obscure, they would jest, "Well it's just the difference between **B** and **H**." Then all would chime in with a resounding "Ahh!"

People began to murmur about a second Nobel for Bardeen. Nobel prizes had been awarded for lesser achievements that BCS, but no one had ever received two in the same field. Bardeen was concerned that, if the Swedish Academy of Sciences held to tradition, his first Nobel Prize would rob Schrieffer and Cooper of theirs.

Bardeen was pleased to learn in November 1967 that Schrieffer was to be awarded the American Physical Society's Buckley Prize. This was not technically an award for BCS but for work done in the last five years. "There can be little doubt, however, that Schrieffer's contribution to the microscopic theory was a significant factor in the award," Bardeen wrote to Charles Kittel.

Bardeen raised the idea of awarding the Comstock Prize of the National Academy of Sciences jointly to Cooper and Schrieffer. "The Comstock Prize is more significant than the Buckley Prize, so that a joint award should make up for any slight Cooper might justifiably feel in regard to the award of the latter to Schrieffer alone." Bardeen's prodding worked. The 1968 Comstock Prize was awarded jointly to Schrieffer and Cooper for "contributions to the theory of superconductivity."

Still Bardeen continued to feel Schrieffer and Cooper deserved a Nobel Prize. He decided to do what he could to arrange a Nobel Prize for BCS. His Illinois colleague Charles Slichter believes he "came to realize that the only way to get a Nobel Prize for that was to arrange a Nobel Prize for something that came after it and depended upon it." Bardeen gave the idea some thought. He "realized that the thing that really appeals to the Nobel Prize committee is to have an international group." Superconducting tunneling was the natural subject. The work depended on BCS, for Josephson's work build directly on the BSC theory; Giaever's work allowed seeing the lattice vibration spectrum; and the Eski diode is based on the tunneling effect. If such a prize were to go to Leo Esaki, Ivar Giaever, and Brian Josephson, the laureates would be Japanese, Norwegian, and British.

In 1967 Bardeen proposed a Nobel Prize for electron tunneling in solids, "an area that has had an extremely rapid development in recent years." He wrote to the Nobel Committee:

> Those I would like to nominate are Leo Esaki for his discovery of interband tunneling in semiconductors and the tunnel diode, Ivar Giaever for his experiments on tunneling through a thin oxide layer separating two metals, particularly when one or both are superconducting, and Brian D. Josephson for his predictions concerning

supercurrent flow through a tunneling junction. All three of these discoveries have opened up new areas of research.

He renewed the nominations in 1971, 1972, and 1973. The tunneling prize was awarded in 1973—to Josephson for his theory of superconductive tunneling, Giaever for his tunneling experiments, and Esaki for the observation of tunneling in p-n junctions. And the prize for BCS was awarded a year earlier, in 1972.

"John was a very politically savvy person," Slichter pointed out. "People have a concept of John Bardeen as being kind of an innocent unworldly person. I don't think that is correct. I'm pretty sure that John Bardeen figured out how to get the Nobel Prize awarded for superconductivity."

Bardeen first learned that he had won a second Nobel Prize in physics from a Swedish journalist who called his home from New York on Thursday, October 19, 1972. "I didn't quite believe him," Bardeen said. "These things are often false rumors." But around dawn the following morning he was awakened by the official call from Sweden.

Once again great excitement ensued in Champaign-Urbana, whose physics Nobelist was now a double laureate! A stream of reporters questioned Bardeen about superconductivity and everything else. When he tried to leave for work that morning, his electric garage door opener failed. *Newsweek* and other magazines and newspapers focused on the irony of the transistorized apparatus failing to work for the man who had invented the transistor. Bardeen hastened to explain that what had failed was not the transistor but the switch outside the door. "I've got my wife working on getting it fixed." (He had, in fact, disconnected the transistor because it was so sensitive that airplanes passing overhead were causing the door to open.) "We've had that door for years, and it never failed us before."

When Bardeen finally arrived at the university, he attended a lunchtime seminar about David Lee's study of superfluids at Cornell. The only difference from the usual brown-bag affair was the champagne ordered by David Pines and Nick Holonyak.

A reporter followed Bardeen to one of his classes. A number of colleagues and other students also attended, for word had gotten around that Bardeen would speak on superconductivity. "At a few

minutes before the hour, Bardeen entered the room, head down. . . .
He glanced at his watch as he reached the front of the room and
began. 'I think this would be a good time to discuss superconductivity,' he said with a small grin." The reporter also attended a
press conference at which Barden "spoke softly, slowly, almost reluctantly. It was almost as though he wondered what all the fuss
was about. But he had the look of a man who is so happy he can
hardly stand it."

On the Saturday following the announcement, John attended
the Illinois home football game. "Bardeen put on his overcoat and
hat, wrapped an apple in a paper towel and drove to Memorial
Stadium to watch the Fighting Illini (0–5 at the time) play
Michigan." It was reported, "Bardeen does not miss home games"
and, according to colleagues, "he roots like hell."

The usual round of congratulations followed. John smiled
broadly when Nick Holonyak teased him that all he needed now
was "another Nobel Prize with two other guys and you've got a
whole Nobel Prize." Hundreds of good wishes flooded in from
friends, colleagues, and people Bardeen had never met. Brattain
wrote, "Sorry you could not get your garage door open!" He said
that he and his new wife Emma Jane would be thinking about him
on December 10 and added, "Be sure you don't need to buy or borrow a dress shirt at the last minute." John replied, "Jane and I have
been going over the souvenirs of the 1956 trip which reminded us
of the marvelous experience we had. It is a surprise and somewhat
overwhelming to be going again after 16 years."

The board of directors of the Champaign Country Club, led by
George Russell, one of Bardeen's golf partners and that year's president, wrote: "I know we are expressing the feeling of all club
members when we say that such a richly deserved reward couldn't
happen to a nicer person." Another admirer complained in a
tongue-in-cheek editorial sent to a local paper about the amount of
publicity John received for the second Nobel, compared with the
publicity garnered by other university employees. "Let's keep these
things in perspective. A Nobel Prize winner comes along every day,
but a new football coach or athletic director is a once in a lifetime
story."

Bardeen tried to respond to all the letters, but in the end ran
out of time and patience. He took special care to answer letters
from young people, such as one sent by eighth-graders in the school

where his nephew taught art. "Being eighth-graders," they told him, "we are really interested in your first Nobel Prize, which dealt with the transistor, since most of us use transistor radios." A nun in Brazil shared her fuzzy but fond memories of working with a scientist named Bardeen in Pittsburgh during World War II. "I would be so pleased if you turned out to be that same gentleman!" Bardeen wrote back that she had probably worked with his brother Tom, who lived in Pittsburgh at that time and had children like those she described.

Jane again chose a blue formal gown to wear at the ceremony, this one of crepe with princess lines and "very pretty beading and embroidery." Recalling King Gustav's scolding in 1956 about leaving the children home, she insisted that this time they all come along. By now the three children had spouses, and there were two grandchildren. It was a great family reunion.

The trip began on December 4 with a minor crisis. Due to icy conditions, their flight from Champaign was cancelled. Pines saved the day by suggesting that the university supply a car and driver to get them to O'Hare Airport in Chicago. "The roads were good shortly north of Champaign-Urbana," Jane scribbled in her diary, "so we arrived at O'Hare in good time."

The next day Jim and his wife Nancy met John and Jane in Copenhagen. They gave themselves two nights and a day "to catch our breath and get used to the change in clock before the swirl of activities in Stockholm." The rest of the family arrived in bunches throughout the day. On December 6 the entire clan had lunch with Aage Bohr and his wife Marietta. Later, John and Jane paid a visit for tea to Aage's mother, Margrethe Bohr, Niels Bohr's widow.

On arrival in Stockholm on December 7, they were immediately sucked into the Nobel activities. Jane enjoyed sharing them with their children, Jane's brother Jim Maxwell and his wife, and their friends the Gaylords, the parents of Bill's wife Marge. Jane was delighted to have so much family present, but wrote in her diary that it left "almost no time to shop or just wander thru the city as I did in 1956." On December 10, the day of the presentation ceremony, Bardeen led the procession with Stig Lundquist, a physicist at the university in Göteborg, Sweden, who would later become chair of the Nobel Committee. As they proceeded into the hall, Jane noticed that "John's shyness caused him to hang back so that L had to grab his hand and pull him forward." Gazing into the audience, John focused on his family—Jane, three children, two daughters-in-law, a son-in-law, and two grandchildren. "We were quite a crew," he told reporters, chuckling softly. "They took up quite a bit of the front row at the presentation."

Bardeen handled part of the huge job of answering congratulation letters while they were in Stockholm. Jane had made up a combined thank-you and season's greetings card, which they both signed. When John later wrote to thank Carl Vernersson of Rank Xerox Sweden, who contributed a secretary to help the Bardeens mail cards, he added, "If your profits show a sharp drop in December, you will know the reason why."

On the way home to Illinois, the Bardeens stopped in Switzerland, where the children skied and John and Jane enjoyed the mountains. They also stopped off at Murray Hill to participate in a small Bell Labs reunion dinner in honor of the twenty-fifth anniversary of the invention of the transistor. Jim Fisk, now president of Bell Labs, had added a note to the invitations, saying that "John Bardeen's second Nobel Prize for his work in superconductivity provides an additional reason for rejoicing at this time."

When the Bardeens finally returned home, they found a letter waiting from the Wisconsin Telephone Company informing Bardeen of their intention of naming December 23, the anniversary of the day the transistor was first demonstrated to Bell Labs executives, a statewide "John Bardeen Day." "Naturally, all of us at Wisconsin Telephone are proud of the research contributions you have made for the good of mankind." Bardeen responded, "I have always had a very warm feeling for my home state, so that this very special honor is greatly appreciated."

The many-body groups at Illinois and Bell Labs continued to compete with one another. Bardeen had always enjoyed the competitive spirit inherent in cutting-edge research, but the quality of the exchanges sometimes disturbed him.

The Allender, Bray, and Bardeen paper on excitonic superconductivity gave rise to another challenge from Jim Phillips, who in 1968 had left the University of Chicago to join the many-body group at Bell Labs. He and Bardeen had exchanged several heated letters about the work during September 1972. They were unable to agree. Phillips then wrote a highly critical response and sent it to

Physical Review Letters, which, because of an editorial blunder, was published on December 4, nearly two months before the Allender, Bray, and Bardeen paper that Phillips was objecting to. Phillips's letter undercut the paper's reception.

Bardeen found the letter waiting when he returned from Stockholm. He was livid at this "inexcusable set of blunders," Bray later recalled. It was "the only time I've ever seen him actually say mad cross words out loud." Early in January 1973 Bardeen wrote an angry note to Samuel Goudsmit, the editor of the *Physical Review*: "An author certainly has a right to see his article appear before a distorted and misleading account of it is given elsewhere." Goudsmit wrote back that the editors presumed "the author must have been in contact with the authors of the cited paper and gotten permission to quote from it." Responding to Bardeen's suggestion that the journal "establish procedures such that cases similar to this one cannot occur," Goudsmit replied, "I do not know how this can be done without the help of the F.B.I."

Karl Hess remembered eating lunch with Bardeen and Bill McMillan shortly afterwards. McMillan remarked that things around the department had been a little dull of late. Unaware of Bardeen's most recent exchange with Phillips, he said, "Shouldn't we invite an interesting person?" He innocently suggested Phillips. Hess watched a red wave of anger travel up Bardeen's neck and into his face at the mere thought of bringing Phillips to Illinois. "Poor McMillan never knew what he did."

The Bray-Allender-Bardeen paper on excitonic superconductivity caused a few more waves later that year when Phil Anderson and John Inkson wrote a critical comment on it for *Physical Review Letters.* Allender, Bray, and Bardeen replied that Inkson and Anderson had misunderstood the paper and had dealt with "semiconductors with different physical properties." The comment and reply were printed back to back. The two groups continued to argue without resolution. Anderson felt that Bardeen was not reading his and Inkson's comment accurately and that he had grown too stubborn to consider that someone else might be right. They finally "agreed to disagree."

Such controversies were tempered by activities such as golf. Allender, Bray, and Bardeen often played golf together. "I got better while I was his graduate student," Bray reminisced. "If a real nice day popped up we would all look out the window," and if the weather was right, "We would just migrate as a whole" to the university golf course near the airport. "It was fun playing with him. He was the usual gentleman there that he was everywhere else. Not terribly talkative there like anywhere else. He was a very good golfer. He usually would beat us."

When Bardeen learned that Allender's father was a golfer, he suggested that when the senior Allender visited Champaign they make a foursome with Bray. The older players scored much higher than the younger ones. Back at 55 Greencroft, Bardeen took out his Nobel medals to show Allender's father. "It was something my father would remember all of his life."

The hours of physical activity in the outdoors helped Bardeen relax and reenergize. He rarely used a cart, preferring to walk the roughly four miles of an eighteen-hole course, and managed to avoid becoming as rotund as his father Charles had been in his later years. Golf provided Bardeen with an opportunity to express some of the attributes that also made him an excellent swimmer, bowler, billiards player—and great scientist. The golf course was another domain for mastering challenging problems and developing the "mental toughness" ascribed to the best golfers. A nongolfer might imagine the game of golf as a leisurely stroll around the greens, but it was nothing like that for Bardeen. Every stroke demanded control.

Golf also required focus for long periods, both within a single game that might last four hours or more. Over the years Bardeen pursued the sport at every opportunity. Like scientists, most golfers experience months or years of consistent and relatively uneventful practice, punctuated by a few brilliant successes. The better golfers develop a core of realistic confidence and optimism that allows them to execute complex skills consistently and keep them in the game over the long term. In the short term, reinforcement (both positive and negative) is immediate, for a ball either does or does not do what is wanted. A player's game suffers if he or she becomes discouraged over a series of less successful shots or overly excited by any one thrilling moment of success.

Although Bardeen liked pitting himself against others, the important competition was against his own skill level and previous scores. He was a favorite golf partner at both the country club and university golf course. He had a slight tremor in his hands (possibly the result of a high fever in his youth) that worsened with age, but it had surprisingly little effect on his golf game. Jim Bray, who played golf with Bardeen regularly in the 1970s said: "He'd get out there and start swinging his golf club as if nothing were wrong at all. He would swing very smoothly and very well." Bardeen, said another golf partner, "never rushed things, not even his golf swings." In his middle years, Bardeen shot in the high 70s, a good score for a man with a fairly sedentary lifestyle. Even in his sixties and seventies, his scores fell respectably in the low 80s.

Bardeen enjoyed the social aspects of golf almost as much as its mental and physical activity. He cultivated numerous golfing friendships at the country club. One foursome consisted of Bardeen; Charles Keck, a local merchant and amateur cartoonist (among other talents); Myron Kabel, an agricultural products sales representative; and King McCristal, a professor of physical education. Esther Werstler, whose husband was another regular partner, appreciated the fact that John never failed to ask about her work at the hospital. John "never gave me the feeling" that he did not want to include her in a game because she was a woman.

In December 1969 Kabel had a heart attack (from which he fully recovered). For a get well card Keck drew a cartoon of the other three golf buddies shivering on a golf cart at the first tee of the country club while waiting impatiently for their friend. John is saying, "Call him again!" Keck is wishing he had a bourbon, while McCristal merely says "Br-r-r-r."

For his golf partners, John was just another good golfer. His scientific renown was not an issue. Charles Slichter treasured the story about when one of Bardeen's longtime golf partners at the Champaign Country Club turned to him and said, "Say John, you know I've been meaning to ask you. Just what is it you *do* for a living?" Slichter commented, "I think if I had won two Nobel Prizes like John has, I would manage to work it into the conversation somehow!"

Several of Bardeen's golfing friends complemented John's reticence with their ebullience, in the same way that John's brother Bill, his friend Walter Osterhoudt, and his colleagues Brattain and Holonyak did. Paul Coleman was highly entertained one day watching the large and garrulous Louis Burtis energetically manage a golf game while Bardeen cheerfully acquiesced.

Bardeen continued to derive an element of fun from golf, even when not on the greens. One spring day in the late 1960s Bardeen and his colleague Harry Drickamer were chatting at a reception. It appeared that the world was flying apart. Students everywhere were protesting U.S. involvement in Vietnam. The National Guard had been called into Urbana-Champaign to maintain control. Jane Bardeen had broken a vertebra in her neck and was wearing a brace, and Drickamer's daughter had recently had orthopedic surgery. "Boy, things aren't very good, are they?" Harry said. "Noooo, not so good," answered John. Then, after one of John's characteristic long pauses, "a little smile comes to the corner of his mouth. And his eyes turned with them, 'You know, Harry,' he says, 'I'm shanking my iron shots and when that happens nothing goes right.'"

After his second Nobel Prize, Bardeen began to prepare for his retirement, set for March 1975. Having taught for more than twenty years at Illinois, he looked forward to spending all his time on research. When new students asked to work with him, he politely explained that he would not be able to see them through to the end of their work and did not want to leave them hanging.

Ravin Bhatt was one of those students. He came to Illinois from India in 1972 hoping to work with Bardeen, but Bardeen was no longer accepting students by the time Bhatt had completed his early course work. Although disappointed, he felt better after Bardeen referred him to McMillan. "If it couldn't be John, Bill was about as good an advisor as I could get." McMillan directed Bhatt to problems of superfluid helium. When Bhatt finished his Ph.D. he went to work at Bell Labs, just as McMillan had done a decade earlier and Bardeen had done after World War II. Bhatt felt that by working with McMillan he "inherited some of the style of physics that John initiated."

Bhatt took the last course that Bardeen taught, Physics 463, an advanced graduate-level course in quantum fluids, one of Bardeen's favorite subjects. It was an intensely demanding course in a tightly circumscribed area of current interest—the kind of advanced course for which the Illinois physics department was famous. Students were expected to immerse themselves in the literature and practice problem solving with minimal guidance. Hoping to take advantage of the last opportunity to take a course with the double Nobel laureate, about sixty students signed up. Under normal circumstances the course would have drawn fewer than ten. As most of them were ill prepared for the work, Bardeen's teaching method quickly reduced the class to about the usual number.

Bardeen's retirement called for a celebration, but Ralph Simmons, then head of the physics department, had to take into account Bardeen's "great distaste for any formal retirement festivities." He compromised by turning Bardeen's retirement party into a symposium on "Frontiers in Condensed Matter, held on October 15, 1976." The menu of the symposium dinner listed: "Poulet Polyphase, Champignons Champaignoise, Boulettes au jus Stockholm, Crabe Etat Solide, Consomme Jeanne, Bouef Bardeen, Pomme de terre Laureate, Tomate grillee transistor, Salad Plusiers corps, Petits Pain PNP, Sorbet Superconductif, Beverage aux choix une ou plus dimensions." The program also included the texts of several songs composed by David Lazarus. One verse, sung to the tune of the "Battle Hymn of the Republic," went:

> There's been BCS and APS
> And PSAC too
> Pals like old TI and Xerox.
> (Don't you wish that it were you?)
> P-N junctions made transistors
> Wigner's training sure came through.
> Our John keeps marching on!
> (*Chorus*):
> Place your bets on the electron.
> Pairs and singlets you'll conject on.
> Nobel Prizes you'll collect on.
> Our John keeps marching on!

After becoming a professor emeritus, Bardeen continued to work in his office almost every day, spending some time in the physics library studying recent articles. He still left his office door open to invite colleagues and students to call on him with their problems. John Miller, a graduate student of John Tucker, Bardeen's collaborator during the mid-1980s, remembers that Bardeen was "amazingly up to date" on the latest theoretical and experimental work on charge density waves and high-temperature superconductivity. When experimenters asked Bardeen for help with interpretation or design of their experiments, he brought to bear his "really encyclopedic knowledge of solid state physics."

Among the few changes that colleagues noticed in Bardeen's patterns after retirement was that he stopped regularly wearing a tie to the university. After the second Nobel Prize, he seemed a bit more relaxed, somewhat more outgoing. "You feel more self-confidence having received a Nobel Prize," he told a reporter. He was more likely to spend sunny afternoons on the golf course. And one could more often find him sitting with his feet up on his desk in a pose that recalled his days at the Naval Ordnance Laboratory more than those as a distinguished university professor. He more readily indulged in relaxed chat with colleagues. In fact, it became increasingly difficult for his colleagues to extricate themselves from such conversations, even to get to their classes on time, for Bardeen would sometimes discourse at length. In his seventies, the man who had spoken so little throughout most of his career became, now and again, almost garrulous.

13

A Hand in Industry

One day in the late 1970s, Bardeen came by the office of Karl Hess, a young postdoc who had recently come from Austria to study with Bardeen. Hess had studied under Karlheinz Seeger, who had worked with Bardeen some years earlier. Hess asked Bardeen what he looked for in choosing a good physics problem.

Speaking slowly and using his fingers to count off his points, Bardeen told Hess that there were three important requirements. "First of all," look at "whether there is a technological basis" for the work. "If you think of something in some theory but it can never be realized because there is no technology there, you are working in empty space." Second, the problem needed to be challenging, "because if it's so simple that you can do it on the back of an envelope, well, the project is over." Third, the research should have applications potential. "That's what most people in basic science overlook. If you do something and you want it to have importance, it has to mean something to the people." Asked once what he did with a problem that showed no promise of having practical applications, Bardeen smiled and replied, "I choose another problem. You don't want to pick one that's too difficult, but you must pick one that will lead to significant results."

With such a high value on the practical side of physics, it was natural for Bardeen to want to maintain strong ties to industry, even after his return to academia. His friend Nelson Leonard, a chemist, said that Bardeen's "face always lit up" when he spoke about the interplay of science, technology, and industry. Charles Gallo, a scientist at 3M, recalled that when Bardeen would speak about high-temperature superconductors at Minnesota Mining and Manufacturing, he "never once mentioned a theoretical, abstract or philosophical motivation. He only stressed the myriad practical applications and the economic viability."

Bardeen's longest industrial association began in 1952, when he was contacted by a small company based in Rochester, New York. The Haloid Company had invested in an electrophotographic copying process to become known as xerography. The concept was still new and relatively undeveloped. Its scientific and engineering problems were daunting. No one could have predicted then how far the commitment to xerography would take Haloid.

In 1906, at the time Haloid was founded, the company specialized in high-quality photographic paper. Its business existed in the margins of the markets occupied by its giant competitor down the road, Eastman Kodak. In the mid-1930s, Haloid acquired the Rectigraph Company, and with it the copying process known as the photostat, which used Haloid's photographic paper. From Rectigraph came John Dessauer, who became Haloid's director of research.

In 1945 Dessauer noticed a brief reference in Kodak's *Monthly Abstract Bulletin* to a process called electrophotography. Immediately struck by the commercial potential of this process, Dessauer brought the work to the attention of Haloid's president, Joseph Wilson.

The concept of electrophotography was then less than a decade old. It had been invented by Chester F. Carlson, a patent lawyer, who wanted to develop a process useful for copying documents. After studying the available photographic methods, including Rectigraph's photostat process and Kodak's verifax process, Carlson came upon an electrographic method that used photoconductivity, a process pioneered by Paul Selenyi. The Hungarian physicist had been dissuaded from pursuing the industrial application of his idea. On the basis of Selenyi's description, Carlson began to investigate how to form electrostatic images on photoconducting insulating layers. The work led him to invent electrostatic electrophotography in 1937. His application for a patent was accepted in 1943.

With the hope of developing his idea into a marketable product, Carlson contacted various corporations, including IBM, A. B. Dick, and RCA. None of the corporations were interested, for the working model that Carlson had constructed failed as often as it worked. But the idea interested Russell Dayton at Battelle Memorial Institute, a small nonprofit research organization based in Columbus, Ohio. In the article that Dessauer read in Kodak's *Monthly Abstract Bulletin*, he learned that Battelle wanted to supplement its research into the electrophotography process by contracting with a larger industrial or university laboratory.

Haloid's president, Joe Wilson, agreed with Dessauer that the electrophotography would dovetail with Haloid's product line. Haloid then entered into an agreement with Battelle. A year later, in 1946, Haloid bought the patents for electrophotography from Battelle. In 1947 this process was renamed "xerography," after the Greek words for "dry" (*xeros*) and "writing" or "drawing" (*graphein*). Part of the way in which Wilson supported research on Carlson's electrophotographic process was by securing military contracts. The process could be used in copying microfilm held in military archives.

Bardeen's connection with Haloid arose from a decision that Dessauer made several years later to draw on the expertise of outside scientists. Harold Clark, a member of Haloid's staff, suggested asking Bardeen whether he would be interested in consulting for the firm. Clark had taken a course from Bardeen at the University of Minnesota and thought of him when he heard that Bardeen was leaving Bell Labs. Dessauer considered it unlikely that a scientist of such high stature would want to associate with such a small company, but felt it would not hurt to ask. Intrigued by the concept of an electrostatic copying process, Bardeen readily agreed to meet with Clark and Dessauer in Cambridge, Massachusetts, where he was giving a seminar.

As the three men strolled along the banks of the Charles River, they talked about the relationship between science and industry. Bardeen may also have recalled the exciting years he had spent in Cambridge as a Harvard junior fellow. As he listened to the Haloid scientists explain the copier project, he may well have recalled the monumental task he had faced making the two copies that

Princeton had required of his doctoral dissertation. He was undoubtedly also intrigued by the vast potential social and economic value of such a technology. In later years he would add that he was immediately attracted to the philosophy of the company's president, Joseph Wilson, who considered research "a key to our development."

Bardeen admired the young company's self-image as an institution willing to take a risk that would "challenge its short-run position in order to buttress the long years ahead." He accepted Dessauer's offer. His agreement, formalized in October 1952, called for six to eight visits per year to Haloid's research laboratory, which at that time was based "in a few laboratory rooms in a converted frame residence." In 1961, as Haloid poised itself to become an international giant, the company would change its name to Xerox Corporation. Its Xerox 914 copier would be hailed by *Fortune* magazine as "the most successful product ever marketed in America."

In the years when Haloid was growing into a Fortune 500 company, Bardeen consulted with its researchers on many scientific issues. For example, after selenium was selected as the photoconductor material, Bardeen made useful suggestions regarding particular design processes, such as reducing the time the photoconductor plate rests between copies (the "fatigue delay") by doping the selenium to better control its electrical properties. This innovation, for which he filed a patent in February 1958, was an important contribution to the research process but was never directly incorporated into commercial Xerox machines.

On a typical consulting visit Bardeen would deliver a lecture in the morning. After lunch, he would visit several research teams in their labs or offices. Between visits he wrote letters and made phone calls about areas in which he felt the company ought to be up-to-date. He commented on possible scientific hires and gave his opinions about who to appoint to positions of leadership. In its early days Haloid had found it difficult to attract the best Ph.D.s. Bardeen's presence as a consultant, and later as a member of the board of directors, helped to allay job candidates' fears that they might not have the freedom or the resources to engage in cutting-edge research.

Xerox scientists found that Bardeen "would listen intensely to what we were saying," giving the researchers "a chance to talk things out." The frugality of his verbal responses led some to "wonder whether he was seriously interested" or possibly so alarmed at the poor quality of work "that he was being kind by not probing too deeply with his questions." But even the scientists who had been nonplussed by Bardeen's unresponsiveness found that "at the end of the day, he provided critical guidance."

Contact with one of the most prominent solid-state physicists in the world also boosted morale at the laboratory. Xerox scientist Charles Duke said that Bardeen projected a philosophy that "the business of research was the discovery of new insights rather than the refinement of the old" and that the best approach was to "hire the best, equip them to the cutting edge of the state-of-the-art, and stay out of their way." In addition, because Bardeen was "unfailingly kind, considerate, and courteous," he contributed toward setting "a standard of humane behavior in fiercely competitive and arrogant surroundings."

Bardeen's reports to Xerox offered guidelines for successful industrial research. For example, in 1978 he specified that only two "levels of effort" were viable: "a major effort at the forefront that is recognized worldwide" or "a minor effort sufficient to keep in close contact with developments made elsewhere." For Xerox's facility in Webster, New York, he advocated the latter, encouraging its scientists to choose "side problems," such as photoelectron spectroscopy, that would bring Xerox into cooperation with other major research centers, such as the surface program of the University of Pennsylvania. He explained that such a strategy would make Xerox "part of a leading research group at a fraction of the cost of the total program."

Bardeen's talks at Xerox often conveyed his philosophy that "invention does not occur in a vacuum." Most advances "are made in response to a need, so that it is necessary to have some sort of practical goal in mind while the basic research is being done; otherwise, it may be of little value." He recalled the period at Bell Labs before the invention of the transistor as a productive time, citing teamwork in the semiconductor subgroup as "the sort of basic research which should be done in industry." Referring to the Bell Labs mission of improving communication and to Mervin Kelly's goal in the late 1930s of replacing the relay and vacuum tube, Bardeen said, "Those doing basic research should be well aware of the long-range goals of the company and of the specific research programs in which they are involved." In 1973 he worried that budgetary requirements imposed by Xerox's "bean-counters" were dictating "too much opportunism, jumping on the latest bandwagon." He felt that the lack of long-range planning created "much of the uncertainty over missions and goals as emphases are changed from year to year."

The most fruitful problems, Bardeen would stress repeatedly, lie at the intersection of science and engineering. The transistor was an example. When Bell Labs engineers were frustrated in the 1930s by the inadequacies of vacuum tubes, their scientist colleagues suggested that it was "possible, theoretically at least, to control the flow of electrons in a semiconductor," even though at that time "no one knew how to do it." For a few years, "semiconductors became one of the most popular fields of physics." The invention of the transistor then "opened up an intensive period of device development and of basic research on semiconductors." The example illustrated an important feedback loop in which product development feeds basic research and vice versa.

Bardeen's many talks and writings about the relationship between science and industrial research express his commitment to the idea that the highest use of science is for the public good. He believed that it made sense to look first at the technological base and then work on developing the corresponding science, rather than "finding something in science and then looking around for applications." He emphasized the difficulty of transferring a scientific discovery that had not been made in the context of any technological mission "into useful products, particularly in competition with products which already exist." In the case of the transistor, where the technological need led the science, it was possible to reach social usefulness directly.

Bardeen disagreed with the usual distinction made between basic and applied research. "Basic research is defined by the National Science Foundation as that directed toward fuller knowledge and understanding rather than toward practical application. I prefer not to stress the last phrase of this definition, since I believe that much good basic research is done with applications in mind." He would remind his audiences that the research program resulting in the transistor "was basic in that it was directed toward understanding the electrical properties of semiconductors, but everyone working on the program was aware of the long-range goal and of its importance." He believed fundamentally that "there is really no sharp dividing line between basic and applied research."

Bardeen also stressed the important role of basic research in the advancement of technology. During one Xerox symposium that he helped to organize, he listed four interrelated reasons why companies should support basic research. (1) Research makes industrial research less a function of cut-and-try by suggesting "areas where significant advances are possible." (2) It can provide the background for innovations. (3) It can help develop personnel in such a way that specialists will be available to consult on thorny problems. And (4) it ensures that industry is in communication with science. In this way, by supporting fundamental research, corporations could avoid the "scientific dust bowl" that Brian Pippard alluded to in his famous 1961 talk "The Cat and the Cream."

Above all, Bardeen emphasized the human aspects of science in industry, saying that "the most difficult to find are the people—the required leadership and qualified scientists." He encouraged industries to offer their best scientists freedoms comparable to those in academia. Bardeen pointed to Bell Labs' "enlightened research philosophy" as a desirable model. Within the confines of the organization's mission, he believed, scientists ought to be free to select their own research problems and "to publish their results, to attend scientific meetings, to visit and give lectures at universities, and, perhaps, occasionally have a sabbatical to get refreshed. In other words, they should participate fully in the scientific community." Bardeen also thought that industrial scientists ought to have the opportunity to see their work come to fruition in marketable products. "Creative work is difficult," he pointed out, "and motivation is required for sustained activity."

In 1961 Bardeen was elected to Haloid's board of directors. This was also the year Haloid changed its name to Xerox and was first listed on the New York Stock Exchange. As a board member he continued to offer advice not only on existing research programs but also on long-term goals, including management policies. That same year he pointed out to Dessauer that "the outstanding problem in computers is a good rapid-access memory," an area to which "Xerox might contribute in the future." Moreover, the company's increasing size meant that its electronics research program "will depend more and more on overall competence rather than patents for protection."

In the 1970s Bardeen also served on Xerox's Technical Advisory Panel (TAP), a committee of scientists that met twice a year to review research activities of the different Xerox laboratories and to advise senior management on the overall quality of research efforts. Charles Duke believed that Bardeen's advice had helped Xerox "build one of the finest industrial laboratories in the world in the fields of organic and disordered solids during the late 1970s."

When Dessauer retired in 1968 Bardeen pushed Xerox to bring in Jacob (Jack) Goldman as the company's chief of research. Bardeen and Brattain used to play poker with him at American Physical Society meetings. Bardeen had admired Goldman's work during his thirteen-year stint as the director of Ford Motor Company's scientific research laboratory, especially his ability to attract top-notch researchers and keep them productive.

Bardeen supported Goldman when he became the driving force behind the creation of Xerox's Palo Alto Research Center (PARC). From his position on the board of directors and as chair of TAP, Bardeen vigorously argued both for PARC and for the continued development of the Webster, New York, facility.

PARC had barely opened when the company faced new challenges. Xerox's key patent on selenium had expired, and the company's near monopoly on copy machines was threatened by the entry into the copier field of several other companies, including IBM and Eastman Kodak, who boasted new technologies that Xerox had yet to integrate. With the impending expiration of other important patents, it was clear that Xerox, now a huge corporation, would shortly be plagued by antitrust suits. The fact that the company had not reached its profit targets in November and December of 1970 added to the atmosphere of crisis at Xerox in the early 1970s. Another problem came from the acquisition, against Bardeen's advice, of the computer company Scientific Data Systems (SDS). Xerox lost well over a billion dollars when SDS had to be closed in 1975.

Xerox's research program came under threat from the "bean-counters," accountants and MBAs who seemed more interested in improving the bottom line in the short term than in supporting scientists whose research was a long-term investment. Goldman sat fuming at a 1971 board meeting while the rest of the members debated whether to abandon PARC. It seemed a reasonable step, given the company's small fixed investment in the new center and its questionable profitability.

Bardeen and another scientific consultant, Robert L. Sproull of Cornell, dissented; they argued that PARC's $1.7 million budget was but a drop in the bucket of Xerox's overall spending. It would be shortsighted and irresponsible to cut off the company's best chance of conducting innovative research that might lead to new products and markets. Opinion turned, and PARC was saved. In the subsequent two decades the computer scientists at PARC developed much of the foundation for the coming computer revolution, including the Alto computer, the first graphics-oriented monitor, the first simple handheld "mouse" inputting device, the first home word-processing program, one of the first local area communications networks, the first object-oriented programming language, and the first laser printer. Why Xerox did not then succeed in leading the production and marketing of computers in the 1980s is an important story that is yet to be fully analyzed by historians of science and technology.

The 1973 decision to relocate its office products and information system development facility in Dallas limited the company's ability to bring digital office technology to market. Texas Instruments, the leader in transistor development and production, was in Dallas, but that did not help Xerox. "Dallas turned out to grow a culture that was completely orthogonal to and independent of the digital world in general and PARC in particular," said Goldman, who considers the move to have been the greatest error in Xerox's history. He and most of the scientific and technical staff at Xerox fought the decision, but executives overruled them, citing cost considerations such as labor, taxes, and transportation.

Bardeen was among the last to learn about the decision on the new Dallas facility. He dashed off a handwritten postscript to a letter he was about to send to Xerox:

> It seems to me that it would be a great mistake to have at least the advanced development activities so far removed from the centers of excellence in this area both in Xerox and the outside technical community. Texas Instruments was successful only because it was able to draw heavily on technology developed at Bell Labs, but we have no similar resource.

Bardeen said he hoped that such an important decision might be postponed until it could be discussed at the upcoming board meeting, scheduled in a little over a week. He was so concerned that he drafted a personal and confidential letter later that day to

Peter McColough, chairman of the board of Xerox, detailing his reasons for opposing the scheme. Bardeen said it would "be difficult to attract to that area the key innovative people required for a successful operation." Moreover, Dallas was "far removed from the major centers of activity which are in the East Coast Boston-Washington complex and on the West Coast."

It was too late. By the time McColough received Bardeen's first letter, the decision had been finalized. Bardeen was deeply troubled that Xerox had made a major decision on the basis of short-term cost-effectiveness without consulting its board of directors or its senior scientific and technical consultants.

Bardeen retired from Xerox's board of directors in 1974, but he continued to consult for the company for another six years. He retired from consulting as well in 1980, after almost thirty years with the company. He wrote, "The problem is not so much the time spent on visits but to try to keep up with the various areas of science so that I can talk intelligently about the problems." Bardeen maintained lifelong friendships with Dessauer and Goldman. In 1976 he received word that Dessauer and his wife were anonymously donating money to fund scholarships at Clarkson College. They hoped Bardeen would consent to have the scholarships named after him. Bardeen wrote Dessauer, "It is your name that should be commemorated. Your great contributions in opening up and developing from scratch a whole new area of technology are not well enough appreciated." Dessauer, however, insisted that the awards be named after Bardeen. John responded that he would "be proud to have the Scholarships named after him in recognition of our long years of close association and friendship."

Bardeen consulted for relatively brief periods for a few other firms. In 1954 he began consulting for General Electric (GE), visiting its labs as often as once a month in the late 1950s and early 1960s. He met there primarily with scientists in areas related to his own interests. When Nick Holonyak left the army in 1959 and went on to work at GE's research center in Syracuse, New York, he occasionally ran into Bardeen there when both were visiting GE's Schenectady laboratory. Holonyak noticed his mentor's name appeared on the same third-floor office door as the director of the semiconductor group. Bardeen resigned from his GE consulting position in 1961 when he joined the Xerox board of directors.

Bardeen also consulted for and sat on the board of Supertex, a company founded in 1976 by one of Bardeen's former students, Henry C. Pao. Supertex specialized in the design and manufacture of integrated circuits and electronic components. Viewing this consulting as part of mentoring a former student, Bardeen accepted only token compensation for it.

On at least two occasions Bardeen's advice saved Supertex from jumping into risky ventures. He advised the firm against accepting incentives to set up a plant in Indiana to provide semiconductors for the automotive industry. On a recent visit to a General Motors facility in Indiana, he had found the employees both extravagantly paid and resentful of working conditions created for their own safety, such as wearing protective gear. In contrast, Silicon Valley workers accepted the protective gear without complaint. Bardeen warned Supertex that shortcomings in worker relationships would last for the life of the plant. Pao never regretted turning down the Indiana offer. Similarly, Supertex heeded Bardeen's advice against accepting what on the surface appeared to be an attractive opportunity to establish a plant in Vancouver. Bardeen thought that Canada's social democratic government might make operations more expensive.

Also on the board of Supertex was Pao's father, a successful manufacturer. The older Pao and Bardeen began to follow every board meeting with a game of golf. Pao himself took up golf so that he could play with them, but never managed to beat either of the two. In playing, Pao noticed that while John's shaky hands made it difficult for him to hold a fork at lunch, on the links the palsy seemed to disappear.

Bardeen never officially consulted for Sony Corporation, but he enjoyed a warm relationship with the company over the years. On October 6, 1989, Sony paid tribute to Bardeen and his scientific contributions to the electronics industry by endowing a $3 million chair at the University of Illinois in his name. Bardeen's old friend Makoto Kikuchi, the research director of the corporation, told reporters that the transistor was responsible for igniting modern technology. "John Bardeen is a great figure in science, and this is our way of honoring him. Because this (university) has been such a good place for him, we hope (the fellowship) will attract others here."

After a long deliberation, the search committee named Nick Holonyak the first recipient of the Sony professorship: the John

Bardeen Chair of Electrical and Computer Engineering and Physics. "Dr. Bardeen once told me (in a low voice) that he wanted to have Nick on the Bardeen Chair," Kikuchi told Jane. "He was such a humble person that he added, 'but I am not related to the Search Committee so this is just my hope.'"

Bardeen's last visit to Japan, in May 1990, was made in connection with Sony's contribution to his chair at the University of Illinois. Sadao Nakajima recalled that on Bardeen's "last day in Tokyo, he invited a number of Japanese friends, and he asked me and my wife to join him. The atmosphere at this dinner was just like a family gathering. Just very quiet and peaceful talking. I think he loved to meet other people in that way, very informal."

To foster relationships between local industry and the University of Illinois, Bardeen helped to establish the Midwest Electronics Research Center. The goal was to assist industries in developing their research capabilities and bring them in touch with research at universities. As Bardeen explained to the Illinois Governor's Office, the center sponsored symposia on various topics and an industrial research visitors program in which scientists encouraged university relations with midwestern firms such as 3M, Honeywell, and Delco.

"When I took John along," joked Paul Coleman, a colleague of Bardeen's in the electrical engineering department, on business trips "we'd go in and talk to the president. If I went to the company [without him], we'd talk to the janitor." Coleman was well aware that other scientific luminaries would be "too busy with their Nobel Prize work than to get in an airplane and fly through a thunderstorm over to Company X, to help Paul and his colleagues to interact with industry."

A typical visit began with a tour of the facility. Then Bardeen would give a short talk, after which the group discussed the firm's problems and goals, often with company scientists and engineers present. The visitors and company executives would then have lunch and usually head off to the golf course. On these trips, as elsewhere, Bardeen was "low keyed, a little on the shy side, and spoke very slowly. But on the golf course, my gosh, that guy was competitive. . . . He would try his hardest to win." Sometimes the company would engrave a plaque with the date and names of the players to commemorate a Bardeen visit and its inevitable game of golf. George Russell, who had been a colleague of Bardeen's at the University of Illinois, kept one of the plaques on his wall even after he became chancellor of the University of Missouri.

Bardeen and his engineering colleagues would advise the companies in planning their research and development strategies. After a visit to 3M in 1985, Chuck Gallo wrote Bardeen a letter of thanks. "The fact that you think the 'search for high temperature superconductivity is worth pursuing' carries great weight with all of us at 3M."

The university visitors could also help the companies hire promising students. And if the company was having problems with its computers or other equipment, Coleman would bring along a technical expert from the university. The university, in turn, gained jobs for its students and some incidental benefits, such as the occasional donation of a perfectly good piece of research equipment, abandoned because the company was updating its technology.

Bardeen never stopped believing that free communication between academic and industrial scientists would result in the most rapid advancement of science and the greatest benefit to society. He carried this conviction not only into his industrial consulting but into the many advisory positions he assumed in government and professional societies.

14

Citizen of Science

B ardeen tried to meet the social obligations that arose from his scientific stature. One of the more fulfilling ones was to serve from 1959 through 1962 on the President's Science Advisory Committee (PSAC), first under President Dwight D. Eisenhower and then under John F. Kennedy. In May 1960 he spoke about this experience to an audience of students when he was delivering the honors convocation to new Phi Beta Kappa graduates at Illinois State Normal University.

Bardeen told his audience that at the root of many social difficulties is "the fact that there are all too few people with a broad understanding of the problems of contemporary life." He urged them to consider jobs in government service or teaching. He went on to explain how PSAC tried to help by analyzing government expenditures on research projects and by recommending how government funds should be distributed among different areas or projects, such as space exploration, missile development, or science education. The committee, he told them, "is doing a very effective job."

The origins of PSAC are tied to Russia's launching of *Sputnik* in October 1957, the first artificial satellite. That Russia had beaten the United States into space stunned Americans. The following month Eisenhower created PSAC in an effort to bring his science advisors closer to the Office of the President. He appointed James

Killian, the president of MIT, to act as the committee's chair, his "special assistant" for science and technology, who represented PSAC in cabinet meetings and in meetings of the National Security Council.

Killian in turn invited eighteen eminent scientists to work on the new committee charged with strengthening American science and bringing it to bear in policymaking. "They looked for generalists," explained Charles Slichter, who served on PSAC in the mid-1960s, "with the idea that for each panel you would put together people with a specific experience and expertise in the area. The idea was to have a broad look and an absolutely fresh look." Later, during Bardeen's first year on PSAC, the eminent chemist, George Kistiakowsky took over Killian's role as chair.

One of the most important duties of PSAC members was to consult with scientific organizations and committees and report back on their findings. PSAC scientists would sit down with groups of experts and ask "all sorts of questions which might open things up in enormously different directions." Bardeen explained that "most of the real work" on PSAC is done by panels composed of two or three members and a few outside experts. The panels prepared reports on a range of important subjects—defense strategy, missiles, nuclear testing, computers, space science, and science and engineering education. The reports were then reviewed by the entire committee, which met twice a month to discuss these reports in the Executive Office Building next to the White House. "I can assure you that there is often heated discussion of various questions at these meetings." Once approved, the report was submitted to the president, then circulated among the appropriate government agencies.

Bardeen applied some of the same skills that he used in solving physics problems to the work of analyzing the government's role in science. For example, in his capacity as PSAC liaison for the atmospheric science panel of the National Academy of Sciences, he suggested that in long-range planning of interagency programs "it might be best to divide up the field into a number of problem areas, and to set up committees of active government and non-government scientists to make projections for each area." The next step might be to put all the pieces back together by having the committees' work "reviewed by ICAS [Interdepartmental Committee for Atmospheric Sciences] as a whole." He told Jerome Wiesner, President Kennedy's science advisor, that "the problems of ICAS are, it seems to me, in large part common to all efforts to get effective interdepartmental planning in science." "What is needed," he had told Kistiakowsky ten months earlier, "is an organization at the policy making level to coordinate the work of the various government agencies engaged in science to review missions and suggest new programs, to review and give support to budgets, and so forth."

Bardeen also spoke about the country's science budget: "There was a time not long ago when science was so starved for funds that one could say almost any increase was desirable, but this is no longer true. We shall have to review our science budgets with particular care to [maintaining] a healthy rate of growth on a broad base and not see our efforts diverted into unprofitable channels."

Bardeen believed that the government should not cover the full cost of research projects and programs. "I feel strongly that [the] costs of basic research should be shared by the institution receiving support." He gave the following reasons for his opinion:

a) Less strict accounting should be necessary if costs are shared.

b) If full costs are paid, including research time of senior faculty, there is no limit to the expansion of large centers. The top few will grow at the expense of less prominent institutions.

c) Encouragement should be given to tapping other sources of revenue.

d) If costs are shared, and there are other sources of revenue, there is less dependence on and thus less danger of control by the federal government.

e) Research is a recognized function of any university involved in graduate education and it should be an obligation of the university to help support it.

Those who served with Bardeen on PSAC included Emmanuel R. Piore, Cyril Stanley Smith, Britton Chance, and Glenn Seaborg. "We thought it was pretty important," said Seaborg, "and we had a very good relationship with President Eisenhower. We met with him and he took us seriously." Slichter, who served on PSAC after Bardeen and Seaborg had left, agreed. "Everyone who was on it was enthusiastic about it, and everyone worked very hard." Although the schedule was grueling, Bardeen found it exhilarating to work on a committee devoted to important social problems and composed of such distinguished members. When President Eisenhower left office, Bardeen wrote "to express my deep gratitude for the opportunity to serve under your leadership."

The work of PSAC enhanced Bardeen's awareness of important issues in science policy, such as arms control, space science, engineering and science education, foreign policy, and atmospheric testing of nuclear weapons. He was encouraged by Eisenhower's and Kennedy's appreciation for the important national role science could play and by their willingness to listen to the advice of top scientists. Eventually, PSAC's accomplishments included laying the groundwork for the National Aeronautics and Space Administration (NASA), encouraging a nuclear test ban treaty (which came to fruition during Kennedy's administration), and strengthening American science education. Bardeen wrote to President Kennedy early in 1963 that "the committee as a whole under Wiesner's outstanding leadership has been a very effective instrument for bringing expert knowledge and wisdom from all parts of the country to bear on these problems."

A memo from PSAC member Harvey Brooks titled "Issues on Research and Development in the Federal Government" elicited from Bardeen a series of handwritten notes that provide insight into his concept of the ideal interplay between government and science. Next to the question, "How should the government attempt to measure the quality and effectiveness of the research performance both of specific agencies and in specific areas and of the government-supported effort as a whole?" Bardeen scribbled a single word: "Experts." The progressive notion of experts as the arbiters of social good was deeply embedded in Bardeen's family history and scientific training. Although he was aware that new ideas for social progress often come from younger members of society, he was convinced that performance is best measured by people with top-notch training and long experience.

Bardeen's position was more democratic in responding to the question "How should the Federal Government decide the magnitude of effort that should be devoted to specific major categories of basic and applied science . . . ?" He wrote: "Role of Congress." It was a given for Bardeen that Congress was entrusted with the responsibility of representing the best interests of citizens when creating and approving budget appropriations.

Beside another question about aligning federally supported projects and problems with "the processes and institutions of formal education," Bardeen wrote: "post-doctoral." He never stopped emphasizing the importance of education. He told Kistiakowsky in 1959 that graduate education in the sciences "is a topic in which I am very much interested," and he saw postdoctoral fellowships as an important way to develop the training of young scientists.

Another question on Brooks's memo asked, "In the support of basic research, how is the government to keep a proper balance between direction from above and initiative from below?" Bardeen wrote: "Multiple sources of support important." He felt that scientists needed the freedom to pursue their research interests, even when those interests might not be compatible with the immediate goals of a particular funding institution, such as the Department of Defense. Although mission-oriented agencies should have their goals and interests taken into consideration by scientists who benefited from their funding, the scientists also had the right and the social obligation to pursue basic research wherever it led them. With multiple sources of funding, one could balance those sometimes competing interests.

In 1969 Senate Majority Leader Mike Mansfield pushed through an amendment to a military authorization bill that prohibited the Defense Department from supporting research not directly related to a military purpose. Strongly opposed to the Mansfield amendment, Bardeen wrote in March 1970 to U.S. Representative Emilio Daddario that it could "in the long run make it impossible for the Defense Department to support basic scientific research in the universities." The consequence would be "drastic cutbacks" in high-quality programs, which, Bardeen judged, could "only be disastrous." He emphasized that military-sponsored research was helping to keep the American economy strong, and develop technology for consumers, both at home and abroad. "Once technological leadership is lost, it would be very hard to regain." He suggested that "the Mansfield amendment should either be deleted in the 1971 budget or reworded" to permit scientific research in areas, such as solid state physics, which are clearly relevant to the military.

Despite Bardeen's belief in government's responsibility for funding science education, he realized that certain types of programs were politically not feasible. Seaborg had chaired a PSAC panel on basic research and graduate education. One theme of the resulting paper, known as the "Seaborg report," was that basic research and graduate education "should be closely coupled." In the committee discussion of the report on April 18, 1960, Bardeen pointed out that however desirable it might be from a science education point of view, the time was probably not right "for direct government support for tenured appointments in universities" and that "there might be political problems involved in direct grants to universities." Instead, he suggested that "research scientists, that is, those without teaching responsibilities at universities, be confined to temporary postdoctoral appointments."

In another note scribbled on a PSAC document, Bardeen again emphasized the importance of having first-rate scientists in many fields. In considering "factors determining the desirable rate of growth in special disciplines," he added "manpower available" to the list of considerations that included "scientific promise, security gains, civilian gains, prestige gains." Given the large federal investment in scientific research, Bardeen thought it would be necessary to devise criteria for evaluating which areas of research to fund and to what extent. "My own feeling," he wrote to I. I. Rabi in 1960, "is that at any given time there is an optimum level of support at which research can be carried out most effectively in the sense of getting maximum productivity per dollar spent. Increasing support beyond this level may be desirable to get results faster, but diminishing returns [set] in very rapidly."

Bardeen consulted for other government agencies before and after his PSAC years. World War II had set a pattern in which scientists routinely entered into agreements with government agencies. Like most American scientists who had worked on defense projects during World War II, Bardeen had maintained ties with the government, serving in various capacities over the years. For instance, in the spring of 1951 he had accepted an advisory post with the Office of Naval Research (ONR) under a personal services contract in which he agreed to consult without compensation. He was part of the Panel on Solid State Physics chaired by Fred Seitz.

During the Cold War such contracts required the preparation of both a Report of Loyalty Data and an affidavit that the scientist was "not a Communist or a Fascist," nor planned to engage in "any strike against the Government of the United States." Despite Bardeen's personal distaste for Red Scare fanaticism, he had no reason not to take loyalty oaths such as the one required for his ONR personal services contract. Similarly, in 1952, when the two-year-old National Science Foundation (NSF) asked Bardeen to serve for a year on an advisory committee in its Division of Mathematical, Physical, and Engineering Sciences, a security investigation was needed to assure the government that giving him access to technical information would not "endanger the common defense and security." Before his appointment was confirmed, he was again required to sign the affidavit swearing that he was neither a Communist nor a fascist and had no intention of participating in any subversive activities. Bardeen subsequently served on various other NSF committees, such as the Special Commission on Weather Modification.

In 1965 Bardeen was appointed to the President's Commission on the Patent System, the only scientist among the commission's fourteen inventors, businessmen, and patent attorneys. He served on this commission for two years. Bardeen wrote to Alfred Marmor (another committee member) in the summer of 1966, "I feel strongly that rapid and free exchange of technical information is one of the main bases for the strength of our economy, and that we differ from Europe in this respect." He felt that the patent system was "a major factor" in America's strong economic position, in part by making that kind of free exchange possible. Not everyone felt that a grace period was necessary, but Bardeen pointed out that "the grace period is valuable in that it allows rapid publication of scientific and technical publications and market test of patentable products prior to application for a U.S. patent."

In 1965 President Lyndon Johnson honored Bardeen with the National Medal of Science, the nation's highest award for scientific achievement. The first medal had been awarded three years earlier to Theodore Von Kármán. The medal recognized both the present implications and the (probable) future influence of a scientist's accumulated work. Bardeen's likeness was also inscribed on a medallion as part of the Franklin Mint's "Medallic History of Science," a series of "100 sterling silver medals recognizing the greatest events in the history of science." The collection included commemoratives for Copernicus, Galileo, Kepler, Newton, Curie, and Einstein.

Bardeen played golf whenever possible on his many travels. One of his favorite courses was the ancient and challenging one in the charming town of St. Andrews, Scotland, which dates back to the Middle Ages. The university at St. Andrews is more than 400 years old; its Royal Golf Club is well over 200 years old. All are situated on a peninsula that juts out into the sea, providing spectacular views all the way around. The town is saturated with golf lore and with shops specializing in golf paraphernalia.

Jack Allen, a discoverer of the superfluidity of liquid ^{4}He and a close friend of Bardeen's, was based at St. Andrews University. Bardeen tried to visit Allen whenever he was in Europe, both to talk physics and to play golf.

When John and Jane visited Allen in 1967, they stayed in an old inn overlooking the beach at a point where the shoreline gives way to jagged rocks. On a Sunday morning, while Jack and John were off playing golf, Jane decided to take a long walk by herself along the shore. Clambering over some boulders, she slipped on seaweed and slammed into the rocks below. Someone on the beach helped her get to the nearest clinic. John rushed to her side when he got the news.

A fractured vertebra in Jane's neck required transporting her to the university hospital at Dundee. According to the orthopedic specialist, she was lucky not to have been killed or paralyzed. "There was hardly any pain," Jane wrote to her sister. "Most of my discomfort was due to getting used to this one position and the traction." The treatment involved wearing a neck brace, and she was confined to bed for six weeks.

Bardeen had to cancel a meeting with John Dessauer at Xerox because of the extended stay in Scotland. As soon as Dessauer heard the news about Jane, he contacted the company's London branch, which "lost no time in locating a projector" to help Jane read while lying flat on her back. But the orthopedist had a better idea. He rigged up a pair of glasses with mirrors at a right angle to the lens, and with this simple arrangement Jane could read with the book propped on her chest. "John was a bit chagrined" when he had to send the Xerox mercy van back to London with its cargo unused.

During the anxious six weeks following Jane's accident, Bardeen stayed in St. Andrews, hitting the links in the morning and visiting Jane in the afternoon. He calculated that buying a set of golf clubs to use during the extended period in St. Andrews would be cheaper than renting. When they went home, the "Bardeen clubs" remained with the St. Andrews physics department, Bardeen's gift to his colleagues at one of his favorite places. Jane told Allen that John had "had the best holiday for years, since he was normally a workaholic."

Allen also shared some of Bardeen's other interests. In a conversation at the 1958 Low Temperature Conference (LT6), held that year in Leiden, Bardeen, Allen, Jan de Boer of Amsterdam, and Sam Collins of MIT discussed the reckless waste of helium in the world. They developed a scheme for separating helium from the natural gas being extracted from the ground in Amarillo, Texas, and then pumping it back into empty wells to await use. The low-temperature physicists believed that "longer range technological developments" could be brought up short if there were not enough helium to keep them going. "An example," Bardeen wrote, "is the use of superconductors in the electric power industry for both generation and transmission. If widely adopted, large amounts of helium would be required." Bardeen and the others contacted some members of Congress, and the idea was put into operation via agreements with private producers.

The contracts were cancelled in 1971, when the Nixon administration abandoned the program. Bardeen wrote to Allen "expressing sorrow and anger that Congress had stopped the process because gas wells in Arizona and Colorado were selling helium and undercutting the Amarillo prices." Trying to save the program, Bardeen wrote to Dr. Hugh Odishaw at the National Research Council of the National Academy of Sciences. "The original basis for the conservation program, that it will pay off economically in a reasonable time, is no longer valid. There is no reason to expect that helium demand will exceed supply in the foreseeable future (say 20 years) even if commercial production no longer goes into storage." Support for the program would thus have to be based on a conservation argument, that of "conserving a unique, irreplaceable resource for future generations." He suggested that perhaps other countries might be willing to help underwrite the cost of the conservation program. "If the program cannot be sustained, future applications of low temperature technology will be greatly hampered." Although they could not save the U.S. helium conservation program, Bardeen and Allen estimated that their plan, while it lasted, had saved about thirty billion cubic feet of helium.

For Bardeen the helium conservation program and its demise symbolized a larger potential problem. He wrote to a colleague in

1974 that "it seems impossible with our political system to do any real long range planning." Furthermore, helium conservation "is just a small part of the overall problem of conserving our resources for future generations." Less than a year earlier, he had told a radio journalist that "the energy crisis has been known to scientists and other people in the field for many years. It should have been planned for." He understood that the measures required to protect the natural basis of the human economy would not be politically popular, but he insisted that it would be necessary. "Our laws are designed to exploit our natural resources. It will have to change. This may mean a complete restructuring of the tax structure, for one thing."

A decade and a half later he had little reason for optimism. "My greatest concern is the environment," he said. "We have to learn to live differently, using less energy and creating far less pollution. I think people will depend more and more on communication in their homes, rather than physically going from place to place." He couldn't guess whether the human species would successfully learn to curb its use of resources but, he said, "we've got to take more responsibility for our actions."

Bardeen also supported such groups as the Population Crisis Committee, a national organization dedicated to stabilizing global population growth. Long before the environmental movement popularized the problem, he joined forty-seven other Nobel laureates in petitioning President Kennedy and United Nations Secretary General U. Thant to make a serious effort to decelerate world population growth. Foreign aid and birth control, Bardeen suggested, were possible means to that end. Closer to home, he supported Planned Parenthood and the Sierra Club.

Several of Bardeen's efforts to offer social service backfired. The honor of being elected president of the American Physical Society (APS) in 1968 turned into an unforeseen burden. The APS had already committed itself to holding its 1970 annual meeting at the Palmer House Hilton in Chicago when violence erupted there on the night of the 1968 Democratic National Convention. Vivid images of police brutality appeared on national television. Journalists filmed police indiscriminately clubbing protestors as they tried to push back thousands of demonstrators outside the Hilton.

A number of professional societies had cancelled their Chicago meetings to protest the police actions and Chicago Mayor Daley's support of them. Many members of the APS wanted their Chicago convention cancelled too, or moved to another location. Other members objected to making any political stand as an organization. Conyers Herring was among those who believed that canceling or relocating the meeting "involves using the APS organization as a weapon on issues that, however important they may be otherwise, do not directly concern the objectives for which APS was set up."

As the new president of the society Bardeen found himself in the unenviable position of mediator. Physicists wrote to him from all over the country, some of them stridently. He took a position of neutrality. Even if the APS council had wanted to move the conference, the logistics would have been nightmarish. No other convention facility capable of handling nearly 10,000 attendees was available at that late date. On November 10, 1968, Bardeen wrote to William C. H. Joiner, an APS member who had wanted to relocate the meeting, to explain why it would be held as originally planned:

> If we were scheduling a meeting from scratch we should certainly take into consideration the feelings of even a minority of members. However, we have a contract with the hotels to meet in Chicago in 1970 and it would be very difficult to find an alternative location at this late date. . . . Holding a meeting because of a prior legal commitment to do so does not argue one way or the other in regard to any political viewpoint.

Bardeen was greatly relieved when his term as APS president ended. The Chicago turmoil had posed such demands on his time that "my physics has been neglected." Years later he just shook his head whenever he heard that someone had agreed to run for the APS presidency.

Bardeen also had the misfortune of chairing the Very Low Temperature Commission of the International Union of Pure and Applied Physics (IUPAP) in 1970. Protests threatened to disrupt the IUPAP's Twelfth International Conference on Low Temperature Physics (LT12) in Kyoto, a meeting for which Bardeen was responsible. The president of the Japanese Physical Society (JPS), Todashi Sugawara, informed Bardeen that some members of the JPS had threatened to make trouble if any physicists associated with the military attended. Bardeen, however, found it unacceptable that a physicist should be excluded from a professional conference for political reasons.

It was in part a question of defining military research. Many American physicists, including Bardeen, were being funded by the military but were not involved in weapons research. He wrote to Sugawara in March that "I regret very much that you have found it necessary to ask representatives of military organizations to withdraw their papers on their own initiative in order to avoid possible trouble at LT12."

After meeting early in June with researchers from military laboratories, both in the United States and Canada, Bardeen wrote again to Sugawara. "Since none of these groups have the remotest connection with weapons research, it does not seem fair to exclude them." He also pointed out that "so few cases are involved, it is hard to see how a big issue can be made. I am sympathetic to the desire for peace and to decrease the power of military institutions, but [the demonstrators] should be able to find much more relevant cases to protest against."

Troubled by the issue, Bardeen contacted Jack Allen in Scotland. Allen, who had preceded Bardeen as chair of the committee, thought that a firm stance might be more palatable if it came from someone other than an American. Allen knew that the IUPAP had already committed itself to a policy of allowing all qualified scientists to attend its conferences, no matter who their sponsors were. The rule had been established by the IUPAP General Assembly in 1957 after the West German contingent tried unsuccessfully to deny access to East German scientists at their Rome meeting. After receiving approval from the IUPAP's president, Allen contacted Sugawara and told him that if anyone from any country was barred from attendance because of their affiliations, the conference would be shut down immediately. Although he was still worried, Sugawara was pleased to have the support of the IUPAP and agreed not to try to screen attendees.

Bardeen had to have emergency prostate surgery that summer, causing him to miss part of the conference. He wrote to Allen to ask whether he would say a few words to open the conference, a ritual that Bardeen normally would have performed as chair. Allen agreed. Despite all the maneuvering, the conference ran without a hitch.

Another incident underscored Bardeen's aversion to political agendas that interfered with the free practice of science. In 1976 when it appeared that two Israeli scientists were being excluded from a conference in Hungary for political reasons, Bardeen wrote to the conference organizer expressing his "grave reservations about attending an international meeting that is not open to all." He felt "strongly that attendance at an international conference should be governed by scientific competence. No one should be excluded because of his country of origin."

Bardeen's ideas about international scientific exchange were similarly liberal. In 1959 he had written to Senator Thomas Hennings, Jr., about the consequences of restricting free exchange of scientific information. He stressed that in the field of solid-state physics "there have to date been few restrictions imposed by other countries, so that the rapid progress made in this field in recent years has been international in scope. In particular, Russian scientists, who have been quite active in the physics of solids, publish their results and give reports of their work at international conferences." For the United States to remain at the front lines in research, Bardeen wrote, American scientists had to be able to "gain from free exchange of scientific information." Twenty-three years later, Bardeen enclosed a copy of that letter in a note to Thomas H. Johnson of the White House Science Council. "I don't remember the reason for writing the letter at that time," he confessed, "but I think the comments are still relevant."

Bardeen's extensive scientific advising over and above his teaching, research, and industrial consulting taxed him physically. In March 1974 Jane wrote to her sister that John was again having trouble with hypertension. It had seemed to be getting better, "but he pushes himself too hard." She worried that he would be "off again tomorrow to D.C., for a meeting with [Senator] Daddario on Science and Government." Just a year earlier Bardeen had collapsed in Washington, D.C., while on a consulting trip for the National Bureau of Standards. John Hoffman, director of the bureau's Institute for Materials Research, sent someone to accompany Bardeen back to Champaign to make sure that he made it home all right. Afterwards, Bardeen wrote to Hoffman to thank him and to let him know that after nearly a week in the hospital for tests, "they were unable to find a specific cause [for the hypertension] but were able to treat it with pills."

PSAC was dismantled in 1973, in a time of rising inflation, the war in Vietnam, a growing deficit in the federal budget, and intensifying political and social conflicts. Policymakers and the public were questioning the role of science in promoting or subverting national values. President Lyndon B. Johnson let PSAC languish. Some members of the committee had become disheartened because of Johnson's policy on Vietnam. Richard Nixon finally abolished PSAC altogether in 1973.

The historian Bruce Smith has divided post-World War II American science policy into three phases: 1945–1966, 1967–1978, and from 1979 on. The first, which began immediately following World War II and lasted through 1966, was an era of hope and confidence that science could exert "a liberating influence on human affairs." That era gave way to one of distrust between scientists and policymakers through the 1970s. It was followed by a rising public skepticism after 1979 about whether science had the power to improve the world. The establishment of PSAC and Bardeen's PSAC work fell within the optimistic first period. The dismantling of PSAC occurred during the antiscience trend of the second period.

The antiscience began to reverse during the third phase, even before Reagan moved into the White House in 1981. Viewing science as a panacea for American weakness in the world's economic and military arenas, Reagan chose a strong science advocate, George Keyworth, as his science advisor. Together Reagan and Keyworth assembled a panel of leading scientists reminiscent of PSAC. Bardeen was invited to serve on the new presidential science committee, called the White House Science Council (WHSC). The WHSC would, however, be nothing like its predecessor.

When Keyworth phoned Bardeen early in 1982 to invite him to serve on the new panel, Bardeen replied, "You don't want me. . . . I am useless in a group larger than two. In a group of two I can hold my own, but otherwise I can't." Bardeen's reticent manner and muffled speech probably did impede his effectiveness in a larger group, but he was hesitant to serve on the new WHSC for other reasons as well. He considered Reagan's policies militaristic and irresponsible. Even so, his colleague Charles Slichter urged Bardeen to join the council "because I felt that they needed really good people with courage and strength." Bardeen accepted the position, "but his heart wasn't in it." He had, according to Slichter, "absolute contempt for Reagan."

In spite of his reservations Bardeen gave the same serious attention to the White House Science Council that he had given to

PSAC twenty years earlier. One of the council's projects that he supported was the creation of multidisciplinary research centers funded by the government. The idea reminded Bardeen of the highly productive groups he had encountered at Bell Labs. Problems could be studied collaboratively by individuals and groups having different sets of skills and knowledge. According to Keyworth, "John was very, very useful in the formation of that." But the White House was less than enthusiastic about the proposal.

Another council project evaluated federal laboratories and suggested ways to improve their effectiveness. Bardeen's long experience with industrial and academic research formed the basis of his strong opinions about America's place in the global scientific and economic community. He was concerned that "while this country is the leader in pure scientific research, it is falling behind in applying science to broaden the technology base for products of the future." The "biggest weakness in this country," he felt, was "in the development of the technology base required for leadership in the period ten or fifteen years out." Long-range planning "should be done by industry in its own self-interest," but he worried that "the pressures for short-term profits and emphasis on return on investment with high interest rates gives priority to near-term product development." Somehow, he believed, government should encourage a long view. "Only a few companies (Bell Labs, IBM, Hewlett-Packard, etc.) have the resources and enlightened management to do the required long-term basic research and technology development. The Federal Labs and to some extent universities help fill in the gap, but there is a weakness in coupling with industry."

Had the White House Science Council confined its attention to issues such as these, Bardeen might have been happy to serve. But less than a year after he joined, Keyworth and Reagan committed the United States to a space-based missile defense program officially called the Strategic Defense Initiative (SDI) and popularly known as "Star Wars." Bardeen strongly opposed the initiative. When the council, chaired by Seitz, met in mid-March of 1983, Keyworth had not raised the issue of SDI for discussion. The committee members were therefore incredulous when on March 23, 1983, Reagan made his famous speech, edited by Keyworth, to the National Association of Evangelicals. Reagan publicly called the Soviet Union an "evil empire" against which the United States would develop an "impenetrable shield." The committee had heard nothing about the idea. "Such a far-reaching proposal," Bardeen later wrote, "should have been reviewed by the Science Council at least, if not by a much wider group of experts."

Two weeks after Reagan's incendiary speech, Bardeen resigned from the council. In a letter to Keyworth he said he needed more time for his own research. His friends and colleagues knew the real reason, that "there was no point in being on a committee which is supposed to give advice when such a truly major scientific and technical decision is made without consulting the body." Bardeen felt that "he was being used for his name," which he didn't want associated with any "committee that blessed the president's program."

Bardeen told Keyworth, "I do not feel that the time I have spent on the WHSC has been as productive as I would like. The decision was triggered by two recent actions by the White House on important issues involving science and technology concerning which the WHSC and other parts of the technical community had very little opportunity for input." One of the "recent actions" was SDI.

The other action concerned a proposal to establish a National Center for Advanced Materials (NCAM) at Berkeley. The issues raised by this proposal related to solid-state physics. At the time Bardeen accepted his position on the council, he had "thought that I might be helpful in areas of my own expertise, particularly materials science and electronics." It was therefore a great frustration, he wrote to Keyworth, that "neither I nor other members of the WHSC were consulted in regard to the NCAM proposal until the November meeting when it was too late for the discussion to have any influence." He added that in discussing the NCAM proposal "with many knowledgeable people from both universities and industry" and even attempting "to put the proposal in the best light possible," he still did not find anyone who considered NCAM a good idea. Bardeen later explained that "when I resigned I was more concerned about the proposed Berkeley National Center for Advanced Materials than I was about Star Wars." Because Congress received so many adverse comments on NCAM from scientists, the proposal was ultimately drastically scaled back.

SDI continued to be a huge worry for many scientists, including Bardeen. The Reagan administration proposed spending billions of dollars on the program, with the aim of making strategic nuclear weapons "impotent and obsolete." The scientific community all but unanimously opposed pouring so much money into a project so unlikely to succeed. As Bardeen saw it, the problem with SDI was much more than the wasted money and poor chance of success. The program would exacerbate the arms race and drain scientific talent away from more socially productive work. He worked closely with some of his Urbana colleagues on a petition opposing Star Wars. Michael Wortis and David Lazarus drew up the proposal; Bardeen carefully reviewed it. He took an active role in "pushing this petition, which was eventually signed by many of the physicists of the world," according to Lazarus. "He was one of the first ones to sign it."

The Strategic Defense Initiative so worried Bardeen that he took the step, unprecedented for him, of writing several short articles and editorials about its dangers. One editorial was sent to the *New York Times,* but he withdrew it when Hans Bethe and Kurt Gottfried asked him to cooperate with their efforts to call attention to the negative impacts that SDI would have on society. Bardeen wrote to Bethe to clarify his position:

> As noted in my letter to *The New York Times,* I am concerned about the effect of a massive build up of R&D for SDI on the consumer economy and on the progress of science and technology in general. The usual argument is about the small positive effect of spin-off rather than the massive negative effect of drawing off funds and talent from more productive areas. One indication of the distortion of priorities is that Keyworth has spent 50% of his time in the past two years on SDI and recently as much as 85%. It should be possible to make a more definitive study of the issue, perhaps with the Apollo program serving as a model.

He reminded Bethe that PSAC's review of the projected Apollo program had discouraged the project, citing "a $25 billion price tag over a ten-year period," but the president and Congress went ahead anyway. "I think that the indirect cost of the programs in diversion of top talent from consumer industries was far greater than the out-of-pocket cost."

Bardeen and Bethe's coauthored editorial opposing SDI was written largely by Gottfried and appeared in the *New York Times* on May 17, 1986. Bardeen approved the text and was happy to lend his name. The piece decried the SDI's lack of scientific basis and pointed out the immense sums of money and scientific talent already pumped into the program, with embarrassingly few results. They argued that SDI would impair the nation's ability to compete in international markets. The article also emphasized the need to reinstitute some sort of science advisory apparatus that would offer independent judgment on scientific matters for the chief executive—even when that judgment was "politically unpalatable." The organization they described resembled the defunct PSAC, on which both Bethe and Bardeen had served during the Eisenhower and Kennedy administrations. That kind of committee, they argued, "even if composed largely of scientists from the weapons labs, would have made it perfectly clear to Mr. Reagan that his vision of defending cities against nuclear attack had no basis in scientific knowledge."

Bardeen blamed much of the downslide of the American economy on shortsighted research programs encouraged by a government intent on escalating the arms race. He came to consider arms control as "the most important issue of our time," one on which the "future of civilization is dependent." In an editorial in *Arms Control Today*, Bardeen focused on problems he associated with SDI, deploring the "militarization of space" because of "staggering" dangers and costs, including the inevitable brain drain. "At a time when our civilian economy needs all the help it can get to remain competitive in world markets, the best scientific and technical brains in the country may be drawn off to work on a project of dubious value."

Aside from his firm stance on arms control and the environment, Bardeen was reluctant to use his position to expound on political issues. Ned Goldwasser said that Bardeen "was very careful about what he endorsed and what he didn't endorse" and generally handled the Nobel laureate's dilemma of being asked to comment on a broad range of subjects by agreeing to speak "only on the things where he had some scientific or technical expertise, knowledge, understanding, that the general public did not have." Bardeen was so cautious about proffering his opinions publicly that people who did not know him well sometimes got the impression that he was not interested in political or social issues.

For example, when asked by the Peace Research Institute to support a statement calling for various measures to prevent escalation during the Cuban Missile Crisis of 1962, he telegrammed a reply:

> In agreement with general principles and would be glad to endorse briefer statement concerning vital interests, extreme danger of col-

> lision course, importance of cooling off period and need to look for
> alternative solutions, but unwilling to endorse entire paper since I
> have no expert knowledge, have had no part in its preparation and
> have objections to some statements therein.

One of the "objections" he referred to concerned using the Pope as
an intermediary. Bardeen scribbled an emphatic "NO!" in the margin next to that suggestion.

On another occasion, the European Committee on Crime Problems asked Bardeen, as a Nobel laureate, to provide his opinion on
whether "the death penalty has a place in contemporary society."
Bardeen responded, "I do not believe that my winning a Nobel Prize
in Physics makes me particularly qualified to give an opinion on
this topic. However, I am personally opposed to the death penalty,
particularly in times of peace." Likewise, to a request to support an
archeological program, he responded, "I do not feel that I should
lend my name to an organization unless I can take an active part."

Bardeen occasionally gave his thoughts away with some small
gesture that his students or other close associates could easily read.
When Jim Bray, who had an office near Bardeen's, heard the news
on the radio in 1972 that the segregationist governor of Alabama,
George Wallace, had been shot, he rushed down the hall to tell
Bardeen, who "didn't say a word. The only thing that happened
was he looked at me for about five seconds, and then cracked the
smallest of smiles and turned his head back to his work. . . . That
told me everything I wanted to know about his opinion of George
Wallace."

Both John and Jane abhorred racism. Jane worked on the problem, along with other social issues, through the local branch of the
League of Women Voters. In the 1950s she found to her dismay that
African Americans were not allowed to sit on the main floor of the
local movie theaters, and "being of the mind of freedom for everybody, I got involved with the whole movement of racial matters."
She helped to organize a league committee consisting of four
women, who coincidentally all had the same given name. Charged
with combating racial discrimination in the Champaign-Urbana
area, the "Committee of Janes" worked with sympathetic religious
groups to urge local retailers to hire African American clerks. One
merchant flatly refused. "I did it for the Jews, but I'm not going to
do it for blacks." It took years of prodding from the committee for
the company to begin hiring a few African Americans. The "Janes"

also designed and executed a campaign that in 1961 succeeded in attracting a chapter of the Urban League—an organization dedicated to the achievement of social and economic equality for African Americans—to Champaign County. Jane believed that "once you break the door open a little, it comes open all the way." Although John shared Jane's convictions about civil and human rights, he was uncomfortable with political activism. He supported her activities from the sidelines, often with his checkbook.

John and Jane were appalled by Bill Shockley's efforts during the 1960s and 1970s to prove "scientifically" that people of color were genetically less intelligent than whites and that the high rates of reproduction of African Americans were causing a decline in the average intelligence of Americans. Bardeen did not publicly denounce his former friend and boss, but he occasionally made comments that revealed his aversion to Shockley's genetic theories. "You can't measure intelligence by IQ or any other single number," he said. "There are many different kinds of intelligence."

Bardeen occasionally participated in political action on behalf of a particular candidate or cause. He served on a local steering committee composed of a group of scientists, both Democrats and Republicans, who united in opposition to Barry Goldwater's extreme conservatism during the 1964 presidential election. The group supported Lyndon B. Johnson and opposed Goldwater for many of the same reasons Bardeen would so adamantly oppose Reagan nearly twenty years later. While they acknowledged the need for military strength, they argued that government was equally responsible for taking the initiative to "eradicate poverty; attack the clusters of urban problems; overcome water and air pollution; upgrade inadequate transportation; promote the prevention of illness; and make the great advances in medicine, hospital design, and health care available to all, regardless of ability to pay."

The committee advocated education as the best defense against mediocrity and closed its "Statement of Principles" with an admonition against intolerance. "In this and all other areas we reject discrimination based on race, creed, or sex as immoral, undemocratic, and savagely wasteful of the nation's human resources." During the campaign, Bardeen participated in the development of advertisements to encourage local citizens to vote for the Johnson-Humphrey ticket.

Bardeen strongly believed that high-quality public education is crucial to the nation's well-being. In 1981 it seemed as if budget cuts would necessitate closing the University of Illinois's Laboratory High School. As an experimental school, Uni High was then an auxiliary project of the College of Education. To protect the college from a major budget cut, Uni was singled out as a project that could be discontinued. Theoretical physicist Shau-Jin Chang in the Department of Physics recalled that both his children, Iris and Michael, who were attending Uni that year, "were really heartbroken when they heard the news. Many of their classmates were crying at school."

After the announcement several physics faculty members had lunch with Bardeen and discussed the closing of Uni High, which all three of Bardeen's children had attended. John told how in 1951, when he was considering whether to accept Illinois's offer, he had spoken about Uni with Philip Anderson at Bell Labs. Anderson, a graduate of Uni High, had told Bardeen that Urbana-Champaign was a great place to raise a family and that Uni High was an outstanding school. That conversation had figured into Bardeen's decision to accept the Illinois offer. He agreed to write a letter to support the effort to save the school. Writing to Goldwasser, then vice-chancellor of the University of Illinois, Bardeen testified that the presence of Uni High had "played an important role in our decision to come to Illinois." The administration kept Uni High as a unit of the university, reporting directly to the chancellor.

Bardeen worried about many educational issues, including how state budget cuts might affect both the quality of education at the University of Illinois and the number of students who would have access to it. "You can't cut back continually and expect to have a first class institution." Recalling that as a college student he had enjoyed a great deal of freedom to study whatever he found interesting and noting that this policy had paid off in his case, Bardeen argued for proposed curriculum changes that would allow students more freedom to follow their own interests. He tried to follow that philosophy in guiding his own graduate students, giving them much leeway.

It seemed to Bardeen that an intermediate advanced degree between the master's and the doctorate might be desirable. He felt that it would benefit "people who do well in course work and have a good understanding of the subject matter but are not the creative sort of person required for a research career." Those people might make outstanding college-level teachers, but their inability to complete a thesis held them back. It would work best, he thought, if it "would not be a consolation degree but an intermediate degree representing a certain level of training and accomplishment."

Bardeen was convinced that the key to increasing popular appreciation of science was "better science education for a larger fraction of the people." Goldwasser recalls that Bardeen was "very supportive of the idea that we have to get more science teaching, better science teaching into the schools as well as the universities." He was proud of the work that his daughter-in-law, Marge Bardeen, Bill's wife, was doing toward that end at Fermilab (Fermi National Accelerator Laboratory). She created programs in which teachers and students could develop a deeper understanding of mathematics and science. She also was active in national attempts to improve science education, working with the U.S. Departments of Education and Energy and undertaking education research of her own with the support of NSF grants.

In 1958 Bardeen participated in a panel at Uni High on education in the natural sciences. He and the other panel members, all University of Illinois professors, had children in Uni High. The group included Seitz, Heinz von Foerster, Frederick Will, and Chalmers Sherwin. They concluded that all students ought to receive a broad education, including English, math, languages, and sciences. Especially talented youth, they thought, should have opportunities for individual development, but not at the expense of learning the fundamentals. "Natural scientists are not all geniuses," their report said. "They are all kinds of people. For every genius there are hundreds less spectacular but still very important. While many show scientific interest at an early age, not all do." In general the group agreed that "up-grading teaching for all will help future scientists, and that high school training in science should be broad." Bardeen argued that we should "develop fundamental thinking instead of memory work," because "fundamental things will stick." Bardeen's own education had followed this model, and he tried to follow a similar philosophy with his students at Illinois.

Scientists in other countries also considered Bardeen a role model. He took his responsibility as scientific ambassador seriously. He was among the American scientists to be invited to visit the Soviet Union in 1960 as part of an exchange program set up by the National Academy of Sciences. Because he was a member of the President's Science Advisory Committee, he would need special clearances in order to go. The clearances proved no problem, but a glitch on the Soviet side caused his trip to be cancelled. While in Europe at another conference that summer, he discussed the impasse with Russian physicists A. F. Ioffe and V. M. Vul. With their intervention, the necessary clearances suddenly came through for his Russian visa, and he left Prague on September 7 for the Hotel Ukraine in Moscow.

The first few days moved "slowly here as far as professional contacts are concerned," John wrote to Jane, but the pace soon picked up. He found himself giving one talk after another about superconductivity or semiconductors. He met numerous scientists whose work he knew well but whom he had never met before because of Cold War restrictions. Among them were Peter Kapitza, Isaac Khalatnikov, Lev Gor'kov, Igor Tamm, Leonid Keldysh, Vitaly Ginzburg, Abram Ioffé, and Vasilii Peshkov. Bardeen had particularly fruitful interactions with the members of Lev Landau's group at the Institute of Physical Problems, some of whom left within the decade to help found the Institute for Theoretical Physics. Although he did not get to meet Landau or E. M. Lifshitz, who were away, he did meet some of their students and colleagues. Young Alexei Alexeivitch Abrikosov, a student of Landau's, came to consider Bardeen a second mentor.

Three years later Bardeen again visited the Soviet Union for a large conference in Moscow and visits to research centers there. David Pines and Leo Kadanoff came along. Bardeen began the first talk at the conference by saying that "we are familiar with the very outstanding work done in solid state physics in the Soviet Union, and know the names of many of your scientists who have contributed so much to our knowledge, but we seldom have a chance to meet with them in person." He called the conference a "welcome opportunity to make new friends and to discuss with them the latest developments in solid state theory." Everyone was "very friendly," he wrote to Jane.

Bardeen found the Landau group impressive. Landau himself had been incapacitated by an accident two years earlier and remained in the hospital. Each of Landau's colleagues, all of whom had previously been Landau's students, was "a distinguished theorist in his own right." Speaking of the group led by Khalatnikov, Bardeen and Pines reported, "Not only are they bright and versatile, they are as well both charming and sophisticated, with a well-developed sense of humor, and a rather broad knowledge of American life and literature." In subsequent years Bardeen saw Russian scientists such as Gor'kov and Abrikosov regularly. "We know the result now," Gor'kov wrote in 1991, of "the cooperation for more than thirty years between two schools which became the base for the wide and fruitful cooperation between physicists of our two countries, in spite of difficult times we have had from time to time." Gor'kov thought that "the giant figure of John was that cornerstone which kept firm the whole construction."

During his last visit to the Soviet Union in 1988, the Soviet Academy of Sciences presented Bardeen with the prestigious Lomonosov Award, the country's highest scientific honor. He called it "one of the greatest distinctions I have received during my career." Previously, in 1982, the academy had honored him with a rare foreign membership.

Bardeen and some of his scientific colleagues were in 1975 among the first to venture into the People's Republic of China in cultural exchange. Bardeen accompanied a group led by Slichter. John wrote Jane from Peking that the team was "going into some real depth in trying to understand [China's] educational and scientific research programs." Most of the Americans became ill at one time or another during their month in China; Bardeen felt lucky to be "one of the few of our group that has managed to stay healthy throughout the trip."

Samuel Chu, a historian who went along to help with cultural and practical questions, returned with a strong impression of Bardeen's character. "I'm certain much of the warmth and comradery of the entire delegation had to do with the presence of John Bardeen. . . . In Xian, where the political types (CCP Party cadres) seemed especially intimidating to their own Chinese physicists, John made sure the latter were treated as intellectual equals to the American physicists, and thereby emboldened them in their research and teaching. . . . John included me as his personal 'China expert.'" Although Bardeen disapproved of the strictly regimented social structure there, he wished that Jane could have been there to "enjoy some of the beauties of China." He thought that she would "find it a fascinating country full of contradictions." She missed him, too. "It will be *so good* to have you home again. This house needs you to make it home."

In 1980 Jane came along on John's second trip to the People's Republic of China. They were "given VIP treatment everywhere." Bardeen's meeting with Vice-Premier Fang Yi at the Great Hall of the People was nationally televised. As a result, over 500 people attended Bardeen's lecture at the University of Nanking, despite scant publicity. Bardeen's birthday, May 23, fell on the same day as the anniversary of the creation of Nanking University, and officials arranged a great celebration to honor both occasions, complete with birthday cake, gifts, and singing students. Bardeen considered it one of the highlights of his trip.

In the same spirit in which Bardeen helped industrial firms determine their future directions, he tried to do his part to help developing nations enhance their programs in science. For example, he suggested that China and the United States establish an exchange program that would "introduce Chinese physicists to a broad range of problems and thus help them achieve a better balance in their own research programs."

On a trip to India in 1977, Bardeen shared with his Indian hosts some of his views on the role of science in society. "Science should set an example of international cooperation rather than competition." Moreover, it should be pursued in the service of improving the quality of life for people of all nations, not just a few of the most wealthy. "My country," he told them, "with a population only one-third of India's was using a large proportion of the world's resources." He hoped that India would, instead of blindly borrowing American technology, develop or adapt technologies that might better "suit Indian conditions."

Bardeen occasionally contributed to projects that had nothing to do with government or industry. In the early 1980s he served on the advisory committee for the International Project on the History of Solid State Physics, which initiated historical study of this major field and yielded the book, *Out of the Crystal Maze*. He helped organize the research and wrote to associates in industry, such as Bob Noyce of Intel, to solicit support. When drafts became available, he critiqued them and offered suggestions to improve the resulting book. Around the same time he agreed to be a science advisor for Elizabeth Ante'bi's *Editions Hologramme* in France, a project on the history of technology.

Sometimes he contributed more than just time. At the end of 1972 he donated 100 shares of Xerox Corporation common stock as seed money to establish a Fritz London Endowment Fund at Duke University. London had taught at Duke from 1939 until his death in 1954. Bardeen eventually contributed to the fund a total of $32,000. "This fund is being established to perpetuate the memory of the late Fritz London," he wrote to the director of development at Duke. "The income is to be used to help support the Fritz London Award for distinguished work in low temperature physics." The fund also supported the Fritz London Memorial Lecture Series at Duke. Bardeen wrote of London, "More than anyone else he pointed out the path that eventually led to the theory of superconductivity for which Leon N. Cooper, J. Robert Schrieffer, and I were awarded the Nobel Prize for Physics in 1972. We are all very grateful to him for the deep insight that helped light the way to understanding."

In April 1973 Edith London, Fritz's widow, wrote a moving letter of thanks to Bardeen. She enclosed a card that London's teacher Arnold Sommerfeld had written to Fritz in 1935 to express his support for his ideas about superconductivity. "Fritz was so happy about this card," Edith London wrote. It was among the first reactions to the Londons' theory published in the *Proceedings of the Royal Society.*

Bardeen wrote back that he had always felt indebted to London "for the basic ideas which led to the theory of superconductivity and also had very high respect for him as a person and as a friend." He added, "I especially appreciate your very generous gift of the card from Sommerfeld who immediately recognized the great insight into superconductivity Fritz and Heinz showed in their 1935 paper." He recalled that he had met Sommerfeld when he was a guest lecturer at the University of Wisconsin in the 1920s. Sommerfeld, "like your husband, had a penetrating mind that could go directly to the heart of a difficult problem. The card is a great treasure to have and I will be sure that it finds its proper place in the history of physics."

In the mid-1980s Goldwasser, who had spent a decade as the associate director of Fermilab in Batavia, Illinois, called on Bardeen for a favor. Goldwasser was gathering support for the planned Superconducting Super Collider, the highest energy particle accelerator project ever attempted (discontinued in 1993 by the Clinton administration). He needed distinguished colleagues to write letters on behalf of the project. Bardeen was "more than willing to provide whatever it was that was needed," for he believed that both large-scale and small-scale science were important. Both needed more support than they were getting. "It wasn't a political gesture," said Goldwasser. "It was a pure belief on his part that that's what was right."

In 1987 Bardeen and Bob Schrieffer wrote a joint letter in response to an editorial that had appeared in the *New York Times* under the title, "Super Hasty on the Super Collider." The article suggested delaying the collider while physicists explored whether the recently discovered phenomenon of high-temperature super-conductivity might drastically reduce its cost. Bardeen and Schrieffer argued against the delay, pointing out that many questions about high-temperature superconductivity remained and that waiting until answers were found could derail the project. "The Super Collider should proceed on its own merits with the option to switch to the new superconductors if they can be developed in time." Bardeen also worried about the waste of talented people whose experience in designing previous accelerators could not be reproduced in decades hence. They stressed that "long-term commitment for increased support of science is needed across the board. It is needed for 'small science' projects such as those that led to the transistor, the laser and high-temperature superconductors as well as for the costly 'large-scale' projects, like the Super Collider, that promise rich dividends in increased understanding of our universe."

Bardeen's extensive travels sometimes offered opportunities to renew old friendships or to see family. At a 1958 conference in Brussels, he met Brattain's new wife Emma Jane. "She is very nice," he reported to Jane on a postcard. "Walter made a good pick." Sometimes Bardeen would become frazzled from all the traveling and lose track of family events. He typically relied on Jane to keep track of them. On the way to the same conference in Brussels where he saw Brattain, he started to worry about missing Bill's graduation from college. He sent a letter from Idlewild (now John F. Kennedy) Airport to Jane asking her to "please send the date of Bill's commencement so I can send him a message."

Bardeen missed the birth of Betsy's first child in Boston when he spent the fall of 1978 visiting the *Institut für Theorie der Kondensierten Materie* at the University of Karlsruhe. There he was consoled by the presence of his son Bill; John wrote Jane that "it has been nice to have his company." Bill was now a high-energy physics theorist who, John reported to Jane, was doing "good work."

Bardeen also tried to connect with more distant family members. During a break between several 1958 conferences, he spent some time touring England, taking long strolls and doing "a little research" on family history. Barden Tower, he discovered, "was named from the district (or chase, as it is called there)." He found no sign of any illustrious Bardens in York, "but there was a John de Barden who paid 3s.4d. for fishing rights on part of the moat in York in 1376." He concluded that "our ancestors must have been just ordinary folk."

Once, while in California, Bardeen convinced his friend and colleague Douglas Scalapino to drive with him to see his relative Douglas Harmer in Santa Barbara. "It was like going back into the early years of Santa Barbara as we sat in the Harmer home and saw the paintings." Harmer's father, an artist, had traveled west as a youth to make his fortune. He was a relative of John's mother Althea (probably her uncle) and had been an inspiration to her. Bardeen appreciated his family's artistic heritage; he had paintings at home that had been part of Althea's collection of Japanese art. Scalapino noted, "John seemed to enjoy all this history immensely, as did we, and I wondered what his relatives thought of the man in their midst who had won two Nobel Prizes in Physics."

John liked traveling best when Jane came along with him, but until the children were grown and off to college she usually preferred to stay home. When he went to China in 1975 he wrote a little mournfully to Jane from his stopover in Tokyo, "It is going to be a long time away from home and I am going to miss you." In later years Jane nearly always accompanied him. They often split up during the day—he played golf and went to meetings while she went shopping, hiking, or sightseeing—then came together in the evenings for dinner and socializing.

When he traveled without Jane, he would miss her efforts to keep him organized. In 1960 he left his raincoat behind at Idlewild Airport in New York and had to buy an expensive replacement in Paris. On the trip to China he forgot his traveler's checks. "Fortunately," he wrote, "I was able to get $500 more checks on a personal check from American Express."

On a short trip in January 1978, the Bardeens had a different kind of ordeal. John and Jane drove from Champaign to Fermilab in Batavia, Illinois, where Bill was employed. It was about a three-hour drive. John gave a talk at the laboratory. Then he and Jane had dinner with Bill and his family. Afterwards, checking the weather report, they learned that a snow storm was predicted in Ohio and the northwest, but the storm was not expected to move into central Illinois very soon. Around 8:00 P.M. they headed south on Interstate 57.

Just past Kankakee, nearly an hour from home, they began to feel the effects of the wind. Whiteouts caused by blowing snow soon made driving nearly impossible. John tried to follow the trucks, but even the huge 18-wheelers were getting bogged down in the drifting snow. Finally, they were forced to concede defeat when their car got stuck in a snow drift. All they could do was wait for help. Periodically John cranked up the engine of their seven-year-old Oldsmobile Cutlass so they could run its heater. Even that small luxury was denied them after about 3:00 A.M. when the car could no longer be coaxed to start. Pulling a few extra items of clothing and a rug from the trunk, which had very nearly frozen shut, they did exercises to try to maintain their body heat. "The only thing that got really cold was my feet," John recalled. "I think if I hadn't sat on them they might have been frostbitten."

At dawn a snowplow and rescue team arrived to pick up travelers who had been trapped on the highway. John and Jane were taken to the Gilman Community High School, where they spent two days and a night. The high school gym and cafeteria had been converted into temporary lodgings for stranded motorists. By the second day Bardeen was back at work, doing some writing in the teacher's lounge. He later told reporters, "I was pretty sure they'd come along and pick us up in the morning." All the same, he added matches and candles to the emergency equipment in the car.

Tributes and recognitions multiplied for Bardeen throughout the 1970s and 1980s. In 1974 he, Brattain, and Shockley were inducted into the National Inventors Hall of Fame, along with Alexander Graham Bell and Eli Whitney. In 1975 he received the prestigious Franklin Medal. He was named the 1976 Scientist of the Year by the international publication *Industrial Research*. In 1977 President Gerald Ford awarded Bardeen the Medal of Freedom, the nation's highest civilian award. His fellow honorees included, among others, the composer Irving Berlin, General Omar Bradley, baseball's Joe DiMaggio, artists Georgia O'Keefe and Norman Rockwell, and former first lady Lady Bird Johnson.

Around the same time, the city of Champaign awarded Bardeen the key to the city, an honor no other University of Illinois faculty member had yet received. By then the Bardeens had been residents for thirty-eight years. Mayor Dannel McCollum stated, "Dr. Bardeen is contributing to the city just with his presence." Bardeen was recognized as a good citizen who voted, served jury duty, and supported local organizations. He set a standard of community involvement that the city fathers wished to acknowledge. He frequented the local country club, supported excellence in the schools, and attended sports events any time his schedule allowed. Bardeen appreciated the honor, for "there is nothing more gratifying than being recognized by your friends and neighbors."

15

Pins and Needles and Waves

And this gray spirit yearning in desire
To follow knowledge like a sinking star,
Beyond the utmost bound of human thought.
        . . .
Old age hath yet his honour and his toil;
Death closes all: but something ere the end,
Some work of noble note, may yet be done,
Not unbecoming men that strove with Gods.
        . . .
Tho' much is taken, much abides; and tho'
We are not now that strength which in old days
Moved earth and heaven; that which we are, we are;
One equal temper of heroic hearts,
Made weak by time and fate, but strong in will
To strive, to seek, to find, and not to yield.
                —Alfred, Lord Tennyson, "Ulysses"

Most physicists accept aging as the ultimate obstacle in research. As their mental agility and speed decline, they shift their efforts to ancillary work, such as teaching or administration. Some change fields. But the few who produce works of genius often try to buck the trend. Expecting their passion and experience to compensate for their aging, they strive to achieve yet one more great work. In this respect Bardeen was typical of the exceptionally creative.

In the 1980s, when Bardeen was in his seventies, his research focused on developing a theory to explain an ordered conduction phenomenon involving charge density waves (CDWs). Several research groups had observed these unusual waves in the early 1970s. The periodic modulations of the electron density and the positions of the lattice atoms occur at low temperatures in certain materials having a quasi-one-dimensional structure—the conduction proceeds inside them along linear chains that hardly interact with one another.

One of the experimentalists working in the field of CDWs was George Grüner at the University of California at Los Angeles (UCLA). Grüner aptly compared the electron density in CDW materials to "the ranks of a marching band." In a 1994 *Scientific American* article coauthored with Stuart Brown, he explained that when there is no voltage the marchers (electrons) stand still. But if a voltage is applied, a new phenomenon occurs above a certain threshold: "the band begins to march" with a definite periodicity that differs from that of the crystal lattice. The large current that forms is nonlinear with the voltage; increases of current are not proportional to increases of voltage, as they are in ordinary conductors. The ground state is no longer static, but takes the form of a sliding wave.

In the early days of the quantum theory of solids, Rudolf Peierls had raised the possibility of creating charge density waves. On purely theoretical grounds, he noted in 1930 that the one-dimensional metal (the linear chain) is mathematically unstable. He demonstrated what became known as the "Peierls instability": when the electron density distorts to form a wave tied to a periodic distortion of the underlying crystal lattice, the energy of the system is lowered. This result suggested that charge density waves might exist in nature, but the waves could not be observed until forty years later when experiments were performed in compounds approximating a one-dimensional metal.

Well before that, in the early 1950s, Bardeen had suspected that CDWs are somehow related to superconductivity. In those years he developed a theory based on the assumption that the CDWs are the mechanism for superconductivity. Unfortunately, the theory worked only in one dimension, and the predicted size of the energy gap did not fit with experimental findings. Still, he continued to believe that the CDWs and superconductivity could somehow be explained in a common theoretical framework. Both are collective states of matter that result from a broken symmetry. In CDWs the symmetry is "translational," that is, the lattice is initially spatially uniform, whereas in superconductors the symmetry is a more mathematical one associated with conservation of particles. In both, the interaction of electrons with the lattice creates an instability that causes energy gaps to form. And both are separated from the normal state by a second-order phase transition (known in the CDW case as the "Peierls transition").

Of particular interest to Bardeen was the fact that defects, or impurities added to the crystal, can "pin" (hold in place) the phase of the charge density waves to the lattice and that the phase can be "depinned" by applying a strong enough electric field.

In 1976, Nai-Phuan Ong and Pierre Monceau in Berkeley demonstrated such pinning and depinning in a classic experimental study involving charge density waves created in the inorganic linear chain compound $NbSe_3$. Bardeen immediately recognized that explaining this phenomenon was a fertile problem. Here was another system to which Fritz London's picture of long-range ordering could apply. Bardeen sensed intuitively that the CDW state is a macroscopic quantum phenomenon, analogous to superconductivity but occurring in a different class of solids. The problem had the added attraction that CDW materials had properties—such as large dielectric constants and nonlinear current-voltage characteristics—of great potential importance to industry.

The pinning and depinning can be imagined in nonquantum-mechanical (classical) terms by assuming the wave resembles a flat solid having a rough surface, like a washboard or sheet of corrugated rubber in whose physical irregularities a defect could be caught. Bardeen's picture of this pinning and depinning process as a quantum-mechanical phenomenon occurring on a macroscopic scale was analogous to Josephson tunneling of condensed electron pairs in a superconductor. Bardeen envisioned a collective tunneling of condensed CDW electrons through a small pinning potential created by the impurities or defects of the material.

"If you have a brick wall," explained Grüner, "and you drive the car into the brick wall, you are more or less sure that you are not going to go through the brick wall to the other side intact." In quantum mechanics this process, known as "tunneling," has a finite probability. The CDW case can be imagined as a sort of generalized Josephson tunneling phenomenon. If one has "say a row of soldiers, a million electrons," the question becomes, "Can they somehow tunnel collectively at the same time?" Bardeen felt sure that it was possible as a consequence of the macroscopic quantum state. "It's a beautiful concept," said Grüner. "And it's something that's revolutionary, new, and has consequences." These were powerful motivations for studying CDWs. According to Grüner, had Bardeen's theory been validated, "without question it would have been a Nobel Prize, a third one."

Data from the early a.c. frequency experiments yielded an exponential dependence of electric field on frequency (a hallmark of tunneling) that could be fitted to Bardeen's quantum-mechanical theory. Applying photon-assisted tunneling theory (PAT), in which absorbed quanta can bring electrons into a state in which tunneling is more likely, Bardeen described his pinning as "weak," because the applied electric field did not need to be very strong.

Initially, the many-body community was intrigued by Bardeen's beautiful theory, but with time more physicists preferred the so-called classical model of an overdamped oscillator motion in a periodic potential. Among the prominent physicists who came to support the classical picture were Grüner, Patrick Lee, T. Maurice Rice, Philip Anderson, Alfred Zawadowski, and Paul M. Chaikin. This opposing model allowed for a quantum-mechanical response to the electric field, but with a considerably larger classical effect. It predicted measurable relationships between the amount of impurities present in the materials and the threshold voltage at which the CDWs would begin to slide. For a number of years, the classical and quantum-mechanical pictures would coexist.

In 1980, while pondering his quantum-mechanical model for the charge density waves, Bardeen recalled the conversation in 1974 in which he had learned about the PAT theory. He had been speaking with John Tucker, a physicist then working at the Aerospace Corporation in El Segundo, California, near Los Angeles. Bardeen was visiting there at the invitation of Ivan Getting, his old friend from the Harvard Society of Fellows, who was now president of the corporation. During the visit, Tucker told Bardeen that he was applying photon-assisted tunneling theory to metal–superconductor tunnel junctions. Later used in the design of astronomical quantum receivers, such junctions could be set up at the interface between two superconductors, so that one electron could tunnel each time a photon was absorbed.

One day in the spring of 1980, the telephone rang in Tucker's office at the NASA Goddard Space Institute in New York City. Tucker heard a soft, raspy voice say, "Hello, this is John Bardeen." At first he thought it was a prank. Bardeen told Tucker that he thought charge density waves were a macroscopic quantum phenomenon. He explained why he suspected that PAT might help to explain their transport. "He was very excited because if this were true, you'd see quantum phenomena in a regime that no one had thought possible."

Tucker was intrigued. "It was only the second example, besides superconductors, where you conduct electricity with a moving quantum ground state. There are other quantum ground states where all of the particles like to get in the same quantum state, such as superfluid liquid helium, but then it doesn't carry a charge." He was also aware that collaborating with Bardeen at this stage would be a gamble. At seventy-two the double Nobel laureate had passed the age at which most physicists contribute to cutting-edge research. On the other hand, if Bardeen's idea were correct, it would be an extraordinary piece of physics. Bardeen was "reaching for something that was exotic, but not wildly implausible. Things that strange had happened before in physics." The work could potentially open an "entirely new realm of quantum physics."

Tucker accepted Bardeen's invitation to come to Urbana in the fall of 1981. Bardeen arranged that Tucker be named an untenured associate professor of electrical engineering, with the possibility of tenure after a few years. The offer also included funds to establish a research laboratory and to support a graduate student. Tucker's first student would be John Miller.

For a few years, Bardeen and Tucker's work attracted favorable attention from the CDW community, which grew rapidly in the 1980s as its members looked ahead to wide use of CDWs in studying chaos and turbulence and in building amplifiers. As in earlier collaborative projects, Bardeen modeled the CDW collaboration on the family. He was the patriarch; Tucker was the adopted son. Jane often invited Tucker and his students to dinner, or to cook for themselves, at the Bardeen home. "We used to hold group meetings over there," Tucker recalled. "And every once in a while we'd cook burgers." Usually they would meet in the large sunroom in the back of the house. "It had the kind of benches that you sit on, but they had drawers underneath." On one occasion, a student opened a drawer and found two Nobel certificates bound in leather. "I said, 'Gee John, you have a better class of junk in your drawers than I do.' Everyone laughed, and he was sort of sitting there kind of scowling, like 'What's so funny?'"

"Throwing theoretical spaghetti at the wall to see what sticks" was how Tucker characterized his early research with Bardeen. The team fitted a rudimentary tunneling hypothesis to the data. Then they adjusted their theory in stages. Miller and other students performed "mixing experiments" on CDWs by injecting signals of different frequency into a sample, holding the direct-current (d.c.) field below the threshold for depinning and adding an alternating-current (a.c.) field of gradually increasing frequency or strength. According to their theory, the tunneling would begin at a specific frequency and exhibit a characteristic dependence on the strength of the a.c. field. When their experiments failed, Bardeen shrugged off his disappointment, muttering, "I guess the system is still noisy."

"Maybe he was just carried away a little by how beautiful this could be," Grüner said. "I think that deep down he thought that there is just one missing link, and that maybe everything is going to fall into place. In superconductivity there was just one missing link for half a century, which was the pairs, and then everything fell into place."

Myron Salamon occasionally dropped by Tucker's lab. Salamon was then director of the National Science Foundation (NSF) program at Illinois' Materials Research Laboratory, which was supporting the CDW work that Bardeen and Tucker were doing. Salamon grew worried when Tucker told him, "I'm really working hard to try to prove John is right." Salamon cautioned, "That's really not the attitude for an experimentalist—you measure what's there and if the theory's right, it's right, and if it's wrong you're even happier because you've got something new."

Bardeen's and Tucker's network of associations with CDW experimenters included Grüner, who recalled, "The first set of experiments that we did really fitted." Bardeen, Tucker, and Grüner

were able to demonstrate that the frequency dependence and voltage dependence had the same functional form. They therefore felt
that they had "strong supporting evidence for the photon assisted
tunneling theory." Experiments in Vienna by Karlheinz Seeger, another collaborator in the CDW work who had been a postdoc at
Illinois in the 1950s also initially appeared to fit Bardeen's theory.
But such fits could also be made to the classical theory. To differen-
tiate between the classical and quantum-mechanical hypotheses,
they needed to demonstrate a threshold energy that depended upon
the absorption of photons. They could not find it.

Little by little, Bardeen and Tucker's theory lost ground. The
Bardeen-Tucker theory had fallen seriously behind the classical one
by the time of the major CDW conference held in Budapest in early
September 1984. At that meeting, Daniel Fisher and Leigh Sneddon
of Bell Labs presented their case for the opposing classical model.
But the CDW controversy was overshadowed by a tragedy that
deeply touched both camps.

Three days before the meeting, William McMillan, Bardeen's
former student and now colleague, was fatally hit by a car as he
was bicycling on a country road near the town of St. Joseph, Illinois,
not far from Urbana. Bill's widow, Joyce McMillan, a crystallogra-
pher at the University of Illinois, said her husband had preferred
running to bicycling but had injured his knees. At the time of the
accident he was bicycling because he needed new running shoes
and "didn't have one minute" to shop for them. At forty-eight,
McMillan was in his prime. "He was building computers and running computers twenty-four hours a day in his basement." By the
time he declined his invitation to the Budapest meeting, his main
interest had shifted to spin glasses, the problem he was modeling
on the computer he had built in his basement. After McMillan's
death Joyce realized he had been working on twelve papers at the
same time, all published in the months before and after his death.

Shocked to learn of the accident, Bardeen immediately sent a
condolence telegram to Joyce from Budapest, explaining that there
would be a time devoted to McMillan at the conference. "I feel a
deep sense of personal loss of a former student and close associate
on the staff of the physics department at the University of Illinois,"
he said at the opening session. He requested a moment of silence.
He also arranged that the conference proceedings be dedicated to
McMillan. The department subsequently established the William

L. McMillan Award, given annually to recognize outstanding contributions to solid-state physics by young researchers within four years of the receipt of their doctorates. Bardeen wrote a short biography of McMillan for the flyers distributed at the award presentation.

Bell Labs continued to raise sharp objections to Bardeen's quantum-mechanical theory. Despite harsh review of the NSF proposals and research papers that Bardeen and Tucker coauthored, Bardeen held fast to his belief that "excellent agreement is found between theory and experiment."

Several younger members of the group surrounding Philip Anderson talked and joked while Bardeen spoke at physics meetings. Tucker recalled that it "greatly upset Bardeen that people at Bell Labs would be calling him crazy and stupid in public, and openly laughing at him." He had never before experienced disrespect from the younger generation.

In the fall of 1984 Joseph Lyding, a new assistant professor in electrical engineering, joined the Bardeen and Tucker team as an experimentalist. Lyding came to Illinois because of "the opportunity to work with someone like John Bardeen. That's a once in a lifetime opportunity." He began to grow high-quality crystals for the CDW experiments. The team had been getting most of their crystals from Grüner. By placing a thermal gradient across the crystals and measuring the effect on narrow band noise. "I could see the frequencies of the narrow band noise peaks shift and split into multiple peaks. It was a nice experiment."

Lyding recalled that it was "really nice working with Bardeen—he was so accessible." When he wrote up his first paper and gave it to Bardeen, "it came back the next day with all sorts of interesting suggestions and comments. It was very positive." Lyding also appreciated Bardeen's frequent visits to the Electrical Engineering Research Laboratory (EERL), at least once or twice a week. "He'd come over to EERL and he would always talk to my students, and John's students. He was very accessible and he never put himself above anybody, which is amazing, all things considered."

In the larger many-body world, however, the physics emerging from the collaboration was not faring well. Bardeen's opponents accused him of blindly censoring their classical theory, rather than responding to its details. As their tone grew harsher, Bardeen's patience wore thin. He returned some of the insults. At times he accused his opponents of supporting the classical theory simply because it was easier to program on computers, tools of the new generation that he himself did not use.

It had never been easy for Bardeen to discard a theory in which he had invested much time and energy, but in his younger years he would do so when necessary. Back then there had been more time left, making it easier to switch commitments. Now, in his late seventies, solving the difficult problems of explaining CDWs was a race against time.

Reluctant to upset him, Bardeen's friends avoided discussing CDWs in his presence. Family members encouraged him to work on his memoirs. He would not. Ignoring the problem was not an option for him, or for Tucker, who pushed Bardeen to be more complete in his explanations.

The mood in Tucker's lab grew tense. "Tucker always seemed on edge," said Fred Lamb, who occasionally stopped by the lab. "And Bardeen would always be scowling." In February 1987 Douglas Scalapino chatted with Bardeen at a conference. "You know, Doug," Bardeen mused, "the establishment's against me." To Scalapino, "it sounded so incongruous that I couldn't help but reply, 'But, John, you *are* the establishment.'" Bardeen "smiled gently as he sometimes did, and we went on talking."

Bardeen's health was noticeably failing. The macular degeneration that impaired his vision was an enormous hindrance to his work. It was not helped much by surgery. His frustration sometimes overwhelmed him. Even playing golf was becoming difficult because of his vision loss and the gout in his legs.

By early 1987, Tucker had to admit to his own serious reservations about Bardeen's theory, particularly its notion of weak pinning. He worried about Bardeen's inability to justify the tunneling expression behind the current-voltage curve. Could the Bell Labs theorists be right? Could the a.c. response of the CDWs be explained as a capacitor effect? Tucker told Lyding about a Bell Labs experiment that disproved Bardeen's tunneling model and "said he actually woke up in the middle of the night realizing that what they'd seen was just some manifestation of this narrow-band noise. He could see his ticket to Stockholm had been ripped in half."

Bardeen and Tucker weathered a storm in the spring of 1987 when Tucker wrote a research proposal that did not oppose the classical model but took an open view. He was trying to reflect his own understanding of the problem and also avoid irritating the opponents. Several of them would surely be called in as reviewers. Outraged, Bardeen refused to sign until Tucker had inserted several sentences favoring the quantum viewpoint. The escalating tension between Tucker and Bardeen concerned Salamon, who worried that their lab had become a poor environment for students.

Just then a new wave of excitement shook the physics community. In late 1987, J. Georg Bednorz and K. Alexander Müller, in Zürich, announced the discovery of a class of ceramics, perovskites that become superconducting at 35 K, a much higher temperature than predicted by the BCS theory. Potential applications boggled the mind. They included magnetic imaging, particle accelerator magnets, electrical generators, and magnetically levitated trains. The superconducting behavior of the perovskites was unexpected because BCS assumed that only metals and alloys could become superconducting. Bardeen and other theorists tried to apply the BCS theory, especially the notion of the Cooper pairs, to the new phenomenon of high-temperature superconductivity. Most were convinced that modifications to the theory would be needed to explain what became known as "high-Tc."

As interest shifted from CDWs to high-Tc, some of the tension around Bardeen relaxed. "All of us were quite excited when high-Tc came out and started working on it immediately, Bardeen among us," recalled Salamon. At the huge "Woodstock of Physics" meeting on high-Tc, held in March 1987 at the New York Hilton, thousands of physicists packed into a huge ballroom to consider the implications of the new kind of superconductivity. Brian Pippard called it a "lemming effect."

"What we found when we took our first data on high heat capacity was that some of the numbers that were being published by Bell Labs didn't hang together," said Salamon. In an effort to input all the data and compare what was known, he started putting together "a spreadsheet with all of the BCS formulas."

Someone at lunch mentioned that "John is trying to do just the same thing; you ought to show it to him." So Salamon took his spreadsheet over to Bardeen, who he found working quietly in his office. "He had his blackboard covered from edge to edge with formulas and expressions and calculations, and he had a little six inch slide ruler he was working with." The numbers did not seem to agree in places, so the two "sat down together, me with my spreadsheet computer output, and he with his things on the black- board, and checked each other's numbers. And the end result was that we agreed on these." Bardeen, "like the rest of us, was trying to figure out how much of the original BCS picture was valid and was going to stay together and how much new was needed."

Ever since the discovery of BCS, Bardeen had been eager to find new mechanisms for superconductivity. He "wasn't at all wedded to the phonon mechanism as being the sole means of doing this." Working together, Salamon and Bardeen became convinced "that the pairing picture was robust and that there may be other attrac- tive interactions that could be at work, but that all of the formulas in the BCS, the pairing mechanism, the opening of the gap, were all going to be robust." In June 1987 they submitted a coauthored ar- ticle on the work to *Physical Review Letters*.

"Boy I'm glad this high-Tc came out," Salamon recalled think- ing, "because it will take John's mind off this big fight he's having with Tucker about the charged density waves." The discovery of high-Tc did "put an end to that fight, or at least the open part." When the NSF granted $350,000 extra to the Materials Research Lab to begin a high-Tc initiative, almost "everybody who was work- ing on these charge-density waves said, 'I want to work on high-Tc anyway.'"

Behind the scenes, Bardeen grew angrier by the day about the opposition to his CDW theory. He could not understand the resis- tance and took every opportunity to express his irritation. In May 1987 he wrote to H. Takeyama and H. Matsukawa, "It is inconceiv- able to me why people do numerical calculations on unrealistic and vastly over-simplified classical models when there is a beauti- ful theory based on quantum transport that is in agreement with experiment and involves only a few measurable parameters." In his younger days only Bardeen's family and close friends were aware of his hot temper because he revealed it so rarely. Typically he would vent against material objects, like golf balls or shoes. Only those close to him knew to interpret as rage the low-pitched hiss or sar- castic laugh that occasionally escaped through his teeth. But as he entered his eighties, cracks formed in his composure, and he was less able to moderate his temper when things really mattered to him.

Tucker tried to avoid Bardeen while he worked out his own

theory of CDWs. By the summer of 1987, Tucker was distinguishing his own "strong pinning" theory from tunneling. Tucker explained recently that "in strong pinning, every impurity stops the CDW from moving within a very small volume at each impurity site. The average phase remains constant, however, over much larger volumes containing a great many individual pinning centers, and phase-slip occurs at each site as the CDW moves along." Hoping to convince Bardeen of the problems he saw in the quantum-mechanical theory of CDWs, Tucker prepared a carefully written account of his strong pinning theory.

As Tucker's theory won support, Bardeen felt even more isolated. "I think he felt betrayed," recalled Salamon. Not since the days at Bell working with Shockley had Bardeen been seriously opposed by one of his collaborators. "I think he felt that he'd brought Tucker here specifically to work on this, to confirm this, and then it turned out that [Tucker had] turned on [him]."

Bardeen tried to help his opponents understand his quantum-mechanical theory. He worked hard to improve his arguments, but that also angered him because he felt he had already offered everything necessary to support the physics. Meanwhile Tucker was growing despondent and drained by his failure to explain his own theory to Bardeen. He worked with Miller to develop more convincing arguments. Bardeen opposed the work as unnecessary. When Tucker called for help from several physics colleagues, they claimed they could not understand the details of the theory and could not therefore pit themselves against Bardeen.

Having difficulty in explaining a theory was not a new thing for Bardeen. More than fifty years earlier he had stumbled in his Princeton prelims when asked to detail how he arrived at the answer to one of Robertson's questions. Bardeen's collaborators, students, and occasionally his teachers had often stepped in to help John resolve his communication problems. But as he neared his eightieth birthday, no one was there to help him explain his theory of charge density waves.

John's easy friendship with Nick Holonyak offered a partial antidote to his intense frustrations in the latter part of 1987. Whenever Nick could see that John was feeling frustrated, he would encourage his friend to take a break and do something for himself. "Don't let people constantly take your time," Nick advised. "Look, if it's a good day and you have such an inclination, go play golf."

From September 21 to 25, 1987, Tucker attended the Second European Workshop on Charge Density Waves. It was held in Aussois, a picturesque ski village in the French Alps. Tucker spoke there about his strong pinning theory at a poster session on non-linear transport. Bardeen was home in Illinois. According to observers, the discussion of Tucker's paper turned into a hot debate about the classical versus quantum-mechanical theories of charge density waves. Seeger reported witnessing "an ugly ceremony" in which a Hungarian physicist forced Tucker to "confess before the full audience that he would no longer support Bardeen's CDW theory."

Also at the meeting was Tucker's student Robert Thorne, who had recently taken a position at Cornell as an assistant professor. In a letter to Bardeen, Thorne described the "general feeling at the conference that, while not dead, the tunneling model seems unnec-essary for interpreting the experiments." He expanded on the "con-sensus that weak pinning does not occur, and that strong pinning and phase slip play a dominant role in CDW transport." He also remarked that Tucker appeared to him to be in a highly emotional state.

Bardeen responded angrily to the reports. He called in Lyding and asked whether he thought Tucker was emotionally unbalanced. "I don't think that's true," Lyding said. Lyding did think that get-ting things out in the open might be helpful. He hoped it "would lead to some kind of calm discussion between Tucker and Bardeen, and they would just resolve it, because they were talking." So he mentioned Thorne's letter to Tucker, who "went straight over to Bardeen and demanded to see the letter, and was very forceful about it." John refused.

Bardeen felt backed into a corner because Tucker was able to "talk fast and think on his feet," Lyding explained. "My impres-sion is that Tucker made him feel very uncomfortable during all of this process when they were dealing with each other in person." It was exceedingly difficult for Bardeen to "respond to that in real time" because he "was a very slow, deep-thinking, methodical type. And to have somebody chewing on your ideas when they're com-ing out of your mouth is very unsettling."

By October 1987 the discussions between Bardeen and Tucker had became so heated that the two resorted to writing letters to each other. In his now shaky script, Bardeen insisted on October

24, 1987 that the weak pinning "is confirmed by many experiments." He was clearly exasperated: "These are simple consequences of quantum physics and the uncertainty principle and do not involve any deep ideas." He suggested to Tucker that he had "developed a mental block on this problem that prevents you from thinking straight." Later in the letter he urged Tucker to "put this problem aside for the time being and work on one that you should be able to make progress on." Tucker was insulted.

Lyding found himself in the role of Bardeen's and Tucker's "messenger boy." He hoped "to provide just enough glue to keep those guys talking." He recalled, "They were both upset. Bardeen would call me over to his office" and soon become "frustrated and angry that I wasn't just accepting his point of view." After a while "we'd get down into a calm discussion, and by the end of the meeting we'd be shaking hands." But when Lyding delivered the message, "Tucker would usually write a letter and send it over to Bardeen." A week or so later Bardeen would "call me over again."

Lyding sometimes scribbled notes while talking on the phone with Bardeen. One note read "diatribe." Another read, "Blamed you for spilling the beans." Still another, "Can't reason with him." Throughout this period Tucker was having back problems, which Bardeen claimed were psychosomatic.

Bardeen tried diplomacy. In a November 6 letter to Tucker, he suggested that they "seem to be narrowing down our points of difference." Tucker did not think so. On November 8 he drafted a letter so emotionally charged that he did not dare to send it for six months. He felt that his reputation was on the line. And he had reached the end of his ability to deal calmly with Bardeen, who insisted that his central argument was "obviously correct as it stood." Tucker admitted that Bardeen's rage had shaken him. "I was actually afraid that you might drop dead in front of me." In this letter, which he filed away but did not destroy, Tucker wrote that he felt, "you and I can no longer have any sort of positive interaction."

Unaware of Tucker's incendiary letter of November 8, Bardeen wrote on November 13, "I hope that we can make progress in narrowing down our points of difference." Tucker wrote back that Bardeen was ignoring available experimental facts. This unproductive and hurtful exchange continued for ten more days. When Bardeen wrote on November 23, "I am glad that we agree that a weak pinning model is required to account for the CDW data," Tucker shot back, "Your statement that we now agree on the requirement for weak pinning could not be further from the truth."

Bardeen let a month go by before making his final move. On January 4 he wrote to the NSF and asked that his name be dropped from the grant with Tucker and Lyding. "Because of the change in emphasis of the work," he explained, as well as "my advancing age and other obligations that I have, I find that I am unable to take an active part as co-director with John R. Tucker and Joseph W. Lyding." By then the NSF had verbally committed to fund Tucker and Lyding alone. The money was subsequently granted, "with no strings attached."

The next day Bardeen informed Tucker that he was withdrawing from the CDW project. He again expressed his doubts about Tucker's physics. Referring to their recent interactions, he wrote to Tucker, "You have by no means relieved my concerns. This I greatly regret because I am very grateful to you for initiating the CDW program here, for supervising the outstanding experimental work of John Miller that helped confirm the validity of your theory of photon-assisted tunneling as well as the beautiful work of Rob Thorne on phase-locking with combined a.c. and d.c. fields." Reviewing some of the physics one last time, he referred to Tucker's strong pinning theory as "a contradiction in terms." He added, "There is nothing personal whatsoever. As I have told you many times, my every wish is to have you succeed. . . . I particularly regret any problems that this may cause the students, but see no other alternative than for us to go our separate ways."

Bardeen continued to work on CDWs alone. His confidence in the validity of his theory never wavered. He gave no ground when submitting his arguments on quantum depinning to the *Physical Review* early in February 1988. "No plausible classical model has been able to account for any aspect of CDW transport, d.c., a.c., or combined d.c. and a.c." He argued that "any approach that does not treat CDW metals as macroscopic quantum systems misses the essential physics of the problem." He called on theorists to pay more attention to experiment. "No sound prediction of classical CDW motion in the charge-compensated sliding regime is borne out by experiment. Experimenters and theorists alike should think of CDW motion as a beautiful example of macroscopic quantum mechanics, with many analogies to superconductivity." The referees called for revision.

About this time Ravin Bhatt was visiting Illinois on a Bell Labs recruiting mission. When Bardeen saw Bhatt pass his office, he hurried out to speak with him. "I know you're recruiting and you've got a lot of things to do and that it's important for Illinois. But I think I finally got the tunneling theory of charge density waves all done. I have a BCS-like wave function for it." Bardeen told Bhatt he was about to write the theory up, "but I want to try it out on you." For roughly an hour the two sat together in Bardeen's office and discussed charge density waves. Bhatt listened carefully while the 80-year-old enthusiastically unfolded his theory. Like most of Bardeen's other colleagues, Bhatt could not understand the argument. In May 1988, the month of Bardeen's eightieth birthday, Tucker sent Bardeen his charged November 8 letter.

The revised manuscript, received on October 24, 1988, was published with some light edits on Valentine's Day 1989. "A considerable body of evidence supports a theory based on quantum tunneling over macroscopic distances," Bardeen wrote. "All evidence indicates that it is necessary to treat CDW metals as macroscopic quantum systems with quantum tunneling as an essential feature."

Several months later Bardeen wrote bitterly to the *Physical Review* that in spite of the success of the tunneling theory of CDWs, "few other theorists have worked on the tunneling theory. It is not only not generally accepted by those in the field, but recent papers by leading experimentalists do not refer to the model as a possible explanation of the data." He acknowledged that he had not expressed his ideas as well as he might have. "While I share part of the blame for this state of affairs in not writing clearly enough," he said, "in large part it is a reflection on the present state of physics. Everyone is so busy doing his own thing that no one puts in the time required to get a real understanding of a difficult subject." He made the same point to Holonyak. "Everyone is so busy writing," he told Holonyak, "and no one reads sufficiently, if at all, and knows little of what is already known."

The last of Bardeen's single-authored papers on CDWs appeared in *Physical Review Letters* in May 1990. He followed up with a more general account in an article on "Superconductivity and Other Macroscopic Quantum Phenomena," published in *Physics Today* in December 1990. That article "mattered a lot to him," said his daughter Betsy. He considered it "in a way a culmination of that work." By then the field had moved in the direction Bardeen had opposed. Virtually everyone in the physics community had decided that CDWs were a "dirt effect."

No one has successfully defended Bardeen's theory, but as Miller later remarked, a few isolated voices remind us that the classical theory does not account for all the observed phenomena. It is not impossible that a complete theory may one day incorporate features that are quantum-mechanical along with others that are classical.

Once or twice in the period from 1980 through 1991, when David Lazarus served as editor in chief for the American Physical Society, he learned that Bardeen had received a bad referee report. Having been Bardeen's colleague since 1951, Lazarus found that surprising. When he investigated the reports, he found that indeed "the referee was totally out of line. I couldn't believe it. I think it was some kid who thought it was sort of a cute thing to cut down an icon." Lazarus acknowledged that "John really did have a hard time with the last few papers and it was not his fault at all. They were important papers, they did get published, but they gave him a harder time than he should have had."

Bardeen continued to fight in the trenches of his field long after he had won two Nobel prizes. Although his quantum-mechanical tunneling theory was not accepted as the explanation for charge density waves, the ideas he developed in his theory may prove applicable in other contexts. Through his excitement and passion about the phenomenon being observed, Bardeen was one of the few who made a huge impact on the CDW field. He suggested crucial experiments and attracted many workers to the study of CDWs.

Only Bardeen's family and closest colleagues understood the force of his passion to do great physics. Perhaps that was because he said so little and appeared so ordinary in his everyday life. To most of Bardeen's associates, and even to casual observers, there seemed no reason for him to keep on struggling so painfully through the rejection of his theory of CDWs.

Psychologists claim such behavior is typical of the highly creative person. The aging Bardeen, remarked Tucker, opened "himself to a lot of criticism and a lot of abuse, which he got in spades, because he thought there was a possibility there was something out there that would really be a big breakthrough. He still had the yearning to do that kind of work."

16

Last Journey

Bardeen sat awkwardly, with a gentle half-smile, as he gazed obligingly into the video camera. His Loomis Hall office was crowded on that day in June 1990, with people and video equipment. It was unusually hot.

Bardeen felt a heavy fatigue. He had not been feeling well since his recent trip to Japan and would not have initiated this filmed interview. But members of NHK, a major Japanese television corporation, had traveled a great distance to record his memories about the invention of the transistor. Also, Bardeen wanted the true story to be known. He had hoped to do the filming in the comfort of his home, but when the interviewer suggested his office instead he had politely acquiesced. Nick Holonyak, John's long-time colleague and friend, was along for support. Holonyak recognized the invention of the transistor as a great transforming moment.

The film captures Bardeen at eighty-two, his well-wrinkled face pleasant and calm. He sits at his desk where a lighted magnifying glass had been mounted on an adjustable arm, a gift lovingly crafted by Holonyak and his students to help alleviate some of the frustration of Bardeen's failing eyesight. The papers and journals stacked on Bardeen's desk and the books arranged on the shelves above were obviously still in use.

Bardeen gave a satisfied grin when the interviewer asked about his daily routine. "I come in most every day, but I don't put in as long hours as I used to. I come in later and leave earlier, but I still come in every day." Asked whether he still was active in research, Bardeen's eyes crinkled. "I'm still active in research," he warbled, his lips curling into a beatific smile. "And the number of papers I've published per year has remained about constant. It hasn't dropped off."

Patiently, in response to the interviewer's questions, Bardeen told the short version of his years at the University of Wisconsin and at Gulf Labs, how he became interested in solid-state physics, and why he came to Bell Labs. Sorting methodically through the pile of papers in front of him, he found the one he wanted to illustrate a particular experiment. He had prepared his notes and illustrations carefully. The paper's rattling betrayed the Parkinsonian symptoms that had been troubling him for years. Ignoring the tremor in his hand, Bardeen pointed slowly to the physics figures and indicated the most important features of each.

Halfway through the taping, Holonyak asked Bardeen whether he was getting tired. John shook his head. He wanted to finish the story. A little later Holonyak interrupted to clarify a point that he believed was historically important. He asked Bardeen on what date he and Brattain understood that holes—the ghostlike positive charge carriers that were in the minority in the semiconductor with respect to the negatively charged electrons—were entering the germanium making the first transistor possible. The date, Bardeen said, was December 16, 1947, one week before the group's demonstration of their invention to the Bell Labs "top brass."

"But John," Holonyak pointed out, "the junction transistor is a bipolar transistor based on the bipolar injection ideas that were yours and Brattain's." He wanted to point out several items to reflect the fact that Shockley's junction transistor, more easily manufactured than the original point-contact transistor, was based on ideas that originated with Bardeen and Brattain. "So it's sort of ironic," Holonyak continued, "that the more manufacturable one for a certain time was the bipolar p-n junction device of Shockley's, but nevertheless based on the injection principle that you and Brattain discovered on December 16, 1947. I think that's key and a very important point."

Bardeen understood well the irony of the patent relations. "Brattain and I have the basic patent on the bipolar principle in which we used metal semiconductor contacts rather than p-n junctions, and Shockley has the basic patent on the junction transistor, which is the common type. And Shockley was the first one, at least at Bell, to suggest the field effect concept. And I have the basic patent on the use of an inversion layer." He glanced at Holonyak and smiled, "which is used in the common type of field effect transistor."

Bardeen's dedication to order and accuracy surfaced when the interviewer asked a question about the use of silicon and germanium. The answer required going backwards in the story to the wartime research. In spite of his declining health, Bardeen had no lapses of memory and did not scramble his statements. After answering the interviewer's query, Bardeen suggested that he edit this part of the account to an earlier spot in the tape. Then he picked up his narrative exactly where he had left off.

Asked about Shockley's reaction to Bardeen and Brattain's discovery, Bardeen murmured, "of course he was very excited about the new principle." Pressed for remarks about Shockley's character, he refused to say anything more revealing than, "Well, I think that his personality changed over the course of time." Shockley eventually "developed problems in relations with others," Bardeen offered. Then he prevaricated, "during the time I was there, our relations were good."

Bardeen had no interest in publicly criticizing Shockley, who had died in August 1989 of prostate cancer. The two men had had their difficulties, but Bardeen genuinely admired Shockley's facile mind. Shockley's more recent behavior baffled, infuriated, and sometimes amused Bardeen. He thought it "a tragedy that he got involved in a field in which he had no expertise—human genetics." Using statistics to bolster his theories, Shockley had argued since the 1960s that supposed social and intellectual deficits of African Americans are racially determined. Shockley thought that improving externalities such as education, housing, or health care would make no difference. Bardeen strongly disagreed. In 1981 he told a newspaper journalist that "you have to judge people individually even if there are statistical differences." Shockley, Bardeen once wrote, "could have accomplished much more in physics and electronics if his interests had not been diverted."

In 1980 Shockley had publicized the fact that he was making contributions to a sperm bank, reckoning that his genetic material might ameliorate "the existence of tragic genetic deficiencies in our society." When the news came out about Shockley's sperm donations, one of Holonyak's West Coast contacts called him with a question for Bardeen. "He wanted to know," Holonyak told Bardeen, "if you've been jacking off for the sperm bank." Bardeen chuckled, then laughed out loud at the absurdity of it. "I'd never seen John laugh harder," Holonyak recalled with a grin.

Brattain had passed away two years before Shockley. He spent his last two decades, after retiring from Bell Labs in 1967, teaching at his alma mater, Whitman College in Walla Walla, Washington. There he became interested in biophysics while continuing to work on solids. Brattain "enjoyed teaching undergraduates," Bardeen fondly recalled, especially in his course, "Physics for Nonscience Majors." After a long battle with Alzheimer's disease, Brattain died in October 1987. Bardeen was therefore not in the mood for celebration two months later at the fortieth anniversary of the transistor. "A part of my life is gone. We worked very closely together."

On the day Bardeen heard about Brattain's passing, he telephoned Holonyak. Nick instantly knew that something was wrong. Usually if Bardeen wanted to talk, he would walk over to Holonyak's lab and sit down for a while. "And I'd drop anything and everything I was doing then, because however long he wanted to stay or talk we would talk. But on this occasion, when Walter died, he called." Nick was aware that his wife Kay was waiting outside, but he let John speak. "Gee," he thought, "this is unlike John to stay on the phone this long." It was obvious to Holonyak that John "was very upset," for Brattain had been "really, truly, his partner." At one point during the conversation, John whispered to Nick, "I'm next." Nick replied, "John, we're all next in a certain sense."

Bardeen missed Brattain. In a tribute to his friend, published in *Physics Today*, he wrote, "While history will remember Walter Brattain for his achievements, I will remember him as a close personal friend, golf and bridge partner, and colleague." Bardeen donated $5,000 to the Walter H. Brattain Lectureship at Whitman College, "in memory of a long-time friend and colleague."

Throughout 1990 Bardeen worked on preparing his scientific papers for deposit in the University of Illinois Archives. He spent many tiring hours organizing his papers with the help of Tonya Lillie, an undergraduate assistant who later helped with the early research for this book. Bardeen was preparing to turn over his accumulated scientific files to the university archives. He knew exactly how he wanted to organize them. By associating his working papers—his notes, reprints, and other materials from other scientists, as well as his own early drafts—with the end products, he hoped to contribute to a better understanding of how he had come to make his scientific breakthroughs. He wanted to demystify the process.

Betsy Bardeen said that her father wanted scholars to be able to trace the evolution of "his scientific ideas and where the different contributions came from." David Pines noted that Bardeen "wanted to emphasize what he'd done in science, how he went about doing it." Equally important to Bardeen was to have his life and work remembered "in a way that would inspire the young."

Everything became more difficult for Bardeen as his eyesight failed. He could no longer read normal print and was dependent on his magnifying tool from Holonyak. As the trip that he and Jane took to Japan in the spring of 1990 had exhausted them both, he declined an invitation to return in the fall. Sony's George Hatoyama wrote in October 1990 to say that he would miss them "coming to Japan this fall, but I believe it is good for you to have rest." Hatoyama sympathized with Bardeen's eye problem. "Although I have not suffered from eye disease, I can imagine how annoying it would be." He added, "You know, you (and I) are not young, and we must rest when we have problems in health."

Painful attacks of gout in Bardeen's legs made walking a challenge, although with the aid of a golf cart he continued to enjoy his favorite sport. He still preferred to walk, but Bill Werstler noticed that he would sometimes "stop, take his shoe off, get his foot back in shape, and go on again." The tremor in his hands had worsened with age. The high blood pressure that he had been fighting for years was controlled with medication. On top of that, he had recently learned that dangerous plaque had begun to build up in his arteries. His failing body had become oppressive to him. In late 1990 he developed a ragged cough that would not go away. Venting his frustration to Holonyak one day, about the gout and his poor eyesight, he gestured tightly toward his chest, "and now *this!*"

In September Bardeen learned that he was to be named one of *Life* magazine's "100 Most Influential Americans of the Twentieth Century." He took the honor in stride. "Probably a few names are on everyone's lists," he said, "but beyond those few, there's lots of room for argument." His colleagues, on the other hand, saw little room for argument about Bardeen's accomplishments. Miles Klein told a journalist that "the invention of the transistor and everything that grew out of that has spawned a second industrial revolution that is at least as important as the first industrial revolution."

In December 1990, Bardeen walked a copy of his *Physics Today* article, which had just appeared, over to Joyce McMillan, Bill McMillan's widow. She was touched by the visit. "He wanted me to be sure and get a copy," she recalled. "They had asked him to do a general overview of the history and the shaping of the superconductivity theory. He said in doing that he had finally really acknowledged Bill McMillan's work on that."

Bardeen made repeated visits to Carle Clinic in Urbana to have his cough checked out. That December, doctors there aspirated nearly a liter of fluid from his chest, a new and frightening development. Chest X rays revealed a mass in his lungs characteristic of cancer. When the local doctors wanted to do a chest biopsy to determine the extent of the tumors, John and Jane balked. Jane did not trust the local doctors. It was time to see a specialist.

Bardeen made an appointment at Boston's Brigham and Women's Hospital which, according to a colleague, had one of the best thoracic surgery groups in the country. Dr. David Sugarbaker was the chief of thoracic surgery there. Bardeen hoped that Sugarbaker's team could remove the mass in his lungs. In Boston he and Jane would also be close to Betsy and her family, a comfort for them both. John and Jane purchased airline tickets but had to cancel their reservation when doctors warned Bardeen not to fly. The opening in the chest wall from the pleural effusions in Urbana had not completely healed, leaving his lungs vulnerable to possible pressure changes in an airplane. They hired a limousine service to take them to Boston.

At Betsy and Tom's house they rested. As always, John enjoyed the company of his grandsons, Andrew and Matthew, who were now 12 and 6. Both children were very attached to their grandfather. Andrew had recently written an essay about his favorite relative. "He learned to swim before he could walk, and he still likes to swim. I do too, and when I was younger, I would ride on his back while he swam the breaststroke. We called these 'loon rides' since baby loons ride on their parents' backs. He still swims, and when he is here, he comes to my swim team practices and meets."

While the Greytaks were away at work or school, Bardeen watched a little television and read as best he could with his failing eyes. He chatted on the phone once or twice with friends and family members. Holonyak recalled that they might have talked a little physics, but mostly Bardeen talked about the upcoming tests and health concerns that were then uppermost on his mind.

The appointment with Sugarbaker fell on Martin Luther King Day, January 21, 1991. Appointments were scheduled lightly, and Bardeen did not have to wait long before seeing the doctor. Jane and Betsy accompanied him. After the examination, Sugarbaker gave Bardeen a thorough briefing on his condition and its possible treatments. John gently quizzed him. Sugarbaker recalled, "I was struck right away with his analytical approach to his problem, his seeking answers. And at the same time he was very much a patient, who knew that he was going to work hard with me to try to make this a successful approach." Together they looked at the X rays. Sugarbaker told his patient everything he wanted to know. Bardeen asked about new innovative approaches to treating his cancer, surgical or otherwise, and participated fully in the process of his own medical care. "He had many questions about his diagnosis and what the potential options would be." They planned to do additional tests to find out whether the cancer had spread via the bloodstream and whether the tumor had metastasized to the lymph nodes.

They agreed that, at the beginning of the following week, Sugarbaker would perform a bronchoscopy and a mediastinoscopy (an examination of the area around the lungs using an endoscope with a light and lenses). The procedures would determine whether the cancer had spread. If it had, surgery would not be an option. The procedure required an overnight stay in the hospital. The night before he went in, Bardeen spent a quiet evening with Jane, Betsy, Tom, and his grandsons.

The surgery went smoothly. John awakened in the recovery room, where he was able to chat drowsily with the nurses. He remained in the hospital that night while the laboratory analyzed the biopsy. Sugarbaker knew what the results were shortly after the surgery, but preferred to give his patient time to recover fully from the anesthesia before discussing them. Jane and Betsy kept John company for a while that evening and promised to come back the next day. They left him watching a basketball game on the television in his room.

Early Wednesday morning, January 30, Bardeen was awake and exchanging pleasantries and small jokes with the nurses. When Sugarbaker came into the room, Bardeen asked how the doctor was doing. "How's your day—are you busy?" he would ask. "We always started by him asking how I was doing." It was, Sugarbaker recalled, "a unique aspect of Mr. Bardeen." Then they got around to discussing Bardeen's diagnosis. The cancer had spread to the point where surgery was not an option. They agreed to go forward with radiation treatments. Sugarbaker was not "overly pessimistic, but Bardeen was a highly intelligent person." The doctor could see Bardeen "knew that this was not good news." At best, they could try to slow the disease process and alleviate some of the symptoms. Bardeen likely had no more than a few months to live.

Sugarbaker checked Bardeen's incision from the day before. It looked fine. He continued his rounds, marveling at Bardeen's calm acceptance of his condition and his dignity in the face of his own mortality. As Sugarbaker was getting ready to step onto the elevator, Bardeen's nurse came running to get him. She had gone in to check on her patient and "found him unresponsive." The record on the cardiac monitor revealed that Bardeen had suffered a massive heart attack. The team could not revive him. He died that morning, January 30, 1991, at 8:45 A.M. Sugarbaker remembered that Bardeen had "expired very quickly, painlessly." Of the thousands of patients he had treated, Bardeen stood out in his memory. For him it was a "very thought-provoking incident—I've never forgotten it." Bardeen knew that he had a "very, very poor prognosis" and "that he had an incurable problem. And he died."

Sugarbaker considered Bardeen to be "a man with a lot of presence." You could tell "there was a lot to this person." Thoracic surgeons often see people when "their backs are up against the wall." At that moment courage is a precious commodity. Bardeen, he felt, had remarkable grace in his final days. "He was rock solid, and very calm" on getting the kind of news that caused others to "shift in their chairs and become uncomfortable."

Bardeen's body was cremated at the Mt. Auburn Crematory in Cambridge. Jane brought his ashes back to Champaign. In the spring the family would bury him in Madison, the city of his youth.

On Friday afternoon, February 8, several hundred friends and colleagues honored Bardeen's memory at a simple memorial service held at the Illinois Student Union of the University of Illinois. The service was followed by a reception at the university's Krannert Center for the Performing Arts. Betsy, Bill, and Jim shared memories of their father; a few of his friends and colleagues paid tribute to his scientific genius and to his humanity.

Bill told the group it was no wonder the "children have all ended up in or close to physics." Part of the reason "was the fact that we all felt the enthusiasm and intensity that Dad gave to physics." The Bardeen children learned by example. "In our house, physics was considered the noblest of professions. We were not told this but just seemed to know it anyway. It was clearly indicated by Dad's own approach to physics and the standards he set in his own life. Dad did not try to teach us physics but encouraged us to find our own way." John didn't mind teaching some physics at home when the occasion called for it. Bill recalled the time when "I was in danger of failing in a course on electricity and magnetism in college." His father spent the Christmas vacation working on problems with him and Bill ended up passing the course.

"I suppose you would not consider my father to be the typical American father," Bill said. "However, for us kids he was always first and foremost our father. Of the three children, I must have been the one who helped Dad learn what it really meant to be a father." He explained that Jim had been "calm and peaceful as a baby and was content to explore the world within his reach. However, I was always the active one, always one step from disaster and one step ahead of anybody trying to catch me." Jim told the group that when the children were young, "he rough-housed with us on the living room floor and later spent many hours playing catch and shagging flies with Bill and me."

"Most of all, my father enjoyed his grandchildren," Betsy said. He "felt completely comfortable playing on the floor with little children, tussling with them, building with blocks, and entering their imaginative play." She told one of her favorite stories, the one "when Andrew was about three, when I heard strange noises coming from Andrew's room at our cabin in New Hampshire. When I looked in, I found them pretending to be loons, diving under the quilt to catch fish and coming up to call to one another. When Matthew came along, he joined in, and the three of them would roll around on the floor in the loft." Recalling her own childhood, she said, "Although we often sat together in a comfortable silence, he was very interested in everything I did, and could be counted upon to come up with odd bits of knowledge to fit in with my interests. When I learned to play jacks and challenged him to a match, he went right through the twenties, all in order and with no mistakes."

David Pines would profoundly miss his friend's daily presence. He had returned to Urbana in 1959 because Bardeen was there. At the memorial he described John as "quiet, genuinely modest, instead of basking in the glory which a continuing series of seminal contributions to science brought his way, he preferred always to work on the *next* scientific challenge. He possessed a rare ability to see into the heart of a problem, to pluck out the key phenomena from a myriad of often conflicting experimental results and to isolate the elements required to develop a coherent description. Nowhere was this more evident than in his work on superconductivity."

"John was a wonderful colleague, and a devoted and caring friend," Pines continued. He recalled a trip that he, his wife Suzy, and their infant daughter once made "by car from Paris to Leiden—John at his request in the back seat with $2^1/_2$-year-old Catherine Pines, who found him so accessible she talked to him, non-stop, for $6^1/_2$ hours."

Nick Holonyak recalled how John's letter of introduction to Makoto Kikuchi had influenced his experience in Japan during the mid-1950s. He told the story about receiving John's card from Stockholm when he was working in Yokohama for the army, and his commanding officer's reaction: "Holonyak! Who in the hell do you know by the name of John in Sweden?" Nick also told the others his theory about why John had misspelled the word "existence" in the card—John had skipped the grades in primary school when he should have learned to spell!

Charles Slichter recalled the moment when he learned from John that the mystery of superconductivity had finally been solved. "I was walking down the hallway of the physics building and John stopped me," he recounted. "It seemed that we stood there for five minutes while I was waiting for him to speak and then he said, 'Well, I think we've figured out superconductivity.'" For Slichter, "this was the most exciting moment in science that I've ever experienced." He fondly recalled how John brought his medal "around to the lab so the rest of us could see it."

Bob Schrieffer reminisced about the sixtieth birthday gift that he and his wife Anne had constructed for Bardeen, the collage of John's students and violins. "And there was John in a set of tails with a red sash as the superconductor in his grand canonical ensemble." Schrieffer told how he had "miscounted the number of students so everything was perfectly arranged as 27, but there were only 26 heads." He decided to correct his error by adding a photo of Niels Bohr as a student, inserting it "right off center, so John wouldn't notice it." Bardeen instantly saw the ringer and said, dryly: "He was not my student. That's Niels Bohr."

Letters from all over the world poured into 55 Greencroft and the Illinois Physics Department expressing heartfelt condolences and sharing warm memories of John. Images of Bardeen helping along a younger colleague or playing with children were familiar themes. His golfing buddies Bill Werstler and Lou Burtis planted a tree on the fifteenth tee of the Champaign Country Club's golf course in memory of their friend.

Fred Seitz wrote Jane about the last time he saw Bardeen, when he visited Urbana in October 1990. The two men had joined some physics colleagues for lunch.

> As the luncheon ended, it became clear that John wanted me to come to his office to talk. We must have spent one and a half hours discussing many things from the past as well as issues concerning the state of his health. He also told me that he was getting his papers in final order since he expected to lose his vision soon. Looking back, it is now clear that he was trying in his way to tell me that this would probably be our last meeting. At the time, that seemed unimaginable.

During the visit, John and Fred had reminisced about the extraordinary period they had lived through at the University of Illinois. Bardeen said, "Well we owe a lot to Louis Ridenour." They agreed that Ridenour, in collaboration with President George Stoddard and Provost Coleman Griffith, had been able to turn the place around, "taking advantage of the opportunities of the period."

Dutch Osterhoudt wrote to Jane from his wife Gretchen's room at Durango Mercy Hospital, where she had undergone surgery three days earlier. "John was so quiet and modest, but the scientific world learned about him from his achievements. Gretchen and I miss him already. We had hoped you and he would come down and stay with us this coming summer." Their grandson Curtis had taken a liking to Bardeen.

John Tucker wrote, "I will always remember the good times."

In May when the first flowers were in bloom and the trees were coming to life, Bardeen's ashes were buried in Madison. John had asked that his remains be laid in the cemetery where Charles Russell and Althea Harmer Bardeen had been buried. They were interred in a plot adjacent to that of his parents. A place beside John was reserved for Jane. The family held a small service at the graveside. Bardeen's nephew Tom came from California to wish his famous uncle Godspeed. He brought his infant son, John, Bardeen's only namesake.

Jane donated John's medals—the two Nobel Prize medals, the Lomonosov, and the Franklin—as well as his papers to the University of Illinois, as he had specified in his will. Most of the estate was held in trusts to be distributed to family members according to the rules of each trust. The family worked hard to preserve all of John's relevant documents and artifacts. Bill and his wife Marge agreed to preserve important family papers in their home in Warrenville, Illinois. At the university, most of the documents in Bardeen's office were moved to the University of Illinois Archives, where they were carefully arranged, boxed, and catalogued for research use according to archival standards.

Jane carried on with her life, but it became difficult for her to get out as much as she would have liked. It was not long before the house on Greencroft became too much for her to look after. She moved to Clark-Lindsey Village, an assisted-living apartment complex in Urbana where some of her friends already lived.

Gretchen Osterhoudt wrote Jane in May 1994 with the sad news that Walter, John's old friend "Dutch" from the University of Wisconsin and Gulf Labs, had died. She, like Jane, was "learning to be a widow."

Declining health eventually forced Jane to move to the skilled care unit at Clark-Lindsey Village. She suffered intensely from the pain in her back. Even as her condition deteriorated, she continued to make friends. She often talked about her long-ago days teaching biology, about her family and her travels. One night she told one of the staff members, "I'm going to die." The woman lingered with Jane in her room for a while. Jane died that morning, March 31, 1997, at 4:00 A.M., seven days short of her ninetieth birthday.

"Every time we attend a funeral service," Jane had once told her sister Betty, "we decide again that we want no such ceremony when we die." She and John agreed that the family could, if they wanted to, have a memorial service conducted by friends and family, "but not a sermon by a stranger, who, if a minister, is bound to dwell on life after death and other religious ideas in which we have no faith." Jane's family and friends honored her memory with a small memorial service on April 5, 1997.

Bill Bardeen designed a single low stone monument for the graves of both his parents. The design includes two N's to represent his father's two Nobel prizes and two flower sprigs to symbolize his mother's interest in the natural world. The stone marks the spot where Jane's ashes now lie with John's, on a low hill just west of Madison.

17

Epilogue:
True Genius and How to
Cultivate It

For centuries the question of genius has been a subject of mystery and debate. Such fundamental questions as "How does genius relate to creativity?," "Is genius hereditary?," or "Can genius be cultivated?" have been surprisingly difficult to answer. Even today there is little agreement in the vast literature that has grown up around such questions of creativity and genius. Yet within that contentious discussion are pockets of agreement, the beginnings of a modern, more realistic picture of genius. John Bardeen's life and work offer a basis for discussing it.

We need to be careful in presenting the results, for the research is incomplete. The picture is still hazy. And Bardeen is only one example. To gain a more complete account will require comparisons across many cases. Even Bardeen's work has not yet been fully mined with respect to questions of genius or creativity, for only in the last years of writing this book did we recognize that it could help to address such questions.

Most scholars who work on creativity agree on three aspects of genius. First, there is a real and substantial distinction between a "true genius" who makes lasting contributions to culture and the fictional archetype of a genius, amply represented in popular culture by such figures as Dr. Frankenstein or Will Hunting. In the group of scholars who had recognized this distinction by the early twentieth century are the biologist, eugenicist, and statistician

Francis Galton, Charles Darwin's first cousin, and the psychologist Lewis Terman, the father of intelligence quotient (IQ) testing.

Galton had hoped to improve human creativity through selective breeding that would enhance intelligence and other "good" characteristics. He initiated the systematic study of human creativity by developing statistical methods for evaluating individual differences. Terman built on Galton's work, designing more sophisticated quantitative tests for studying creativity. An outgrowth of his famous tests conducted at Stanford University in the early 1920s was the extensive study conducted by Terman's student Catherine Cox on the future success of three hundred gifted children having superior IQ scores.

A second general agreement among scholars of creativity is that genius and creativity must refer to a particular domain having a well-defined knowledge structure. Thus Mozart was a genius with respect to music, Newton a genius with respect to physics. This association with a domain brings genius and creativity into direct contact with culture and society, for a group of experts in the domain must agree that an achievement adds something new and important to it. If not, the idea or act can have no lasting impact.

Such "gatekeepers" of knowledge are united by their acceptance of the "paradigms" of their domain, a term that Thomas S. Kuhn introduced into the study of the history of science about four decades ago when he pointed out that in any given period scientific fields are characterized by a set of agreements about how knowledge is produced in that field. This description of a field in terms of its paradigms subsequently entered many fields of knowledge and allows us to define what a highly creative act is, namely one that creates, replaces, or restructures a field's paradigms. The degree of restructuring can vary a great deal. Bardeen's creativity in physics was far more limited in scope than that of Newton or Einstein, but it nevertheless brought significant paradigm shifts into the existing structure of knowledge in solid-state physics, the domain in which Bardeen worked for almost his entire career.

The third agreement in the field of creativity, not yet as widespread as the other two, describes a genius in terms of a profile, a configuration of heterogeneous features that support the highest levels of creative work in a field. Unlike the popular stereotype, with its familiar features (e.g., otherworldly, alone, charismatic, eccentric, unbalanced, self-taught, etc.), the profile of the true

geniuses who live and work in the real world includes features such
as intelligence, passion, confidence, focus, perseverence, and the
habit of breaking down problems into smaller parts. While many of
the features are similar when we compare individuals—just as different human noses or lips are similar to one another—the constel-
lations they form are as individual as a human face.

This notion of a genius profile idea is not a new one. Even Cox,
whose work stressed genetic giftedness, conceded that "high intellectual traits" are not alone responsible for the eminence of her
subjects. They achieve success "also by persistence of motive and
effort, confidence in their abilities, and great strength or force of
character." What is new about the genius profile is the wide agreement the idea is coming to enjoy. As this conception offers fertile
research opportunities, we will examine this profile in more detail,
drawing on Bardeen's life and science.

Genius and exceptional creativity can be represented in three qualitatively different dimensions: (1) *an individual (or personal)
dimension*, which includes features such as talent and motivation
that refer to the particular domain in which the person worked; (2)
a methodological dimension, which includes the variety of
problem-solving approaches such people employ in their domain;
and (3) *a contextual dimension*, which includes environmental
factors, such as family and education. These dimensions are not
independent of one another. For example, Bardeen's family strongly
influenced his motivation and values. But examining the dimensions separately gives us the basis for studying them more closely.

Most of the features in the personal dimension have been hotly
debated over the years. For instance, while almost everyone
acknowledges the importance of talent and intelligence, there is no
general agreement about whether these traits are genetically determined. By any accounting Bardeen was talented in the area of mathematics; his outstanding ability in mathematics was apparent to
his parents and teachers when he was a child. But it is not yet
possible to say whether this talent derived solely, or even primarily, from genetics, as Galton, Terman, and Cox would have argued,
or from his having been exposed to the field of mathematics over a
long period of time.

Today most scholars would agree that such a talent is probably based at least partially on genetic ability. Yet startling experiments have shown that people with rather ordinary ability in areas such as chess or music can be trained to achieve master-level performance by subjecting them to the right kind and amount of training over a long period of time, usually about ten years. There is indeed some evidence that Bardeen exposed himself massively to mathematics from his young years, or perhaps as a result of his enthusiasm for the subject. Talent appears to be based on both genetics and training.

While early in the twentieth century Terman and others assumed that IQ was the necessary prerequisite for creativity, much of the current literature about creativity raises questions about the role of inherited intelligence, at least as measured by standard tests. Feynman, who was obviously highly creative, is said to have had an IQ of 129, which, while well above average, is not in the high range of most recognized geniuses. Some psychologists have suggested that creativity has more to do with having the right kind of intelligence for creative work in a particular domain. For instance, the kind of intelligence that Bardeen had made it possible for him to "see" the electrons in solids.

In the same camp with those who believe that talent is largely a function of exposure are those who emphasize a person's passion or motivation for work in a domain. The great teacher Confucius emphasized passion: "Only one who bursts with eagerness do I instruct; only one who bubbles with excitement, do I enlighten," he wrote. The psychologist Teresa Amabile and her students insist that "intrinsic motivation" (a quality referred to even by members of the opposite camp, including Galton) fuels the massive investment in learning about a given domain, an exposure they believe is the true basis for talent. From this point of view, self-help gurus who advise following one's passions are not wrong, because passion inspires exposure and the drive for excellence, which, in turn, often sends a person looking for supportive environments. From the fact that motivation can be cultivated through positive reinforcement, it follows that talent can too. Although this argument opposes genetic determinism, it does not rule out the possibility that a genetic component figures importantly in an individual's potential abilities.

Passion is a compelling candidate for the source of creativity because it can be viewed as the source of most other qualities in the individual dimension of the genius profile, such as the will to learn, perseverance, focus, confidence, or the drive to achieve mastery. Bardeen had strong passion for physics and mathematics, and he also possessed the complex of traits that one can expect passion to bring. He learned early on that persevering and maintaining his focus on difficult problems, even in the face of frustration, often brings enormous rewards. Through his experience he learned that struggling with difficult mathematics problems builds the intellectual muscles and confidence for doing creative work in mathematics. He probably did the most to develop himself into a world-class physicist during his time at Harvard when he struggled almost constantly with problems that were then intractable. The confidence and the intellectual strength he developed as a physicist and mathematician also supported his ability to take measured risks.

Bardeen's intense focus on the problems he engaged with would sometimes unnerve his associates, especially when his attention wandered off into what appeared a meditative trance—sometimes right in the middle of a conversation. Schrieffer thought that at the times when Bardeen was on "cloud nine," he was letting nature "flow into his being," saturating his consciousness with information about the system he was studying. By coupling this intense focus with persistence and his years of training and experience, Bardeen nurtured his legendary "intuition" for physics.

Bardeen's communion with physics problems was not unlike the riveted attention a young child focuses on something of interest. Alison Gopnik and others have compared scientists to grown-up children. Howard Gardner has correlated the single-mindedness required in creative work with a retention of childish ways of focusing on how the world works, a notion that Einstein also touched on in his autobiography. Bardeen's rapport with children may well have been related to his ability to attach his intense child-like focus to physics problems.

Some research suggests that childhood traumas, even such drastic ones as the death of a parent, can actually benefit creativity. Bardeen was twelve when his mother, Althea, died. Rather than crushing him, confronting that painful experience appears to have enhanced his inner strength. Similarly, the marginalization he experienced in both grade school and high school may have helped him learn how to stand apart from the crowd. In times of emotional stress, Bardeen often turned to mathematics and physics, which offered controllable worlds in which to escape from the pain in his personal or social life.

Sports offered other controllable worlds—as well as a multitude of additional benefits, including stress relief, physical exercise, social interaction, entertainment, and a model for teamwork. The sports that most appealed to Bardeen were ones in which he could practice mastery and engage in constructive competition. Bardeen was intensely competitive in both work and sports, but his kind of competition seems to have been driven less by a desire to win over another person than a wish to triumph over nature's hardest problems. Schrieffer said that Bardeen was "in competition basically with himself or with nature."

Studies of the personal dimension of genius suggest that creative people often continue to work actively into their senior years. Bardeen was no exception. His passion or ability for research hardly diminished with age, although his mental flexibility appears to have declined somewhat. In his seventies Bardeen took on a problem as difficult as any he had attempted earlier in life. His bold and controversial theory of charge density waves might have earned him a third Nobel Prize had it hit pay dirt. Despite his inability to convince his colleagues of the merits of his work on charge density waves, he remained confident that his theory was correct. He never lost his excitement for the problem. A colleague said, "He's the only man I ever knew who in his later years was still actively excited, working like crazy about physics, and not just working but writing papers that were right at the forefront and controversial." We find a similar pattern in Einstein's struggles late in life to develop a unified field theory and in Dirac's last theoretical work on the magnetic monopole.

Attempts to describe the second, i.e., methodological, dimension of the genius profile have benefited from extensive experimental research conducted by cognitive psychologists. For over two decades their dominant metaphor for human problem solving has been information processing by computer. The pioneering work on scientific thinking by Herbert Simon and his collaborators has described problem solving as a search through a "problem space."

Among the many intriguing results of experimental studies in this field are those designed to explore the differences between novice and expert problem solvers. In exploring the differences in the recall that master and novice chess players have of the arrangement of pieces on a chess board, psychologists have found evidence to support the theory that master players have at their disposal more complex ("chunked") mental structures than do novice players. These structures were created over the course of their long periods of strenuous training and experience. There is reason to believe that such chunking of knowledge also occurs in the sciences. The development of the expert-level structures necessary for creative work appears to require about ten years of intense study and practice to develop. Thus highly creative individuals often produce major achievements about every ten years—as did Bardeen in his work on the transistor and superconductivity, which culminated in 1947 and 1957. The resulting master-level cognitive apparatus makes the search through problem space much faster. According to this notion, then, the meaning of a person's "intuition" in a domain is strongly correlated with the amount and intensity of their experience in their domains.

Bardeen's notebooks and letters document many problem-solving approaches that have been studied experimentally. A frequent method of Bardeen's is one that psychologists sometimes call "problem decomposition," another name for the methodology that Bardeen said he learned from Wigner. Bardeen's scientific notes include many lists of "subproblems," smaller problems to be solved along the way to solving larger problems. The subproblems he laid out in his study of superconductivity in 1951 included deriving the London equations for a multiply connected body and proving the current and effective mass theories for one-electron wave functions. Colleagues have reported that Bardeen gave every problem, no matter how small, his full attention. While Wigner may have urged Bardeen to use this analytic approach, Althea Bardeen probably preconditioned him to decompose problems, just as she had encouraged her students at the Dewey School to break down problems.

Bardeen's own references to breaking down problems conflate two kinds of decomposition, both of which he routinely attempted. In one approach he would simply break a larger problem down into smaller parts and, when possible, delegate responsibility for solving the parts to other members of his team. Thus in the push to complete the BCS theory, Bardeen asked Schrieffer to work on the thermodynamic properties and Cooper to handle electrodynamics, while he himself worked on transport and nonequilibrium properties.

In the other kind of problem decomposition, Bardeen would reduce the problem to be solved to the simplest model that contained the basic features of the larger problem. In his words, "You reduce a problem to its bare essentials, so that it contains just as much of the physics as necessary." Thus in 1952, in studying the coupling of electrons in a superconductor with the lattice vibrations, Bardeen asked Pines to examine the problem of the polaron, a simpler problem containing some of the important features of superconductivity. After that, he could add in more features of the larger problem.

Bardeen was not concerned about elegance in his approach to problem solving. He preferred, as Seitz said, to "bully through" to the answer. He was usually willing to try a range of approaches. For instance, in the weeks before inventing the transistor, Bardeen and Brattain studied a series of systems different from one another in material, geometry, or an aspect of structure. To keep control, they tried to make only one or very few changes in moving from one system to the next. It was a kind of experimental tinkering, useful in the absence of a complete theory.

Bardeen would also at times employ the theoretical counterpart of such experimental tinkering. A kind of "brainstorming," such theoretical tinkering resembles techniques that were widely publicized during the 1950s and 1960s as a means for inspiring creativity. It was a time when creativity was being recognized as a tool useful in fighting the Cold War.

In those days, creativity was often described as the result of "divergent" (or "lateral") thinking, an alternative to "convergent" (logical) thinking. In this picture, creativity was possible when the mind moved off its "beaten track." Such studies of divergent thinking were officially sanctioned by the academic world. In 1950 the psychologist Joy P. Guilford, then president of the American Psychological Association, called for a scientific study of divergent thinking. He encouraged psychologists to work on identifying cognitive operations involved in creative problem solving. The resulting research motivated many variations on what came to be called the "psychometric approach," for example, the work of Raymond Cattell in the 1960s and 1970s on fluid intelligence and that of Gardner in the 1980s on multiple intelligences.

The same basic approach was publicized by the creativity gurus of the 1950s and 1960s, such as Edward de Bono, whose seminars were widely attended by corporate executives. Some of the brainstorming (or "blockbusting") exercises followed lines pioneered by Alex Osborn in the 1950s, with tasks such as listing as quickly as possible all words one can associate with some central element of a problem.

While the fervor of the divergent thinking movement of the fifties and sixties has long since died down, we can recognize in the work of Bardeen and other creative people a tendency to engage in a kind of brainstorming reminiscent of de Bono and Osborn at moments in their research when they feel "stuck." Consider the variety of approaches Bardeen and Brattain used twelve days before their great discovery of the transistor. They had achieved a structure that yielded some current amplification, but they did not understand why. Struggling to advance, they tried many variations—of material, geometry, or technique—carefully observing what improved or diminished the effect. A kind of brainstorming also showed up when Schrieffer tinkered in January 1957 with a range of theoretical ideas at the time he wrote down the expression for the BCS ground-state wave function. Countless examples of this approach can be found in the history of technology, for instance, in the time-pressured work on implosion during the Manhattan Project, in Fermilab's building of the first large superconducting magnet accelerator, and in the Wright brothers' experiments on flight.

Such work—and much of Bardeen's problem solving—is heavily grounded in experimentation. A trained engineer as well as a physicist, Bardeen believed that "experiment came first, experiment came second, experiment came third," in the words of Pines. The theoretical physicist Alexei Alexeivich Abrikosov explained that, unlike many theorists who wish "to create laws and to impose [them] on nature," Bardeen would always begin with the results from experiments. "He felt," wrote Schrieffer, "that formalism could lead one astray, unless it was closely tied to experiment and physical intuition. He also felt that mathematics had not yet matured to the level required to attack straightforwardly many of the frontier problems." On one occasion a vice-president at Sharp Corporation asked Bardeen about the other theorists with whom he worked. Bardeen responded, "I don't generally work with theoreticians. I like to examine the data myself."

Within experimental research programs are those rare moments when anomalies signal new insight. The physicist Leon Lederman called such moments "epiphanies" because they result in "the sudden understanding of something new and important, something beautiful." Of course, epiphanies can only happen when anomalies are recognized and acted upon. In Bardeen's work on the transistor, unexpected experimental results brought some of the most dramatic progress. For example, the first observation of a transistor effect in December 1947 was the result of accidental electrical shorting owing to the fact that a germanium oxide they thought was present had actually washed off. Germanium's material property of not sustaining such an oxide changed the structure of the experiment in an unexpected way allowing holes to enter the inversion layer on the germanium slab. Bardeen and Brattain were brought to that realization when they noticed the most obvious consequence of this "accident," that the current was flowing in the opposite direction to what they expected.

There is ample data in the historical papers of Bardeen to fuel study of the roles of analogy and metaphor in his scientific thinking. Finding the best way to represent a problem was a continuing concern for Bardeen. For instance, after Cooper's suggestion of electron pairing, Bardeen, Cooper, and Schrieffer were stuck for over a year because they did not know how to represent mathematically the situation in which many electron pairs overlap physically with many other pairs. At one point, following in the footsteps of Fritz London, Bardeen suggested changing what he referred to as the "language" of the representation, from ordinary three-dimensional position space to a space in which the electrons are specified by their momentum coordinates. Almost a year after BCS was invented, Schrieffer developed a clever dance analogy in an effort to explain the representation problem to colleagues and students. Comparing the electron pairs in a superconductor to couples dancing the Frug on a crowded floor allowed him to portray the nature of the interactions between the pairs.

Psychologists describe analogy as a mapping from a more familiar domain (the "base") to the less familiar one being studied (the "target"). By explicitly conveying relationships from the base into the target domain, researchers, including Dedre Gentner, have explained how scientists can set the stage for developing a new picture. Bardeen's entry on December 24, 1947, in his Bell Labs notebook, reproduced in part in Figure 17-1, illustrates how he and

DATE Dec 24, 1947
CASE No. 38739-7

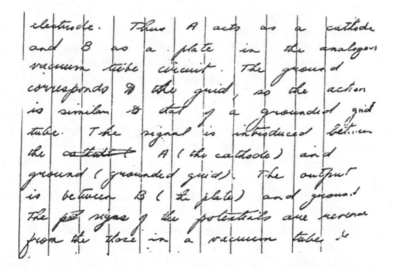

FIGURE 8 Analogy of transistor to triode system (Bardeen notebook, Dec. 24, 1947.

Brattain mapped well-studied features (grid and plate) of the familiar system of the triode onto the unknown semiconductor system they had built, in an attempt to understand how the semiconductor system could in fact amplify. A deeper analysis of how such analogies brought Bardeen and other scientists to their major advances could shed new light on our understanding of creative discovery in the sciences.

In using metaphor, one is also (as in the case of analogy) comparing things that are not yet understood with things that are better understood, but the link between the two is more direct: one describes the unknown *in terms of* the known, even when the details don't match up perfectly. For example, when Bardeen suggested that Gerald Pearson conduct a low-temperature experiment in which the electrons would be "frozen" into the surface states, he meant that the electrons act *as though* the crystal were frozen like ice. Drawing explicitly on a series of examples from the history of physics, Arthur I. Miller has shown how metaphors can themselves act as models in research and thus become powerful instruments of discovery and invention.

For cases in which the base and the target domains overlap, Bardeen would sometimes employ metaphors as bridging principles in which part of the known domain can point the way to understanding an unknown domain of physics. For instance, he specified, in a manner resembling the way Niels Bohr used his Correspondence Principle, that the unknown superconducting energy states must correspond one to one with free electron states in the presence of a weak residual interaction inducing superconductivity.

Whenever possible, Bardeen distributed his work among the members of his team in such a way that coordinating their particular experiences and talents could help him make a more effective attack on the problem at hand. Some of his best work occurred in collaborations in which the other members offered expertise that differed from his own. Bardeen's and Brattain's notebooks from 1946 and 1947 document their rich interplay in the context of the Bell Labs semiconductor group, which had been, in turn, structured in a way likely to motivate interdisciplinary collaboration.

In later years Bardeen consciously built such interdisciplinary collaboration into his problem-solving methodology, for instance, adding Cooper to the superconductivity team to gain access to his expertise in quantum field theory. Bardeen's network of collaboration also extended outwards from his immediate team into the larger physics community. In one crucial example, a phone call on May 15, 1950, from Bernard Serin offered Bardeen the clue "that electron-lattice interactions are important in determining superconductivity."

Yet another strategy in Bardeen's methodological profile was to limit the focus of his research to only a few questions. Psychologists conceive of an expert as one who "knows a lot about a small number of things." By staying in the same area of physics for many years, Bardeen developed deep expertise and a profound base of knowledge with which he could probe research problems in the domain of solids. Wigner, whose own professional interests shifted several times, told an interviewer that Bardeen "had one wonderful quality—he didn't change the subject of his interest too often and therefore he could make very important contributions."

Bardeen's regular and systematic updating of his knowledge base should not be overlooked as an important and powerful methodology. Almost daily he took the time to visit the library and read publications of interest to his work. When starting a new problem, he always began with a literature review. By widening and deepening his base of knowledge, such library research often gave Bardeen the advantage over other workers.

These overlapping approaches to problem solving, clearly illustrated in the documentation of Bardeen's work, help us model one physicist's creative methodology. To generalize, or perhaps to develop a typology of alternative models, it is crucial for historians, psychologists, and other scholars of creativity to study many cases. In the area of the history of science, this program of research is well underway in the research of scholars such as Nancy Nersessian, Ryan Tweney, Frank Sulloway, Howard Gruber, Arthur I. Miller, Gerald Holton, Keith Simonton, Dedre Gentner, and Pat Langley. But the community of scholars conducting studies of this kind is still too small to accomplish the larger task.

The third dimension of the genius profile deals with context, the many environments, such as family or workplace, that can nurture the other dimensions, for instance, by instilling values, offering connections, and providing moral or financial support. Bardeen's parents and grandfather all stressed learning, creativity, diligence, and the joy that comes from work in the interest of social betterment. John's parents strongly supported the development of qualities useful to his education and emerging creativity. For example, they applauded his early tendency to "hang on" to problems stubbornly. Bardeen's colleagues often spoke of his "bulldog tenacity."

Bardeen's mother also set the model of an independent person with the courage to follow her passion. The fact that she could at age twenty-five break from her family to study art cleared the way for John to make a similar move when, at age twenty-five, during the Depression, he quit his secure engineering post so that he could enter the graduate program at Princeton. An equally important model was set by his father, Charles, the academic scientist dedicated to solving problems for the public good. Bardeen's groundedness in the problems of real materials and his commitment to industry and government had their basis in the progressive culture and ideals of his youth in Madison. "The greatest opportunity of all is to serve," his grandfather had written to him.

The fact that Bardeen's parents intervened to support his academic development impacted both his intellectual and personal development. He reflected in a later interview that, although he had been marginalized in his classes with children four or five years his senior, the fact that he had been skipped three grades had generally benefited him. Not only was he challenged, especially in math, but, as he said, he had more time to concentrate on his studies. He considered it valuable to have avoided the boredom that gifted children often experience in school.

In his later years, Bardeen often landed in settings that reproduced his productive early learning environments. Similar in philosophy to the Dewey School, where Althea had taught before marrying Charles, Wisconsin's University High was a progressive school aimed at supporting the creativity of its students. Charles credited Uni High with nurturing John's mathematical talent. As Charles explained to his father, the Wisconsin Uni students were "allowed to go ahead in any line they take up as fast as they show ability and the hard and fast grade system does not exist." The "graduate student's paradise" that Bardeen had encountered at Princeton, the stimulating contexts of the Harvard Society of Fellows, the semiconductor subgroup at Bell Labs in 1945–1947, and the Department of Physics at the University of Illinois all resembled Uni High in supporting Bardeen's creative work in physics.

The Bardeen home was an important factor, too. An anchor offering stability and attachment to reality, home and family were critical, especially in times of stress or when Bardeen ventured off into unknown scientific territory. After John's marriage in 1938, Jane Maxwell assumed the responsibility for maintaining this stability, even while she herself felt separated from her husband's most passionate interests.

Bardeen's mentors played crucial roles in his creativity. Walter W. Hart, Bardeen's seventh- and eighth-grade math teacher, fanned the flames of John's interest in mathematics. When Bardeen was an undergraduate at Wisconsin, Paul Dirac, Peter Debye, and especially John Van Vleck introduced him to modern physics in the period when quantum mechanics was born.

Bardeen's mentors opened doors on his behalf. Van Vleck wrote many letters of support, e.g., to Princeton, Harvard, Minnesota, and the University of Illinois. Consistent with Harriet Zuckerman's thesis about the attraction between "Nobel-bound" mentors and students is the fact that an unusually large number of Bardeen's physics teachers later received Nobel prizes—Dirac, Debye, Van Vleck, Bridgman, and Wigner. Bardeen learned from them what it takes to do world-class research in physics. Although none of these teachers had yet won their Nobel prizes at the time Bardeen studied with them, Bardeen and these mentors recognized each other, somehow.

Bardeen studied with Wigner during the two or three years when Wigner was passionately interested in solids. In 1945 he worked at Bell Labs, by all accounts then one of the best places in the world to study semiconductors. In 1951, when Bardeen came to the University of Illinois, Seitz was building there one of the first large American solid-state departments in academia. How did it happen that Bardeen and other creative individuals so often worked in environments and with mentors who supported their creative work?

Bardeen himself claimed that "accomplishments are a good bit of luck—being in the right place at the right time—and having the right associates." He considered himself "lucky" to have been "on the ground floor of solid-state physics." But serendipity cannot explain why Bardeen so *often* found himself in the right place at the right time to support his creativity. For any major invention to occur, a large number of contingencies need to mesh. How could such a highly improbable act occur twice for Bardeen? Why do such "epiphanies" typically occur more than once in the careers of other creative people, often with a frequency of about ten years?

It appears that creative people actively seek or create the environments that support their intensely dedicated work in an area. They create their "historical junctures" by placing themselves in the "right place at the right time" and among the "right associates," whom they are able to recognize because of their own experiences, interests, and passions. Recall the care with which Bardeen selected his thesis advisor and made his various career moves.

It seems clear that exceptional creativity is the product of many factors, some individual, some methodological, others contextual. Establishing these more rigorously and determining how they interact forms the central problem for the next stage of understanding genius and creativity. Much progress has been made, but the research is still at an early stage.

Different approaches are needed, for each method has its advantages and its disadvantages. In experiments the parameters can be altered and controls are possible, but experiments can rarely be performed on the most creative individuals, nor can they usually run over the decade-long periods required for major creative work. Surveys and interviews allow people to be questioned about their work, but they are unreliable, biased, and, in the case of surveys, inflexible. Individuals do not know themselves all that was involved in their creative work. Computer simulations offer unlimited possibilities for analyzing a great variety of parameters, but so far at least computers are not able to handle nonrational variables. Videotapes of research in the making offer the researcher a ringside seat at the moment of creativity, but like experiments, their use is limited to short time periods.

Much of what we know must come from biographies that examine the lives and works of creative people. These too have disadvantages, for biographies, and histories in general, are only as good as the sources they are based on. Creative people never record every step of their work and human memory distorts the path of discovery. Frans de Waal aptly wrote of scientific research, "In hindsight, the path taken may look straight, running from ignorance to profound insight, but only because our memory for dead ends is so much worse than that of a rat in a maze." Most of the legendary stories about individual acts of creativity—such as Fredrich August von Kekule's famous dream of the snake biting its tail, which is said to have led to his discovery of the benzene ring, or Samuel Taylor Coleridge's account of conceiving his poem Kubla Khan in an opium dream—have been disproved.

Even the documents that people leave behind can be misleading, because individuals have a tendency to shape the evidence so as to make their discovery paths appear straighter than they actually were. Well-meaning family members, eager to shrink a massive collection of sources, often toss out the documentation of false starts. Even archivists occasionally destroy historical data, if only by rearranging papers for more convenient retrieval.

The road ahead for scholars of genius and creativity must therefore remain arduous. But the problems they face are not insurmountable because their subjects are not the otherworldly savants that geniuses were once thought to be. They are real people, like John Bardeen, highly motivated to develop the elements of genius that exist potentially in all of us.

Bibliography

The following abbreviations and acronyms are used in the Bibliography and the Notes:

AIP	Niels Bohr Library, American Institute of Physics, College Park, Maryland
AT&T	American Telephone and Telegraph Corporation Archives, Warren, New Jersey
BFC	Bardeen Family Collection, Warrenville, Illinois
BNB	Bell Laboratories Notebook (AT&T)
HARB	Harvard University Archives, Cambridge, Massachusetts
NARA	National Archives and Records Administration, Washington, D.C.
NOL	Naval Ordnance Laboratory (now Naval Surface Weapons Center, Dahlgren, Virginia)
PRIN	Seeley G. Mudd Manuscript Library, Princeton University, Princeton, New Jersey
STAN	Stanford University Archives, Shockley Collection
UIUC-A	John Bardeen Papers, University of Illinois Archives, Urbana
UIUC-P	Department of Physics, University of Illinois, John Bardeen Collection, Urbana
UMP	Department of Physics, University of Minnesota, John Bardeen Collection, Minneapolis

331

USOPM   U.S. Office of Personnel Management, OPF/EMF
           Access Unit, St. Louis, Missouri
UWA   Archives, University of Wisconsin Memorial
           Library, Charles Russell Bardeen Collection,
           Madison
WHIT   Walter Brattain Collection, Whitman College, Walla
           Walla, Washington
WHS   Wisconsin Historical Society, Steinbock Archives,
           Madison

All interviews listed, unless otherwise indicated, are in UIUC-P.

Abrikosov, A. A. 1993. Interview by Vicki Daitch, August 5.
Albert, R. S. 1993. Genius and Eminence: The Social Psychology of
    Creativity and Exceptional Achievement. New York: Pergamon
    Press.
Albert, R. S., and M. A. Runco. 1999. A history of research on cre-
    ativity. Pp. 16–31 in Handbook of Creativity, R. Sternberg, ed.
    Cambridge, England: Cambridge University Press.
Allender, D. 1993a. Interview by Vicki Daitch, February 11.
———. 1993b. Interview by Vicki Daitch, February 23.
Allender, D., J. W. Bray, and J. Bardeen. 1973. Model for an exciton
    mechanism of superconductivity. Physical Review B 7(3)(Feb-
    ruary 1):1020–1029.
Alpert, D. 1992. Interview by Vicki Daitch, March 5.
Amabile, T. M. 1996. Creativity in Context. New York: Westview
    Press.
Anderson, A. 1992. Interview by Fernando Irving Elichirigoity,
    January 13.
Anderson, P. 1958a. Coherent excited states in the theory of super-
    conductivity: Gauge invariance and the Meissner effect. Physi-
    cal Review 110:827–835.
———. 1958b. Random-phase approximations in the theory of su-
    perconductivity. Physical Review 112:1900–1916.
———. 1961. Tunneling from a many-particle point of view. Physi-
    cal Review Letters 6(2):57–59.
———. 1970. How Josephson discovered his effect. Physics Today
    (November):23–29.
———. 1987. Interview by Lillian Hoddeson, July 13.
———. 1992a. Interview by Lillian Hoddeson, Vicki Daitch, and
    Michael Riordan, March 17.
———. 1992b. Tunneling into superconductors. Physical Review
    Letters 9(4):147–149.

———. 1998. Take some germanium and two bits of gold foil. Review of Crystal Fire: The Birth of the Information Age in the Times Higher Education Supplement (March 13):1.

Anderson, P. W., and J. M. Rowell. 1963. Probable observation of Josephson superconducting tunneling effect. Physical Review Letters 10(6):230–232.

Anspacher, W. 1992. Interview by Vicki Daitch, July 30.

Baker, W. 1992. Interview by Lillian Hoddeson and Michael Riordan, September 25.

Baker, W., and P. Miller. 1984. A Commemorative History of Champaign County, Illinois: 1833–1983. Champaign: Illinois Heritage Association.

Bardeen, C. R. 1907. Abnormal development of toad ova fertilized by spermatozoa exposed to the roentgen rays. Journal of Experimental Zoology 4(1).

———. 1909. Variations in susceptibility of amphibian ova to the X rays at different stages of development. Anatomical Record 3(163).

———. 1918a. Determination of the size of the heart by means of the X-ray. American Journal of Anatomy 23.

———. 1918b. The value of the roentgen ray and the living model in teaching and research in human anatomy. Anatomical Record 14(337).

Bardeen, C. W. 1910. A Little Fifer's War Diary. Syracuse, N.Y.: Charles William Bardeen.

Bardeen, Jane. 1991a. Interview by Lillian Hoddeson and Fernando Irving Elichirogoity. June 6.

———. 1991b. Interview by Vicki Daitch, September 29.

———. 1991c. Group interview with Brian Pippard, David Pines, Lev Gor'kov, Ansel Anderson, Gordon Baym, and Charles Slichter by Lillian Hoddeson, October 9.

———. 1993. Interview by Vicki Daitch, November 23.

———. 1994. Interview by Vicki Daitch, December 2.

Bardeen, J. 1936. Theory of the work function: II. The surface double layer. Physical Review 49(May 1):653–666.

———. 1937. Conductivity in monovalent metal. Physical Review 52(October 1):688–697.

———. 1938. An improved calculation of the energies of metallic Li and Na. Journal of Chemical Physics 6(July):367–371.

———. 1940. Electrical conductivity of metals. Journal of Applied Physics 11:88–111.

———. 1941. Theory of superconductivity. Abstract. Physical Review 59:928.

———. 1942. Analysis of pressure data. NOL Report 45. Washington, D.C.: United States Navy Yard, February 24, NOL.

———. 1947. BNB 20780, AT&T.

———. 1950a. Handwritten notes on superconductivity, May 15–18, UIUC-A.

———. 1950b. Zero-point vibrations and superconductivity. Physical Review 79:167–168.

———. 1956. Theory of superconductivity. Theoretical part. Pp. 274–369 in Handbuch der Physik. Vol. 15, Sigfried Flügge, ed. Berlin: Springer.

———. 1957. Semiconductor research leading to the point-contact transistor. Pp. 77–99 in Les Prix Nobel en 1956, K. M. Siegbahn et al., eds. Stockholm: P. A. Nordstet & Sons.

———. 1960. Talk at Illinois State Normal University, May. UIUC-A.

———. 1961. Tunneling from a many-particle point of view. Physical Review Letters 6(2):57–59.

———. 1962. Tunneling into superconductors. Physical Review Letters 9(4):147–149.

———. 1963. Interview by Lincoln Barnett, May 21, AT&T.

———. 1965a. Basic research in the industrial laboratory. Pp. 56–70 in Science and Society: A Symposium. New York: W. A. Benjamin.

———. 1965b. Interview by Maynard Brichford, February 8, 1965, UIUC-A.

———. 1965c. 1968. Let's not overlook basic research. Undated manuscript, Xerox files, 1968–1969, UIUC-A.

———. 1973a. History of superconductivity research. Pp. 15–57 in Impact of Basic Research on Technology, B. Kursunoglu, and A. Perlmutter, eds. New York: Plenum.

———. 1973b. Interview by Louise Geislers, WILL Public Radio, Urbana, Illinois, November 6, UIUC-A.

———. 1977a. Interview by Lillian Hoddeson, May 12.

———. 1977b. Interview by Lillian Hoddeson, May 16.

———. 1977c. Interview by Lillian Hoddeson, December 1.

———. 1977d. Interview by Lillian Hoddeson, December 22.

———. 1978. Interview by Lillian Hoddeson with Gordon Baym, April 14.

———. 1980a. Tunneling theory of charge-density-wave depinning. Physical Review Letters 45:1978.

———. 1980b. Reminiscences of the early days in solid state physics. Proceedings of the Royal Society of London A371:77–83.

———. 1980c. Interview by Lillian Hoddeson, February 13.

———. 1984a. Beginnings of solid state physics and engineering. The Bridge (National Academy of Engineering) 14(4):8–12.

———. 1984b. Interview by William Aspray, May 29.

———. c. 1984. Interview by Lillian Hoddeson.

———. 1985a. Macroscopic quantum tunneling in quasi one-dimensional metals. II. Theory. Physical Review Letters 55:1010.

———. 1985b. William L. McMillan. In Lecture Notes in Physics: Charge Density Waves in Solids, Proceedings of the International Conference Held in Budapest, Hungary, September 3–7, 1984, Gy Hutiray and J. Sólyom, eds. Berlin: Springer-Verlag.

———. 1985c. "A Meeting to Honor William L. McMillan," Department of Physics, University of Illinois at Urbana-Champaign.

———. 1986a. Consultation on Star Wars. Letter to the editor. Science. January 17.

———. 1986b. The origins of Star Wars. Arms Control Today (July 1), clipping, UIUC-P.

———. 1987. Solid-state physics—1947. Solid State Technology (December):69–71.

———. 1988. Walter H. Brattain. Physics Today (April):116–118.

———. 1989a. Classical versus quantum models of charge-density-wave depinning in quasi-one-dimensional metals. Physical Review B 39(6):3528–3532.

———. 1989b. Depinning of charge-density-waves by quantum tunneling. Physica Scripta T27:136–143.

———. 1990a. Interview by NHK television, Japan, June, courtesy of Nick Holonyak.

———. 1990b. Superconductivity and other macroscopic quantum phenomena. Physics Today 43(12):25–31.

———. 1990c. Theory of size effects in depinning of charge-density waves. Physical Review Letters 64(19):2297–2298.

———. 1994. Walter Hauser Brattain. Memoirs of the National Academy of Sciences 63:68–87, Washington, D.C., National Academy of Sciences.

Bardeen, J., and W. Brattain. 1948. The transistor: A semi-conductor triode. Letter to the editor. Physical Review 74(2):230–231.

———. 1949. Physical principles involved in transistor action. Physical Review 74(April 15):1208–1225.

Bardeen, J., and J. Keithley. 1942. Proposed design of a mine utilizing the decrease in water pressure created by the passage of a ship. Naval Ordnance Laboratory Report 498. Washington, D.C.: U.S. Navy Yard. February, NOL.

Bardeen, J., and J. M. Kendall. 1941. Gravitational fields. Naval Ordnance Laboratory Report 316. Washington, D.C.: U.S. Naval Gun Factory. September 11, NOL.

Bardeen, J., and A. Nier. 1941. The production of concentrated carbon 13 by thermal diffusion. Journal of Chemical Physics 9:690–692.

Bardeen, J., and D. Pines. 1955. Electron-phonon interaction in metals. Physical Review 99:1140–1150.

Bardeen, J., and J. R. Schrieffer. 1961. Recent developments in superconductivity. Progress in Low Temperature Physics 3: 170–287.

Bardeen, J., and G. H. Shortly. 1942. Firing areas for Mark 13 mine with M4 Mod 1 mechanism. Naval Ordnance Laboratory Report 605. Washington, D.C.: U.S. Navy Yard. May 28, NOL.

Bardeen, J. et al. 1942. Target areas of submarines, Naval Ordnance Laboratory Report 528. Washington, D.C.: U.S. Navy Yard. May 28, NOL.

Bardeen, J., and J. H. Van Vleck. 1939. Current in the Bloch approximation of "tight binding" for metallic electrons. Proceedings of the National Academy of Sciences (February):82–86.

Bardeen, J., L. N. Cooper, and J. R. Schrieffer. 1957a. Microscopic theory of superconductivity. Physical Review 106:162–164.

———. 1957b. Theory of superconductivity. Physical Review 108:1175–1204.

Bardeen, J., G. Baym, and D. Pines. 1962. Interactions between ^{3}He atoms in dilute solutions of ^{3}He and superfluid ^{4}He. Physical Review Letters 17(7):372–375.

Bardeen, J., W. Shockley, and W. Brattain. 1972. Interview by John L. Gregory, April 24, AT&T.

Bardeen family interview. 1992. Interview by Lillian Hoddeson, Michael Riordan, and Fernando Irving Elichirigoity, March 15.

Bardeen, W. 1993. The Barden-Bardeen Genealogy: A Revision of "The Barden Genealogy," a Manuscript by Leon R. Brown of Rochester, New York. Baltimore: Gateway Press.

———. 1995. Interview by Lillian Hoddeson, March 23.

———. 2000. Interview by Vicki Daitch, May 30.

Baym, G. 1992. Interview by Fernando Irving Elichirigoity, March 4.

Baym, G., and L. Kadanoff. 1962. Quantum Statistical Mechanics: Green's Function Methods in Equilibrium and Nonequilibrium Problems. New York: Benjamin.

Baym, G., C. J. Pethick, and D. Pines.1969. Superfluidity in neutron stars. Nature 224:673–674.

Belyaev, S. T. 1959. Effect of Pairing Correlations on Nuclear Properties. *Matematish-Fysiske Meddelelser Konglige Danske Videnskabernes Selskab* 31(11):1–35.

Benjamin, A. C. 1955. Operationalism. Springfield, Ill.: Charles C. Thomas.

Bennett, R. D. 1987. Putting science to sea in World War II: The development of the modern Naval Ordnance Laboratory. On the Surface (July 10), NOL.

Bernstein, J. 1984. Three Degrees Above Zero. New York: Charles Scribner's Sons.

Bethe, H. 1972. Interview by Charles Weiner, October 1966 to May 1972, AIP.

———. 1981. Interview by Lillian Hoddeson, April 29.

Bethe, H., and J. Bardeen. 1986. Back to Science Advisors. New York Times, Op-Ed page, May 17.

Bethell, J. T. 1998. Harvard Observed: An Illustrated History of the University in the Twentieth Century. Cambridge, Mass.: Harvard University Press.

Bhatt, R. 1992. Interview by Vicki Daitch, March 17.

Bleaney, B. 1982. John Hasbrouck Van Vleck. Pp. 627–665 in Biographical Memoirs of Fellows of the Royal Society, 28. London: The Royal Society.

Boden, M. A., ed. 1996. Dimensions of Creativity. Cambridge, Mass.: MIT Press.

Bohm, D. 1981. Interview by Lillian Hoddeson, May 8.

Bohr, A., B. R. Mottelson, and D. Pines. 1958. Possible analogy between the excitation spectrum of nuclei and those of the superconducting metallic state. Physical Review 110:936–998.

Bondyopadhyay, P. K. 1998. In the beginning. Proceedings of the IEEE 86(1):63–77.

Bone, D. 1989. The Emptiness of Genius: Aspects of Romanticism. Pp. 113–127 in P. Murray, ed. Oxford: Basil Blackwell.

Booth, E., ed. 1969. The Beginnings of the Nuclear Age. New York: Newcomen Society in North America.

Bradley, H. C. 1992. Charles R. Bardeen. Wisconsin Medical Alumni Magazine Quarterly 32(4):2–18.

Bradshaw, G. 1992. The airplane and logic of invention. Pp. 239–250 in Cognitive Models of Science, R. N. Giere, ed. Minneapolis: University of Minnesota Press.

Brattain, R. 1993. Interview by Lillian Hoddeson and Michael Riordan, February 20.

Brattain, W. n.d. How the transistor was named. Unpublished document, WHIT.

———. n.d. The saga of an expedition to Stockholm, Sweden, December, unpublished, WHIT.

———. 1947. BNB 18194, AT&T.

———. 1963. Interview by Lincoln Barnett, February 14, AT&T.

———. 1964. Interview by A. N. Holden and W. J. King, January, AIP.

———. 1974. Interview by Charles Weiner, May 28, AT&T.

Bray, J. 1993a. Interview by Vicki Daitch, February 8.

———. 1993b. Interview by Vicki Daitch, February 11.

Bridgman, P. 1927. The Logic of Modern Physics. New York: MacMillan.

———. 1936. The Nature of Physical Theory. Princeton, N.J.: Princeton University Press.

———. 1980. Reflections of a Physicist. New York: Arno Press.

Brinkley, D. 1988. Washington Goes to War. New York: Alfred Knopf.

Brinton, C. 1959. The Society of Fellows. Cambridge: Society for Fellows of Harvard University, distributed by Harvard University Press.

Brown, L. M., and L. Hoddeson, eds. 1983. The Birth of Particle Physics. New York: Cambridge University Press.

Brown, L., R. Brout, T. Y. Cao, P. Higgs, and Y. Nambu. 1997. Panel session: Spontaneous breaking of symmetry. Pp. 478–522 in The Rise of the Standard Model: Particle Physics in the 1960s and 70s, L. Hoddeson, L. M. Brown, M. Dresden, and M. Riordan, eds. New York: Cambridge University Press.

Brown, S., and G. Grüner. 1994. Charge and spin density waves. Scientific American (April):51–56.

Buderi, R. 1996. The Invention That Changed the World: How a Small Group of Radar Pioneers Won the Second World War. New York: Simon and Schuster.

Burnham, J. C., J. D. Buenker, and R. M. Crowden. 1976. Progressivism. Cambridge, Mass.: Schenkman Publishing.

Burtis, L. 1995. Interview by Vicki Daitch, April 21.

Campbell, D. 1998. Interview by Lillian Hoddeson, March 27.

Chandler, A. D., Jr. 1977. The Visible Hand. Cambridge, Mass.: Belknap Press of Harvard.

Chang, I. 1988. To scientists he's an Einstein. To the public he's—John Who? Chicago Tribune. January 3, clipping, BFC.

Chase, W. G., and H. Simon 1973. Perception in chess. Cognitive Psychology 4:55–81.

Chi, M. T. H., P. J. Feltovich, and R. Glaser. 1981. Categorization and representation of physics problems by experts and novices. Cognitive Science 5:121–152.

Clark, J. n.d. Wisconsin High's 'thorough courses' Made it Model School. Newspaper clipping, WHS.

Clark, P. F. 1967. The University of Wisconsin Medical School: A Chronicle, 1848–1948. Madison: University of Wisconsin Press.

Cohen, M. H., L. M. Falicov, and J. C. Phillips. 1962. Superconducting tunneling. Physical Review Letters 8(8):316–318.

Coleman, P. 1992. Interview by Vicki Daitch, February 7.

Condon, E. U., and G. H. Shortley. 1935. The Theory of Atomic Spectra. Cambridge, England: Cambridge University Press.

Cooper, L. 1956. Bound electron pairs in a degenerate Fermi gas. Physical Review 104:1189–1190.

———. 1987. Origins of the theory of superconductivity. IEEE Transactions on Magnetics MAG-23(2)(March 1987):376–379.

———. 2000. Interview by Lillian Hoddeson, August 4.

Crosby, A. 1976. Epidemic and Peace, 1918. Westport, Conn.: Greenwood Press.

———. 1977. The influenza pandemic of 1918. Pp. 5–13 in Influenza in America, 1918–1976: History, Science and Politics, J. Osborn, ed. New York: Prodist.

Csikszentmihalyi, M. 1990. Flow: The Psychology of Optimal Experience. New York: HarperCollins.

———. 1996. Creativity: Flow and the Psychology of Discovery and Invention. New York: HarperCollins.

Curti, M., and V. Carstensen. 1949. The University of Wisconsin, a History: 1848–1925, Vol. 2. Madison: University of Wisconsin Press.

Dahl, P. F. 1992. Superconductivity: Its Historical Roots and Development from Mercury to the Ceramic Oxides. New York: American Institute of Physics.

Davies, M. 1970. Peter Joseph Wilhelm Debye. Pp. 175–232 in Biographical Memoirs of Fellows of the Royal Society, 16. London: The Royal Society.

De Bono, E. 1992. Serious Creativity: Using the Power of Lateral Thinking to Create New Ideas. New York: HarperCollins.

DeBrunner, P. 1992. Interview by Vicki Daitch, February 12.

Del Guercio, G., R. Hudson, I. Flatow, and E. Augenbraun. 1998. Transistorized! KTCA-ScienCentral. Public Broadcasting System, Special Broadcast, November 1999.

Dessauer, J. 1971. My Years With Xerox: The Billions Nobody Wanted. Garden City, N.J.: Doubleday.

———. 1992. Interview by Lillian Hoddeson, July 28.

Dessauer, J., and E. Pell. 1992. Interview by Lillian Hoddeson, July 28.

de Wette, F. 1993. Interview by Vicki Daitch, November 12.

Dirac, P. A. M. 1930. The Principles of Quantum Mechanics. Oxford: Clarendon Press.

Dribin, A. 1997. John Bardeen and Xerox: Bridging the gap between science and the empirical art. Senior thesis for History 296. University of Illinois, Urbana.

Drickamer, H. 1992. Interview by Fernando Irving Elichirigoity, January 10.

Duke, C. B. 1992. Consultant to industry, advisor to government. Physics Today 45(April):56–62.

———. 1993a. Interview by Vicki Daitch, November 9.

———. 1993b. Interview by Vicki Daitch, December 2.

Dunbar, K. 1997. How scientists think: On-line creativity and conceptual change in science. Pp. 461–494 in Creative Thought: An Investigation of Conceptual Structures and Processes, T. B. Ward, S. M. Smith, and J. Vaid, eds. Washington, D.C.: American Psychological Association.

Eckert, M., H. Schubert, and G. Torkar, with C. Blondel and P. Quédec. 1992. The roots of solid-state physics. Pp. 3-87 in Out of the Crystal Maze, L. Hoddeson, E. Braun, J. Teichmann, and W. Weart, eds. New York: Oxford University Press.

Eddington, A. S. 1923. The Mathematical Theory of Relativity. Cambridge, England: Cambridge University Press.

Einstein, A. 1949. Autobiographical notes. Pp. 3–95 in Albert Einstein: Philosopher-Scientist, P. A. Schilpp, ed. New York: HarperTorch-book.

Elkins, T. A. n.d. A brief history of Gulf's geophysical prospecting. Gulf Research and Development Corporation Archives, Pamphlet, Senator John Heinz Pittsburgh Regional History Center, Pittsburgh, Pennsylvania.

Ericsson, K. A., R. T. Krampe, and T. R. Clemens. 1993. The role of deliberate practice in expert performance. Psychological Review 100:363–406.

Fagen, M. D., ed. 1975. A History of Science and Engineering in the Bell System: The Early Years (1875–1925). Murray Hill, N.J.: Bell Telephone Laboratories.

———. 1978. A History of Science and Engineering in the Bell System: National Security in War and Peace (1925–1975). Murray Hill, N.J.: Bell Telephone Laboratories.

Falk, V. S. 1992. Dean Bardeen. The Wisconsin Medical Alumni Magazine Quarterly. 32(4):19, 20, 22–23.

Feist, G. J. 1999. The Influence of Personality on Artistic and Scientific Creativity, in Handbook of Creativity, R. Sternberg, ed. Cambridge, England: Cambridge University Press.

Feist, G. J., and M. E. Gorman. 1998. The psychology of science: Review and integration of a nascent discipline. Review of General Psychology 2(1):3–47.

Feynman, R. 1957. Superfluidity and superconductivity. Reviews of Modern Physics 29(2)(April):205–212.

———. 1985. "Surely You're Joking, Mr. Feynman": Adventures of a Curious Character. New York: W. W. Norton.

Filene, P. 1970. An obituary for 'the progressive movement.' American Quarterly 22(Spring):20–34.

Fisk, James. 1976. Interview by Lillian Hoddeson and Alan Holden, June 24, AIP.

Fleming, D., and B. Bailyn, eds. 1969. The Intellectual Migration: Europe and America, 1930–1960. Cambridge, Mass.: Harvard University Press, Belknap Press.

Flexner, A. 1960. An Autobiography. New York: Simon and Schuster.

Forman, P. 1969. The discovery of the diffraction of X rays by crystals: A critique of the myths. Archive for History of Exact Sciences 6:38–71.

Frauenfelder, H. 1992. Interview by Fernando Irving Elichirigoity, February 28.

Friedman, A. J., and C. C. Donley. 1985. Einstein as Myth and Muse. Cambridge, England: Cambridge University Press.

Fröhlich, H. 1950. Isotope effect in superconductivity. Letter. Proceedings of the Physical Society (London) A63:778.

———. 1954. On the theory of superconductivity: The one- dimensional case. Proceedings of the Royal Society of London A223:296–304.

Gallo, C. 1994a. Interview by Vicki Daitch, April 24.

———. 1994b. Interview by Vicki Daitch, May 3.

Galton, F. 1869. Hereditary Genius: An Inquiry into Its Laws and Consequences. London: Macmillan.

———. 1874. English Men of Science. London: Macmillan.

———. 1883. Inquiries into Human Faculties. London: Macmillan.

Gardner, H. 1985. Frames of Mind: The Theory of Multiple Intelligences. New York: Basic Books.

———. 1993. Creating Minds: An Anatomy of Creativity Seen Through the Lives of Freud, Einstein, Picasso, Stravinsky, Eliot, Graham, and Gandhi. New York: Basic Books.

Gavroglu, K. 1995. Fritz London: A Scientific Biography. New York: Cambridge University Press.

Geison, G. L. 1995. The Private Science of Louis Pasteur. Princeton, N.J.: Princeton University Press.

Gentner, D. 1989. The mechanisms of analogical learning. Pp. 199–241 in Similarity and Analogical Reasoning,S. Vosniadou and A. Ortony, eds. New York: Cambridge University Press.

Ghiselin, B., ed. 1952. The Creative Process. New York: Mentor.

Gholson, B., W. R. Shadish, R. A. Neimeyer, and A. C. Houts. 1989. Psychology of Science: Contributions to Metascience. New York: Cambridge University Press.

Giaver, I. 1974. Electron tunneling and superconductivity. Reviews of Modern Physics 46(2):245–250.

Giaver, I., and C. Bean. 1995. Interview by Vicki Daitch, May 5.

Giere, R. N., ed. 1992. Cognitive Models of Science. Minnesota Studies in the Philosophy of Science. Minneapolis: University of Minnesota Press.

Gilbarg, D. 1992. Interview by Vicki Daitch, July 14.

Gilbert, M. 1989. The Second World War: A Complete History. New York: Henry Holt.

Ginsberg, D. 1993a. Interview by Vicki Daitch and Lillian Hoddeson, December 13.

———. 1993b. Interview by Vicki Daitch and Lillian Hoddeson, December 16.

Ginzburg, V. L., and D. A. Kirzhnits. 1964. On the superfluidity of neutron stars. Zhurnal Eksperimental'noi i Teoreticheskoi Fiziki 47:2006–2007. (Translated into English in 1965. Sov. Phys. JETP 20:1346–1348).

Gleick, J. 1992. Genius: The Life and Science of Richard Feynman. New York: Pantheon Books.

Glover, J. A., R. R. Ronning, and C. R. Reynolds, eds. 1989. Handbook of Creativity. New York: Plenum.

Goldman, J. 2000. Interview by Vicki Daitch, February 8.

Goldstein, A. 1993. Finding the right material: Gordon Teal as inventor and manager. Pp. 93–126 in Sparks of Genius: Portraits of Electrical Engineering Excellence, F. Nebeker, ed. New York: IEEE Press.

Goldwasser, N. 1992. Interview by Vicki Daitch, Febuary 13.

Gopnik, A., A. N. Meltzoff, and P. K. Kuhl. 1999. The Scientist in the Crib: What Early Learning Tells Us About the Mind. New York: HarperCollins.

Gorman, M. E. 1992. Simulating Science: Heuristics, Mental Models and Technoscientific Thinking. Bloomington: Indiana University Press.

Graham, D., with G. Yocom. 1990. Mental Toughness Training for Golf. New York: Steven Green Press.

Graham, O., Jr. 1971. The Great Campaigns: Reform and War in America, 1900–1928. Englewood Cliffs, N.J.: Prentice-Hall.

Gray, J. 1951. The University of Minnesota, 1851–1951. Minneapolis: University of Minnesota Press.

Greytak, B. 2000a. Interview by Vicki Daitch, May 10.

———. 2000b. Interview by Vicki Daitch, May 24.

Greytak, T. 1993. Interview by Vicki Daitch, August 6.

Gruber, H. 1981. Darwin on Man: A Psychological Study of Scientific Creativity. Chicago: University of Chicago Press.

Grüner, G. 1983. Charge density wave transport in linear chain compounds. On Solid State Physics. Comments on Modern Physics: Part B 10(6):183–197.

———. 2000. Interview by Lillian Hoddeson, October 3.

Guerlac, H. E. 1987. Radar in World War II. New York: American Institute of Physics/Tomash.

Guilford, J. P. 1950. Creativity. American Psychologist 5:444–454.

Hackett, D. 1989. Sony gift honors Nobel laureate who's still working quietly at UI. Clipping. Champaign-Urbana News Gazette. October 6, BFC.

Hadamard, J. 1945. An Essay on the Psychology of Invention in the Mathematical Field. Princeton: Princeton University Press.

Hames, J. 1992. Interview by Vicki Daitch, January 13.

Handel, K. 1998. The uses and limits of theory: From radar research to the invention of the transistor. Paper presented at the annual meeting of the History of Science Society, Kansas City, Missouri, October 21–25, 1998.

Handler, P. 1993. Interview by Vicki Daitch, December 22.

Hansen, H., ed. 1974. Illinois: A Descriptive and Historical Guide. New York: Hastings House.

Harmer, A. 1936. Teaching Report. Pp. 335–336 in The Dewey School: The Laboratory School of the University of Chicago, 1896–1903, K. C. Mayhew and A. C. Edwards, eds. New York: D. Appleton-Century.

Hart, D. M. 1998. Forged Consensus: Science, Technology, and Economic Policy in the United States, 1921–1953. Princeton, N.J.: Princeton University Press.

Hart, W. W., C. Gregory, and V. Schult. 1945. Mathematics in Daily Use. Boston: D. C. Heath.

Hartmann, G. 1992. Interview by Vicki Daitch, July 28.

Harvard Society of Fellows. 1987. Videotaped conversation, John Bardeen and Ed Purcell, with Carrol Williams. Cambridge, Mass.: Harvard Society of Fellows. Videotape. May.

Hays, S. P. 1959. Conservation and the Gospel of Efficiency: The Progressive Conservation Movement. Cambridge, Mass.: Harvard University Press.

Henschel, A. B. 1992a. Interview by Vicki Daitch, January 13.

———. 1992b. At home with Charles Bardeen. The Wisconsin Alumni Magazine Quarterly 32(4):19–20.

———. 1995. Interview by Vicki Daitch, April 17.

Herken, G. 1992. Cardinal Choices: Presidential Science Advising from the Atomic Bomb to SDI. New York: Oxford University Press.

Herring, C. n.d. On the wisdom of cancelling a meeting in a "disapproved" city. Unpublished, UIUC-A.

———. 1974. Interview by Lillian Hoddeson, July 23, July 29, October 31.

———. 1992a. Interview by Vicki Daitch, March 16.

———. 1992b. Recollections from the early years of solid-state physics. Physics Today (April):26–33.

Herring, L., and C. Herring. 2000. Interview by Lillian Hoddeson, July 12.

Hess, K. 1991. Interview by Fernando Irving Elichirigoity and Vicki Daitch, October 1.

Hiltzik, M. 1999. Dealers of Lightning: Xerox PARC and the Dawn of the Computer Age. New York: HarperCollins.

Hoch, P. K. 1993. The reception of Central European refugee physicists of the 1930s: USSR, UK, USA. Annals of Science 40:217–246.

Hoddeson, L. 1981a. The discovery of the point-contact transistor. Historical Studies in the Physical Sciences 12(1):41–76.

———. 1981b. The entry of the quantum theory of solids into Bell Telephone Laboratories, 1925–40: A case-study of the industrial application of fundamental science. Minerva 18(3):422–447.

———. 1981c. The emergence of basic research in the Bell Telephone system, 1875–1915. Technology and Culture 22:512–544.

———. 1987. The first large-scale application of superconductivity: The Fermilab energy doubler, 1972–1983. Historical Studies in the Physical and Biological Sciences 18(1):25–54.

———. 1994. Research on crystal rectifiers during World War II, and the invention of the transistor. History and Technology. 11:121–130.

———. 2001. Toward a history-based model for scientific invention: Problem-solving practices in the invention of the transistor and the development of the theory of superconductivity. Paper presented at Model-Based Reasoning: Scientific Discovery, Technological Innovation, Values Conference, University of Pavia, Italy, May 17.

Hoddeson, L., and G. Baym. 1980. The development of the quantum mechanical electron theory of metals: 1900–1928. Proceedings of the Royal Society of London A371:8–23.

Hoddeson, L., and M. Riordan. 1997. Minority carriers and the first two transistors. Pp. 1–33 in Facets: New Perspectives on the History of Semiconductors, A. Goldstein and W. Aspray, eds. New Brunswick, N.J.: IEEE Center for the History of Electrical Engineering.

———. 2001. The electron, the hole, and the transistor. Pp. 327–338 in Histories of the Electron: The Birth of Microphysics, J. Z. Buchwald and A. Warwick, eds. Cambridge, Mass.: MIT Press.

Hoddeson, L., G. Baym, and M. Eckert. 1987. The development of the quantum-mechanical theory of metals, 1928–1933. Reviews of Modern Physics 59(1):287–327.

Hoddeson, L., E. Braun, J. Teichmann, and S. Weart, eds. 1992. Out of the Crystal Maze. New York: Oxford University Press.

Hoddeson, L., P. Henriksen, R. Meade, and C. Westfall. 1993. Critical Assembly: A History of Los Alamos During the Oppenheimer Years, 1943–1945. New York: Cambridge University Press.

Hoddeson, L., L. M. Brown, M. Riordan, and M. Dresden, eds. 1997. The Rise of the Standard Model: Particle Physics in the 1960s and 1970s. New York: Cambridge University Press.

Hofstadter, R. 1955. The Age of Reform: From Bryan to FDR. New York: Vintage Books.

Holmes, F. L. 1985. Lavoisier and the Chemistry of Life: An Exploration of Scientific Creativity. Madison: University of Wisconsin Press.

Holonyak, N. 1991a. Interview by Susan Stamberg. Morning Edition. National Public Radio, January 31.

———. 1991b. Interview by Lillian Hoddeson and Fernando Irving Elichirigoity, May 29.

———. 1992. John Bardeen and the point-contact transistor. Physics Today 45(4):36–43.

———. 1993a. Interview by Vicki Daitch, January 21.

———. 1993b. Interview by Lillian Hoddeson and Michael Riordan, July 30.

———. 1993c. John Bardeen: 1908–1991. Pp. 3–12, Memorial Tributes, National Academy of Engineering of the United States of America, vol. 6. Washington, D.C.: National Academy Press.

———. 1998a. Interview by Lillian Hoddeson, June 12.

———. 1998b. Interview by Lillian Hoddeson and Vicki Daitch, August 6.

————. 2000. Interview by Vicki Daitch, December 6.

————. 2001. Interview by Vicki Daitch, February 4.

Holton, G. 1973. Thematic Origins of Scientific Thought. Cambridge, Mass.: Harvard University Press.

————. 1978. The Scientific Imagination: Case Studies. Cambridge, England: Cambridge University Press.

Homans, G. C., and O. T. Bailey. 1959. The Society of Fellows, Harvard University, 1933–1947. Pp. 1–37 in The Society of Fellows, C. Brinton, ed. Cambridge, Mass.: The Society of Fellows of Harvard University.

Hone, D. 1993. Interview by Vicki Daitch, March 16.

Howe, M. J. A. 1999. Genius Explained. Cambridge, England: Cambridge University Press.

Hoxie, R. G. 1990. Eisenhower and "my scientists." National Forum (Fall):9–12.

Huff, H. R. 2001. John Bardeen and Transistor Physics. In Characterization and Metrology for ULSI Technology, A.C. Diebold et al., eds. College Park, MD: American Institute of Physics.

Hughes, T. P. 1987. American Genesis: A Century of Invention and Technological Enthusiasm, 1870–1970. New York: Viking.

Ihde, A. 1990. Chemistry as Viewed from Bascom's Hill: A History of the Chemistry Department at the University of Wisconsin in Madison. Madison: Wisconsin University Press.

Inkson, J. C., and P. W. Anderson. 1973. Comment on "Model for an exciton mechanism of superconductivity." Physical Review B 8(9):4433–4434.

Jones, H., N. F. Mott, and H. W. B. Skinner. 1934. A theory of the form of the X-ray emission bands and metals. Physical Review 45:379–384.

Josephson, B. D. 1962. Possible new effects in superconductive tunneling. Physics Letters 1(7):251.

————. 1966. Relation between the superfluid density and order parameter for superfluid He near T. Physics Letters 21:608–609.

————. 1974. The discovery of tunneling supercurrents. Reviews of Modern Physics 46(2):251–254.

Kanigal, R. 1991. The Man Who Knew Infinity: A Life of the Genius Ramanujan. New York: Scribner's Sons.

Kaplan, L. 1946. The History of the Naval Ordnance Laboratory, 1918–1945, Part 1, Administrative History. Naval Ordnance Laboratory Report 1000. Washington, D.C.: U.S. Navy.

Keegan, John. 2000. His finest hour. U.S. News and World Report (May 29):50–57.

Keller, E. F. 1983. A Feeling for the Organism. San Francisco: Freeman.

Kelly, M. J. 1943. A first record of thoughts concerning an important postwar problem of the Bell Telephone Laboratories and Western Electric Company, May 1, 1943, AT&T.

———. 1950. The Bell Telephone Laboratories—An Example of an Institute of Creative Technology. Proceedings of the Royal Society of London 203a:287–301.

Kevles, B. H. 1997. Naked to the Bone: Medical Imaging in the Twentieth Century. Reading, Mass.: Helix Books, Addison-Wesley.

Kevles, D. J. 1978. The Physicists: The History of a Scientific Community in Modern America. New York: Knopf.

Keyworth, G. 2000. Interview by Lillian Hoddeson, March 12.

Kingery, R. A., R. D. Berg, and E. H. Schillinger. 1967. Men and Ideas in Engineering. Urbana: University of Illinois Press.

Kleppner, P. 1970. The Cross of Culture: Social Analysis of Midwestern Politics, 1850–1900. New York: Free Press.

Kragh, H. 1990. Dirac: A Scientific Biography. New York: Cambridge University Press.

Kuhn, T. S. 1959. Energy conservation as an example of simultaneous discoveries. Pp. 321–356 in Critical Problems in the History of Science, M. Clagett, ed. Madison: University of Wisconsin Press.

———. 1962 and 1970 (2nd edition). The Structure of Scientific Revolutions. Chicago: University of Chicago Press.

Kulkarni, D., and H. A. Simon. 1988. The processes of scientific discovery: The strategies of experimentation. Cognitive Science 12:139–175.

Lakoff, G., and M. Johnson. 1980. Metaphors We Live By. Chicago: University of Chicago Press.

Landau, L. D. 1957. The Theory of a Fermi Liquid. *Zhurnal Eksperimental'noi i Teoreticheskoi Fiziki* 30(1956):1058–1064 (trans. Soviet Physics JETP 3[1957]:920–925).

Langley, P., H. Simon, G. Bradshaw, and J. Zytkow. 1987. Scientific Discovery: Computational Explorations of the Creative Processes. Cambridge, Mass.: MIT Press.

Lazarus, D. 1987. The Loomis Legacy. Urbana: Department of Physics, University of Illinois, UIUC-P.

———. 1992. Interview by Fernando Irving Elichirigoity, February 24.

———. 1993. Interview by Fernando Irving Elichirigoity, February 11.

Lederman, L., with D. Teresi. 1993. The God Particle: If the Universe Is the Answer, What Is the Question? New York: Houghton Mifflin.

Lee, T. D., F. E. Low, and D. Pines. 1953. The motion of slow electrons in a polar crystal. Physical Review 90:297–302.

Leggett, A. J. 1975. Theoretical description of the new phases of liquid ^{3}He. Reviews of Modern Physics 47:331–414.

———. 1993. Interview by Vicki Daitch and Lillian Hoddeson, November 30.

Leonard, N. 1992. Interview by Vicki Daitch, April 17.

Lewis, S. M. 1991. Xerox Corporation in International Directory of Company Histories Chicago: St. James Press, pp. 171–173.

Link, A. 1954. Woodrow Wilson and the Progressive Era, 1910–1917. New York: Harper & Bros.

Link, A. S., and R. L. McCormick. 1983. Progressivism. Arlington Heights, Illinois: Harlan Davidson.

Lombroso, C. 1891. The Man of Genius. London: Walter Scott.

London, F. 1935. Macroscopic interpretation of superconductivity. Proceedings of the Royal Society of London A152:24–54.

———. 1950. Superfluids, Vol. 1: Macroscopic theory of Superconductivity. New York: John Wiley.

London, F., and H. London. 1933. Supraleitung und diamagnetismus. Physica 2:341–354.

Luttinger, J. M. 1960. Theory of the Fermi surface. Pp. 2–8 in The Fermi Surface, W. A. Harrison and M. B. Webb, eds. New York: John Wiley.

Lyding, J. 2000. Interview by Lillian Hoddeson, October 17.

Mapother, D. 1993. Interview by Vicki Daitch, November 24.

Margenau, H. 1956. Interpretations and misinterpretations of operationalism. In The Validation of Scientific Theories, 3rd ed., P. Frank, ed. Boston: Beacon Press.

Markowitz, J. 1976. The Gulf Oil Corporation—How it all began 75 years ago. Pittsburg Post-Gazette. January 15, p. 4.

Marlowe, D. 1992. Interview by Vicki Daitch, July 26.

Marshall, E. 1985. Keyworth quits White House post. Science (December 13):1250.

Maxwell, E. 1983. Interview by Gordon Baym, March 4.

Mayhew, K. C., and A. C. Edwards. 1936. The Dewey School: The Laboratory School of the University of Chicago, 1896–1903. New York: D. Appleton-Century.

McAfee, J. 1977. From Spindletop to total energy: Gulf and the smaller business community. The Smaller Manufacturer (July/August):12.

McCarthy, C. 1912. The Wisconsin Idea. New York: Macmillan.

McCormick, R. L. 1981. The discovery that business corrupts politics: A reappraisal of the origins of progressivism. American Historical Review 86:247–274.

McDonald, D. G. 2000. The Nobel laureate vs. the graduate student: John Bardeen and Brian Josephson debate superconductive tunneling. Unpublished manuscript, UIUC-P.

———. 2001. The Nobel laureate vs. the graduate student: John Bardeen and Brian Josephson debate superconductive tunneling. Physics Today (July):46–51.

McDonald, D. G., and R. Kautz. 1999. Interview by Lillian Hoddeson, March 24.

McKay, K. 1992. Interview by Lillian Hoddeson and Michael Riordan, September 26.

McMillan, J. 1995. Interview by Vicki Daitch, April 14.

Medin, D. L., and B. H. Ross. 1996. Cognitive Psychology. Fort Worth, Texas: Harcourt Brace College Publishers.

Meek, Walter J. 1935. Charles R. Bardeen obituary. Science (December 27):606–607.

Migdal, A. B. 1959. Superfluidity and the moments of inertia of nuclei. Nuclear Physics 13:655–674. (Also, 1959. Zh. Exp. Theor. Fiz. 37:249.)

Miller, A. 1990. The Untouched Key: Tracing Childhood Trauma in Creativity and Destructiveness. Translated by H. Hannum and H. Hannum. New York: Doubleday.

Miller, A. I. 1984. Imagery in Scientific Thought: Creating Twentieth- Century Physics. Boston: Birkhauser.

———. 2000. Insights of Genius: Imagery and Creativity in Science and Art. New York: Springer-Verlag.

———. 2001. Einstein, Picasso: Space, Time, and the Beauty that Causes Havoc. New York: Basic Books.

Miller, J. 1993. Interview by Vicki Daitch, March 8.

Mitchell, B. 1987. Nobel Prize winner Bardeen keynotes symposium. On the Surface 10(July 10).

Monceau, P., N. D. Ong, and A. M. Portis. 1976. Electric field breakdown of charge-density-wave-induced anomalies in $NbSe^3$. Physical Review Letters 37:(10)(September 6):602 –606.

Montgomery, S. L. 1994. Minds for the Making: The Role of Science in American Education, 1750–1990. New York: Guilford Press.

Morrison, G. B. 1995. Interview by Vicki Daitch, April 13.

Morse, P. M. 1977. In at the Beginnings: A Physicist's Life. Cambridge, Mass.: MIT Press.

Mott, N., ed. 1980. The Beginnings of Solid State Physics. London: The Royal Society.

Mott, N., and H. Jones. 1936. The Theory of the Properties of Metals and Alloys. New York: Dover Publications.

Murray, P., ed. 1989. Genius: The History of an Idea. Oxford: Basil Blackwell.

Nakajima, S. 1993. Interview by Vicki Daitch, August 4.

Nambu, Y., and G. Jona-Lasinio. 1961. Dynamical model of elementary particles based on an analogy with superconductivity, I. Physical Review 122:345–358.

Nersessian, N. J. 1984. Faraday to Einstein: Constructing Meaning in Scientific Theories. Dordrecht, the Netherlands: Martinus Nijhoff.

———1986. The Process of Science: Contemporary Philosophical Approaches to Understanding Scientific Practice. Dordrecht, the Netherlands: Martinus Nijhoff.

Nesbit, R. C., and W. F. Thompson. 1989. Wisconsin: A History. Madison: University of Wisconsin Press.

Nettleton, L. L. 1940. Geophysical Prospecting for Oil. New York: McGraw-Hill.

Newell, A., and H. Simon. 1972. Human Problem-Solving. Englewood Cliffs, N.J.: Prentice-Hall.

Nickles, T., ed. 1980. Scientific Discovery: Case Studies. Boston Studies in the Philosophy of Science, Vol. 60. Dordrecht, the Netherlands: D. Reidel.

Nier, A. 1993. Interview by Vicki Daitch, October 5.

Nitzsche, J. C. 1975. The Genius Figure in Antiquity and the Middle Ages. New York: Columbia University Press.

Osheroff, D., R. C. Richardson, and D. M. Lee. 1972. Evidence for a new phase of solid ^{3}He. Physical Review Letters 28:885–888.

Osterhoudt, W. 1991. Interview by Vicki Daitch, November 16.

Pais, A. 1982. "Subtle is the Lord. . . .": The Science and the Life of Albert Einstein. New York: Oxford University Press.

Pake, G. 1992a. Interview by Fernando Irving Elichirigoity, March 17.

———. 1992b. Consultant to industry, adviser to government. Physics Today (April):56–62.

Pao, H. 1994. Interview by Lillian Hoddeson, January 11.

Patterson, J. 1987. The Dread Disease: Cancer and Modern American Culture. Cambridge, Mass.: Harvard University Press.

Pearson, G. 1976. Interview by Lillian Hoddeson, August 23.

Pearson, R., ed. 1992. Shockley on Eugenics and Race: The Application of Science to Human Problems. Washington, D.C.: Scott Townsend.

Peierls, R. E. 1929. Zur Theorie der galvanomagnetischen Effekte. Zeitschrift für Physik 53:255–266.

———. 1930. *Zur Theorie der elektrischen und thermischen Leit-fähigkeit von Metallen. Annalen der Physik* 4:121–148.

———. 1964. The Quantum Theory of Solids. Oxford: Clarendon Press.

Pell, E. 1992. Interview by Lillian Hoddeson, July 28.

———. 1995. Interview by Vicki Daitch, May 21.

Perkins, D. 1981. The Mind's Best Work. Boston: Harvard University Press.

Peters, L. J. 1949. The direct approach to magnetic interpretation and its practical applications. Geophysics 14(July):29–32.

Peters, L. J., and J. Bardeen. 1930. The solution of some theoretical problems which arise in electrical methods of geophysical exploration. Bulletin of the University of Wisconsin, Engineering Experiment Station Series No. 71, Monograph, Madison: University of Wisconsin.

———. 1932. Some aspects of electrical prospecting applied in locating oil structures. Physics 2(3):103–122.

Peters, P. 1972. Bardeen more than just a scientist. Champaign-Urbana Courier. October 22, clipping. BFC.

Peterson, L. 1981. John Bardeen: Quiet brilliance. Champaign-Urbana News Gazette. February 9, clipping. BFC.

Phillips, H. 1972. Superconductivity mechanisms and covalent instabilities. Physical Review Letters 29:23(December 4):1551–1554.

Pierce, J. 1992. Interview by Lillian Hoddeson and Michael Riordan, June 29.

Pines, D. 1981. Interview by Lillian Hoddeson, August 13 and 16.

———. 1993. Interview by Lillian Hoddeson and Vicki Daitch, December 3.

Pines, D., and J. R. Schrieffer. 1958. Gauge invariance in the theory of superconductivity. *Nuovo Cimento* 10:496–504.

Pines, D., and C. P. Slichter. 1972. The 1972 Nobel Prize for physics. Science (November 3): clipping, BFC.

Pippard, A. B. 1961. The cat and the cream. Physics Today 14(11):38–41.

Polleti, P. 1972. Professor Bardeen—Research top priority. Illinois Technograph Engineering Magazine (October):14–15.

Portnoy, W. 1993. Interview by Vicki Daitch, February 24.

Rabi, I. I. 1980. Dedication of the Loomis Laboratory of Physics. Urbana: Department of Physics, University of Illinois, February 14, UIUC-P.

Reimann, P., and M. Chi. 1989. Human expertise. Pp.161–191 in Human and Machine Problem Solving, K. Gilhooly, ed. New York: Plenum.

Rhodes, R. 1986. The Making of the Atomic Bomb. New York: Simon and Schuster.

Rigden, J. S. 1987. Rabi: Scientist and Statesman. New York: Basic Books.

Riordan, M., and L. Hoddeson. 1997a. Crystal Fire: The Birth of the Information Age. New York: W. W. Norton.

———. 1997b. The origins of the p-n junction. IEEE Spectrum 34(6):46–51.

Riordan, R., L. Hoddeson, and C. Herring. 1999. The invention of the transistor. Reviews of Modern Physics 71:S336–S345.

Riordan, R., L. Hoddeson, A. Kolb, R. Jacobs, and G. Sandiford. Forthcoming. Tunnel Visions: The Rise and Fall of the Superconducting Supercollider.

Robertson, J. K. 1974. St. Andrews: Home of Golf. Fife, Scotland: Citizen Office.

Rodgers, D. T. 1982. In search of progressivism. Reviews in American History 10:113–132.

Rowell, J. M. 1963. Magnetic field dependence of the Josephson tunnel current. Physical Review Letters 11(5):200–201.

Rowland, B., and W. B. Boyd. 1953. U.S. Navy Bureau of Ordnance in World War II. Washington, D.C.: Government Printing Office.

Runco, M. A., and R. S. Albert, eds. 1990. Theories of Creativity. Newberry Park, Calif.: Sage Publications.

Russell, G. 1995. Interview by Vicki Daitch, April 17.

Sah, C. T. 1993. Interview by Vicki Daitch, April 1.

Salamon, M. 2000. Interview by Lillian Hoddeson, October 6.

Salamon, M., and J. Bardeen. 1987. Comment on Bulk Superconductivity at 91K° in Single-Phase Oxygen-Deficient Perovskite $Ba_2YCu_3O_{9-\Delta}$. Physical Review Letters 59(22)(November 30): 2615–2616.

Sapolsky, H. 1990. Science and the Navy: The History of the Office of Naval Research. Princeton, N.J.: Princeton University Press.

Sayen, J. 1985. Einstein in America: The Scientist's Conscience in the Age of Hitler and Hiroshima. New York: Crown Press.

Schaffer, S. 1990. Genius in Romantic natural philosophy. Pp. 82–98 in Romanticism and the Sciences, A. Cunningham and N. Jardine, eds. Cambridge, England: Cambridge University Press.

Schapiro, S. 1963. Josephson currents in superconducting tunneling: The effect of microwaves and other observations. Physical Review Letters 11(2):80–82.

Schmidt, R. 1995. Interview by Vicki Daitch, April 24.

Schrieffer, R. 1974. Interview by Joan Warnow and Robert Williams, September 26, AIP.

———. 1992a. Interview by Vicki Daitch, March 16.

———. 1992b. John Bardeen and the theory of superconductivity. Physics Today 45(4) (April):46–53.

Schweber, S. 1986. The empiricist temper regnant: Theoretical physics in the United States, 1920–1950. Historical Studies in the Physical and Biological Sciences 17(1):55–98.

———. 1994. QED and the Men Who Made It: Dyson, Feynman, Schwinger, and Tomonaga. Princeton, N.J.: Princeton University Press.

Seaborg, G. T. 1995. Interview by Kristine Fowler, May 2.

———. n.d. National service with ten presidents of the United States. Lawrence Berkeley Laboratory Report LBL-704. Berkeley, Calif.: Lawrence Berkeley Laboratory Technical Information Department.

Seitz, F. 1940. The Modern Theory of Solids. New York: McGraw-Hill.

———. 1980. Biographical notes. Pp. 84–99 in The Beginnings of Solid State Physics, N. Mott, ed. London: The Royal Society.

———. 1981a. Interview by Lillian Hoddeson, January 26.

———. 1981b. Interview by Lillian Hoddeson, March 3.

———. 1991. Interview by Lillian Hoddeson, December 28.

———. 1991. Francis Wheeler Loomis. National Academy of Sciences: Biographical Memoirs 60:119.

———. 1992. Interview by Vicki Daitch, March 16.

———. 1992a. The Princeton years and beyond—1930–45. Talk at 1992 March meeting of the American Physical Society, Indianapolis. Also published in Nuovo Cimento della Society Italiana D-Condensed Matter, Atomic Molecular & Chemical Physics Fluids, Plasma, Biophysics, 15D, ser 1(2–3)(Feb.–March 1993):131–138.

———. 1992b. Interview by Lillian Hoddeson and Michael Riordan, September 26.

———. 1993. Interview by Lillian Hoddeson, Vicki Daitch, and Irving Elichirigoity, April 22.

———. 1994. On the Frontier: My Life in Science. New York: AIP Press.

———. 1997. Interview by Gino Del Guercio, c. September. UIUC-P.

———. 1998. Interview by Vicki Daitch and Lillian Hoddeson, March 24.

———. 1999. Letter to Lillian Hoddeson, November 11. UIUC-P.

———. 2000. Interview by Lillian Hoddeson, September 17.

———. 2001. Interview by Lillian Hoddeson, April 10.

Seitz, F., and N. Einspruch. 1998. Electronic Genie: The Tangled History of Silicon. Urbana: University of Illinois Press.

Shapiro, S. 1963. Josephson currents in superconducting tunnelling: The effect of microwaves and other observations. Physical Review Letters 11(2):80–82.

Shockley, E. 1994. Interview by Lillian Hoddeson and Michael Riordan, January 13.

Shockley, W. 1939. On the surface states associated with a periodic potential. Physical Review 56:317–323.

———. 1950. Electrons and Holes in Semiconductors, with Applications to Transistor Electronics. New York: Van Nostrand.

———. 1974. Interview by Lillian Hoddeson, September 10.

———. 1976. The path to the conception of the junction transistor. IEEE Transactions on Electronic Devices ED-23, 7:597–620.

Shoenberg, D. 1938, 1952. Superconductivity. Cambridge, England: Cambridge University Press.

Shortley, G. H. 1935. The Theory of Atomic Spectra. Cambridge, England: Cambridge University Press.

Simmons, R. 1993. Interview by Lillian Hoddeson and Vicki Daitch, December 6.

Simon, H. A. 1978. Information-processing theory of human problem solving. Pp. 271–295 in Handbook of Learning and Cognitive Process, Vol. 5, W. K. Estes, ed. Hillsdale, N.J.: Erlbaum.

Simon, R. 1973. John Bardeen invented the transistor and fathered the theory of superconductivity but he's still not a household word. Chicago Sun Times Midwest Magazine. March 18, clipping, BFC.

Simonton, D. 1988. Scientific Genius: A Psychology of Science. Cambridge, England: Cambridge University Press.

———. 1990. Psychology, Science, and History: A Treatise on Their Convergence. New Haven, Conn.: Yale University Press.

———. 1999. Origins of Genius: Darwinian Perspectives on Creativity. New York: Oxford University Press.

Slater, J. C. 1975. Solid-State and Molecular Theory: A Scientific Biography. New York: John Wiley & Sons.

Slichter, C. 1992. Interview by Fernando Irving Elichirigoity, February 26.

———. 1995a. Interview by Vicki Daitch and Kristine Fowler, April 18.

———. 1995b. Interview by Vicki Daitch and Kristine Fowler, May 2.

———. 1995c. Interview by Vicki Daitch, May 10.

Slossum, E. E. 1920. Creative Chemistry: Description of Recent Achievements in the Chemical Industries. New York: Century.

Smaldane, J. P. 1977. History of the White Oak Laboratory, 1945–1975. Silver Spring, Md.: Naval Surface Weapons Center, White Oak.

Smith, B. L. R. 1990. American Science Policy Since World War II. Washington, D.C.: The Brookings Institution.

Smith, C. S. 1965. The prehistory of solid-state physics. Physics Today 18(12)18–30.

Smith, D. K., and R. C. Alexander. 1988. Fumbling the Future; How Xerox Invented, Then Ignored, the First Personal Computer. New York: William Morrow.

Sommerfeld, A., and H. Bethe. 1933. *Elektronentheorie der Metalle*, Pp. 333–622 in *Handbuch der Physik*, Ser. 2, Vol. 24, H. Geiger and K. Sheel, eds. Berlin: Springer.

Sopka, K. R. 1988. Quantum Physics in America: The Years Through 1935. New York: Tomash.

Sparks, M. 1993. Interview by Lillian Hoddeson and Michael Riordan, June 17.

Sparks, M., and B. Sparks. 1992. Interview by Lillian Hoddeson, August 11.

Sternberg, R., ed. 1988. The Nature of Creativity: Contemporary Psychological Perspectives. Cambridge, England: Cambridge University Press.

———. 1999. Handbook of Creativity. Cambridge, England: Cambridge University Press.

Sternberg, R. J., and J. Davidson, eds. 1995. The Nature of Insight. Cambridge, Mass.: MIT Press.

Sternberg, R. J., and N. K. Dess. 2001. Creativity for the New Millennium. American Psychologist 56(4):332–363.

Stevens, A. 1941. Washington: Blight on democracy. Harper's (December):51–52.

Stone, M. 1992. Interview by Vicki Daitch, February 19.

Sugarbaker, D. 2001. Interview by Vicki Daitch, April 6.

Sullivan, W. E. 1936. Charles Russell Bardeen, 1871–1935, in memoriam. The Anatomical Record 65(2). From an address given at the Fifty-second Session of the American Association of Anatomists, Duke University, Durham, North Carolina, April 9–11.

Sulloway, F. J. 1983. Freud: Biologist of the Mind. New York: Basic Books.

———. 1996. Born to Rebel: Birth Order, Family Dynamics, and Creative Lives. New York: Pantheon Books.

Suzuki, S. 1970. Zen Mind, Beginner's Mind. New York: Weatherhill.

Szanton, A. 1992. The Recollections of Eugene P. Wigner, as Told to Andrew Szanton. New York: Plenum Press.

Tucker, J. R., J. H. Miller, K. Seeger, and J. Bardeen. 1982. Tunneling theory of ac-induced dc conductivity for charge-density waves in NbSe. Physical Review B 25(4)(February 15):25.

Tamm, I. 1932. Uber eine mögliche Art der Elektronenbindung an Kristalloberflächern. Physikalische Zeitschrift der Sowjetunion 1:733–746.

Teal, G., and L. Teal. 1993. Interview by Lillian Hoddeson and Michael Riordan, June 19.

Tennyson, A. 1987. Ulysses. Pp. 11–12 in The New Oxford Book of Victorian Verse, C. Ricks, ed. Oxford: Oxford University Press.

Thelen, D. P. 1969. Social tensions and the origins of progressivism. Journal of American History 56:323–341.

Thompson, C. 1951. Since Spindletop: A Human Story of Gulf's First Half-Century. Pittsburgh: Gulf Oil.

Thorne, R. E. 1996. Charge density wave conductors. Physics Today (May):42–47.

Thwaites, R. 1899. The Story of Wisconsin. London: Lothrop.

Tinkham, M. 1993. Interview by Vicki Daitch, August 4.

Torrey, H. 1983. Interview by Lillian Hoddeson, January 19.

Torrey, H. C., and C. A. Whitmer. 1948. Crystal Rectifiers. New York: McGraw-Hill.

Trick, T. 1991. Interview by Fernando Irving Elichirigoity and Vicki Daitch, September 24.

Tsuneto, T. 1993. Interview by Vicki Daitch, August 5 and 6.

———. 1994. Interview by Gordon Baym, July 2.

Tucker, J. 1995. Interview by Lillian Hoddeson and Joseph Tillman, November 14.

Tweney, R. D. 1985. Faraday's discovery of induction: A cognitive approach. In Faraday Rediscovered: Essays on the Life and Work of Michael Faraday: 1791–1867, D. Gooding and F. James, eds. New York: Stockton Press.

Tweney, R. D., M. E. Doherty, and C. R. Mynatt, eds. 1981. On Scientific Thinking. New York: Columbia University Press.

Van Vleck, J. H. 1926. Quantum Principles and Line Spectra. Bulletin No. 54. Washington, D.C.: National Research Council.

———. 1932. The Theory of Electric and Magnetic Susceptibilities. Oxford: Clarendon Press.

———. 1964. American physics comes of age. Physics Today (June):21–26.

Vinen, J. 1993. Interview by Vicki Daitch, August 5.

Waley, A., trans. 1989. The Analects of Confucius. New York: Vintage Books.

Walter, M. L. 1990. Science and Cultural Crisis: An Intellectual Biography of Percy Williams Bridgman (1882–1961). Stanford, Calif.: Stanford University Press.

Ward, T. B., S. M. Smith, and R. A. Finke. 1999. Creative cognition. Pp. 189–212 in Handbook of Creativity, R. Sternberg, ed. Cambridge, England: Cambridge University Press.

Warne, L. 1991. Vignettes from a Century of Service to the University, the State and the Nation, 1890–1990. Urbana-Champaign: University of Illinois, Department of Physics.

Weinberg, G. 1994. A World at Arms: A Global History of World War II. Cambridge, England: Cambridge University Press.

Weiner, C. 1968. A new site for the seminar: The refugees and American physics in the thirties. Perspectives in American History 2:190–234.

Weinstein, J. 1968. The Corporate Ideal in the Liberal State, 1900–1918. Boston: Beacon Press.

Weisberg, R. W. 1986. Creativity: Genius and Other Myths. New York: W. H. Freeman.

———. 1993. Creativity: Beyond the Myth of Genius. New York: W. H. Freeman.

———. 1999. Creativity and knowledge: A challenge to theories. Pp. 226–250 in Handbook of Creativity, R. Sternberg, ed. Cambridge, England: Cambridge University Press.

Weisberger, B. 1994. The LaFollettes of Wisconsin. Madison: University of Wisconsin Press.

Weisskopf, V. 1951. Nuclear models. Science 113(January 26):101–102.

Werstler, B., and E. Werstler. 1992. Interview by Vicki Daitch, February 18.

Wiebe, R. 1967. The Search for Order. New York: Hill and Wang.

Wigner, E. 1963. Interview by Thomas Kuhn, November 21 and December 3, AIP.

———. 1981. Interview by Lillian Hoddeson with Gordon Baym and Frederick Seitz, January 24.

———. 1984. Interview by William Aspray, April 12, IEEE.

Wigner, E., and F. Seitz. 1933. On the constitution of metallic sodium. I. Physical Review 43:804–810.

Wigner, E., and J. Bardeen. 1935. Theory of the work function of monovalent metals. Physical Review 48(July 1):84–88.

Wigner, E., and F. Seitz. 1934. On the constitution of metallic sodium. II. Physical Review 46:509–524.

Wilson, A. 1936. The Theory of Metals. Cambridge, England: Cambridge University Press.

Wilson, A. H. 1931. The theory of electronic semi-conductors. Proceedings of the Royal Society of London A133:458–491; The theory of electronic semi-conductors—II, A134:277–287.

Woodward, A. 1979. A Biographical Sketch of Alexander Francis Harmer. Pamphlet, BFC.

Wooldridge, D. 1976. Interview by Lillian Hoddeson, August 21.

Woolf, H. 1980. Albert Einstein: Encounter with America. Pp. 28–37 in Some Strangeness in the Proportion: A Centennial Symposium to Celebrate the Achievements of Albert Einstein, H. Woolf, ed. Reading, Mass.: Addison-Wesley.

Young, P. 1972. Supercold superprize. National Observer (December 9). Clipping, BFC.

Zachary, G. P. 1997. Endless Frontier: Engineer of the American Century. New York: Free Press.

Zawadowsky, F. 1998. Interview by Lillian Hoddeson, April 2.

Zuckerman, H. 1977. Scientific Elite: Nobel Laureates in the United States. New York: Free Press.

———. 1983. The scientific elite: Nobel laureates' mutual influences. Pp. 157–171 in Genius and Eminence: The Social Psychology of Creativity and Exceptional Achievement. International Series in Experimental Social Psychology, R. S. Albert, ed. Oxford: Pergamon Press.

Acknowledgments

We warmly thank all the individuals and institutions who helped us write this book. For major financial support throughout, especially of student assistants, we are deeply grateful to the Richard Lounsbery Foundation. We thank the Campus Research Board, the Department of Physics, and the Department of Electrical Engineering at the University of Illinois at Urbana-Champaign (UIUC), Texas Instruments Corporation, American Telephone & Telegraph Corporation, and William and Jane Bardeen for generous seed grants that allowed us to formulate this biography project well enough to seek major funding. We thank the Dibner Fund for its support of graduate assistants during the second and third years; the Alfred E. Sloan Foundation and the UIUC Center for Advanced Study for supporting Hoddeson's work in the seventh year; and the John Simon Guggenheim Memorial Foundation, for awarding Hoddeson a fellowship that allowed her to focus entirely on Bardeen's life and science during the book's final writing stage in 2000 and 2001.

A large number of colleagues and friends helped us in crucial ways. At the University of Illinois, we particularly thank the Department of Physics (headed by Ansel Anderson at the time our work began, then David Campbell, and later Jeremiah Sullivan) for providing a home, constant encouragement, constructive criticism, and countless support services. We are most grateful to the members of the Bardeen History Committee (Ansel Anderson, Gordon Baym, Raymond Borelli, David Campbell, David Pines, and Charles Slichter) for offering the encouragement that brought this project to life. We thank Baym for many technical explanations of Bardeen's physics. For administrative help beyond the call of duty, we thank many who are presently or were formerly on the physics department's staff, especially Borelli, Steve Keen, Steve Knell, Joy Kristunas, Greg Larson, Barbara Leisner, Rebecca McDuffee, Mary Kay Newman, Mary Ostendorf, Robert Williams, and Carolyn Wright.

Thanks to the Department of History (chaired by Charles Stewart when the work began, later James Barrett, then Peter Fritzsche) for repeatedly granting Hoddeson released time from teaching to work on this book. Thanks also to the history department's admistrative and business staff, especially Sandy Colclasure, Stanley Hicks, and Aprel Orwick for their patient and tireless assistance.

We thank all the graduate and undergraduate students who helped us over the years with research tasks, such as conducting and transcribing interviews, gathering source materials, cataloging, editing drafts, and locating references: Everitt J. Carter, Andrew Dribin, Fernando Irving Elichirigoity, Kristine Fowler, Tonya Lillie, Cathleen McFarland, James Nelligan, Steven Rouse, Nicole Ryavec, Glenn Sandiford, Derek Shouba, Joseph Tillman, and Patricia Wenzel. Lillie not only helped us gather and process interviews, identify sources, enter data, and much more, she also served as an assistant to John Bardeen in his last year while he was organizing his scientific papers. She later worked with the Department of Physics and the university archives to make his papers available to scholars.

We are grateful to Horace Judson and Michael Riordan for editorial help at pivotal stages of the work, and to William Bardeen, Gordon Baym, John Tucker, George Grüner, Nick Holonyak, Howard R. Huff, Arthur I. Miller, Fred Seitz, and others who read chapter drafts and offered a wealth of useful comments. Thanks to Jenny Barrett and Firmino Pinto for rescuing the electronic manuscript from a frightening computer crash. Thanks to all the archivists who helped us identify sources at Harvard, the University of Wisconsin, the University of Minnesota, American Telephone & Telegraph, the Naval Surface Weapons Center and the University of Illinois. We especially thank University of Illinois archivists Maynard Brichford and William Maher and their assistants for help navigating Bardeen's scientific papers.

Thanks to the many individuals who offered letters, clippings, photographs, and other documentary materials, also who contributed hours of their own time in tape-recorded interviews. Sadly, neither Jane Bardeen nor Betsy Bardeen Greytak lived to see this book published; without their insights and support this biography would have been inconceivable. We wish to extend our deepest appreciation to Frederick Seitz, Nick Holonyak, and Bill Bardeen for repeated interviews and countless historical leads and materials. Thanks to our editor, Jeffrey Robbins of the Joseph Henry Press for his sage advice, his encouragement, and especially, his humor.

Last but not least, heartfelt thanks to our loved ones—Lillian's daughter and son, Carol and Michael Baym, her husband Peter Garrett, and Vicki's husband Hiram Daitch—who offered not only the time and the emotional support needed for the completion of this work but artistic, literary, and computer assistance.

Lillian Hoddeson, Urbana, Illinois
Vicki Daitch, Canterbury, New Hampshire
May 2002

Notes

1
The Question of Genius

p. 2 **popular magazines sometimes credit:** For example, *Time* magazine, December 31, 1999, 57.

p. 2 **A product of Hollywood:** Seitz (1994), 67.

p. 3 **a little embarrassed:** Lazarus (1992).

p. 3 **the "nerve cell":** Riordan and Hoddeson (1997a), 201.

p. 4 **"Every theory of superconductivity":** Hoddeson, et al. (1992), ch. 8.

p. 4 **Nobel Committee awarded:** Marie Curie's two Nobel Prizes had been in different fields—physics in 1903 and chemistry in 1911. Linus C. Pauling, whose first Nobel Prize was for chemistry in 1954, received his second in 1962 for contributions to world peace. In 1980, Frederick Sanger became the second person to receive two Nobel Prizes in the same field when he won his second prize in chemistry.

p. 4 **" . . . John Who?":** Chang (1988).

p. 4 **"father of the Information Age":** Film director David Frankel raised the same question about Bardeen on reading his obituary in 1991. When he shared the question with friends at *The New York Times Magazine*, they responded by including Bardeen in their New Year's feature in 1995, "Lives Well Lived," *The New*

York Times Magazine, January 1, 1995, Section 6. Bardeen, however, remains largely unknown.

p. 5 **typically possessed by:** Ghiselin (1952); Lombroso (1891); Murray (1989); Nitzsche (1975).

p. 5 **these "Illuminati" merged:** Schaffer (1990).

p. 5 **romantic image of the genius:** Bone (1989); Galton (1869, 1874); Weisberg (1993).

p. 5 **the stereotype does not fit:** For scientists and mathematicians see Gleick (1992); Kanigel (1991).

p. 5 **"the history of science":** Gleick (1992), 313–29, quotes on 329.

p. 5 **bongo-drumming, or his entertaining tales:** Feynman (1985).

p. 5 **wild-haired Albert Einstein:** Pais (1982).

p. 6 **" . . . really have to listen":** Herring (1992b), 30.

p. 6 **"Whispering John":** Hess (1991); Holonyak (1993c).

p. 6 **he appeared "flat":** Philip Foy, in del Guercio (1998).

p. 6 **Only close friends:** Holonyak (1993a).

p. 6 **At the Naval Ordnance Laboratory:** Anspacher (1992).

p. 7 **"very easy":** Bray (1993a).

2
Roots

p. 8 **Ten-year-old John Bardeen:** The incident is described in A. Bardeen to C. W. Bardeen, n.d., c. 1918, UWA.

p. 8 **"He did it right after":** A. Bardeen to C. W. Bardeen, n.d.; C. R. Bardeen to C. W. Bardeen, May 5, 1919, UWA.

p. 8 **"John just hangs on":** A. H. Bardeen to C. W. Bardeen, n.d., UWA.

p. 8 **The American roots:** W. Bardeen (1993), i–iv.

p. 9 **what constituted the progressive movement:** The following works represent only a small sampling of the diverse interpretations of the period known as the Progessive Era: Filene (1970); Hofstadter (1955); Link and McCormick (1983); McCormick (1981); Wiebe (1967).

p. 10 **John Bascom:** McCarthy (1912), 28–32.

p. 10 **"Fighting Bob" La Follette:** Some of the earlier sources are Curti and Carstensen (1949); McCarthy (1912); and Thwaites (1899). We are grateful to Joseph Tillman for helping us to establish these connections in John Bardeen's heritage. For a more recent history of Wisconsin, see Nesbit and Thompson (1989).

p. 10 **Another student influenced:** Curti and Carstensen (1949), 1:288, 2:18–19, 311–312; Nesbit and Thompson (1989), 427; Weisberger (1994), 10, 12.

p. 11 **"the boundaries of the University":** Curti and Carstensen (1949), 2:5–9.

p. 11 **"the expert on tap":** Nesbit and Thompson (1989), 426.

p. 11 **he found Charles Russell Bardeen:** "Resolution of the Faculty of the University of Wisconsin on the Death of Dean Bardeen," undated document in UWA.

p. 11 **a poor drummer:** W. Bardeen to L. Hoddeson, February 1, 2002; C. W. Bardeen (1910), BFC.

p. 11 **The School Bulletin:** W. Bardeen (1993), 208; Syracuse, N.Y., Wednesday Evening, August 20, 1924, "C.W. Bardeen, Publisher and Educator Dies,": unattributed newpaper clipping in BFC.

p. 12 **"It is the useless life":** C. W. Bardeen to C. R. Bardeen, February 8, 1916, C. R. Bardeen Papers, UWA.

p. 12 **"a fair scholar":** C. R. Bardeen, "A Journey for Speech, The dairy [*sic*] of one of a party of seven, the five children of which went to Leipsic, Germany to learn German, French, etc.," 1888, UWA.

p. 12 **"the only real happiness":** Quoted in Clark (1967), 41. Original in C. R. Bardeen Papers, UWA.

p. 12 **facilities were inadequate:** Bradley (1992), 2.

p. 12 **first person to graduate:** Clark (1967), 4, 41–42.

p. 12 **"live an effective life":** C. R. Bardeen to C. W. Bardeen, February 9, 1900, BFC.

p. 12 **create a new medical school:** Sullivan (1936).

p. 12 **The two doctors rode:** Falk (1992), 22.

p. 13 **Charles taught anatomy:** Clark (1967), 4.

p. 13 **Her father opposed:** C. R. Bardeen to C. W. Bardeen, January 11, 1920, UWA.

p. 13 **he too had left home:** Alexander Harmer appears likely to have been Althea's uncle, but we have not been able to verify the exact relationship.

p. 13 **"among the greatest":** Montgomery (1994), 137–144.

p. 13 **a history of this Laboratory School:** Mayhew and Edwards (1936), vii.

p. 13 **"to work out with children":** Ibid., v–vi.

p. 13 **"The school whose work":** Ibid., xv–xvi.

p. 13 **Dewey's ideas:** For a modern account of "intrinsic" motivation, see Amabile (1996).

p. 14 **"explorations where":** Montgomery (1994), 137–144. We find almost the same idea in John Bardeen's later writings on industrial research and science education.

p. 14 **"the essence of":** Mayhew and Edwards (1936), 435–436.

p. 14 **"Again this method involves":** Report by Althea Harmor in Mayhew and Edwards (1936), 335.

p. 15 **"calling the constructive":** Ibid.

p. 15 **"She had no capital":** C. W. Bardeen to C. R. Bardeen, June 22, 1905, UWA.

p. 15 **her decorating business:** W. Bardeen (1995).

p. 15 **"The more I see":** C. R. Bardeen to C. W. Bardeen, June 22, 1905, UWA.

p. 15 **after she had given in:** C. R. Bardeen to C. W. Bardeen, June 14, 1905.

p. 15 **They were married:** C. R. Bardeen to C. W. Bardeen, June 1, 1905, UWA.

p. 15 **"an exceptionally beautiful home":** C. R. Bardeen to C. W. Bardeen, June 27, 1905, UWA.

p. 15 **Madison's art world:** "Appreciate Jap Art Exhibit: Many People View Rare Collection" and "Lectures on Japanese Art," two unattributed newspaper clippings in BFC.

p. 15 **Charles continued:** C. R. Bardeen to C. W. Bardeen, June 14, 1905, UWA.

p. 16 **full teaching hospital:** "Obituary: Charles R. Bardeen, 1871–1935," in *Science* 82 (December 27, 1935), 606.

p. 16 **"made it impossible":** C. R. Bardeen to C. W. Bardeen, January 12, no year, UWA.

p. 16 **crass commercialization of medicine:** "Bardeen Warns of Money-Hungry Doctors; 300 at Honor Banquet," *Capital Times*, June 21, 1932; "25 Years of Service," the *Wisconsin State Journal*, June 21, 1932.

p. 16 **" may have been boiling":** Clark (1967), 45.

p. 16 **"in his informal, often blundering":** Harland Mossman to Paul Clark, July 29, 1964, cited in Clark (1967), 47–48.

p. 17 **"... very loving pair":** A. Bardeen to C. W. Bardeen, n.d., UWA.

p. 17 **"Althea is unusually affectionate"**: C. R. Bardeen to C. W. Bardeen, February 12, 1906, UWA.

p. 17 **". . . unnatural for a healthy woman"**: C. R. Bardeen to C. W. Bardeen, April 19, 1905, UWA.

p. 17 **"you were his best friend"**: A. Bardeen to C. W. Bardeen, n.d., UWA.

p. 17 **Althea gave birth:** "I believe in a single Christian name and in having that name distinctive." C. R. Bardeen to C. W. Bardeen, May 31, 1906, UWA.

p. 17 **She was exhausted:** C. R. Bardeen to C. W. Bardeen, April 20, 1906; July 6, 1906, UWA.

p. 17 **"what spare time"**: C. R. Bardeen to C. W. Bardeen, June 26, 1906, UWA.

p. 17 **apartment seemed too small:** C. R. Bardeen to C. W. Bardeen, March 20, 1907, UWA.

p. 18 **"far less of a care"**: C. R. Bardeen to C. W. Bardeen, July 28, 1908, UWA. John Bardeen's birth weight is noted in Jane Bardeen to Mrs. J. R. Maxwell, February 28, 1939, BFC.

p. 18 **"Charles' devotion to John"**: A. Bardeen to C. W. Bardeen, n.d., UWA.

p. 18 **Charles felt himself "blessed"**: C. R. Bardeen to C. W. Bardeen, February 16, 1910, UWA.

p. 18 **The stucco next door:** C. R. Bardeen to C. W. Bardeen, June 23, 1910, UWA.

p. 18 **Charles converted the attic:** A. Bardeen to C. W. Bardeen, n.d., UWA; Henschel (1992).

p. 19 **"After a necessary operation"**: A. Bardeen to C. W. Bardeen, n.d., UWA.

p. 19 **"I was obliged"**: A. Bardeen to C. W. Bardeen, n.d., UWA.

p. 19 **"always indulged John"**: A. Bardeen to C. W. Bardeen, n.d., UWA.

p. 19 **"huggy-kissy"**: Henschel (1995).

p. 19 **". . . not in the concrete"**: Clark (1967), 43–44.

p. 19 **"concentrated essence of brain"**: A. Bardeen to C. W. Bardeen, n.d., UWA.

p. 19 **". . . didn't learn to spell"**: Young (1972), clip in UIUC-P. For example, he misspelled the word "existence" in a card to Nick Holonyak in 1956 describing the experience of his first Nobel Prize. John Bardeen to Nick Holonyak, December 13, 1956, UIUC-A.

p. 19 **"... talent with figures":** C. R Bardeen to C. W. Bardeen, February 12, 1913, UWA.

p. 19 **Finishing third-grade math:** A. Bardeen to C. W. Bardeen, n.d., UWA.

p. 20 **"He is getting away":** C. R. Bardeen to C. W. Bardeen, July 14, 1919, UWA. In the same letter Charles writes, "since John took the scholarship I have become John's father."

p. 20 **"... a genius for mathematics":** A. H. Bardeen to C. W. Bardeen, n.d., UWA.

p. 20 **more time for his studies:** John Bardeen (1977b).

p. 20 **playmates of his own age:** Jane Bardeen (1991b).

p. 20 **Run, Sheep, Run:** Rosemary Royce Bingham to John Bardeen, February 1, 1973, UIUC-A.

p. 20 **Four Lakes Stamp Company:** Henschel (1995).

p. 20 **"socially he is reserved":** A. Bardeen to C. W. Bardeen, n.d., UWA.

p. 20 **interested in chemistry:** Slossum (1920).

p. 20 **Eager to encourage his son's scientific interest:** C. R. Bardeen to National Stain and Reagent Co., Norwood, Ohio, December 27, 1921, UWA.

p. 20 **"I dyed materials":** John Bardeen (1977a).

p. 20 **"dime store wires":** R. Bingham to J. Bardeen, February 1, 1973; C. R. Bardeen to C. W. Bardeen, July 6, 1922, UWA.

p. 21 **"... I never got that far":** John Bardeen (1977b).

p. 21 **"wig-wag" signaling:** John Hames to Vicki Daitch, October 15, 1991; Hames (1992).

p. 21 **"introduce pupils to":** Newspaper article, Jim Clark, "Wisconsin High's 'Thorough Courses' Made it Model School," n.d., Wisconsin Historical Society, misc. Steinbock archives, Box 94/14, 1 of 1.

p. 21 **The fourth quarter:** According to remaining records, John attended the summer session at least once. Permanent record cards of the Wisconsin High School of the University of Wisconsin Archives; Madison Central High School permanent record card, Admission Records, UWA.

p. 21 **"... freedom from John":** A. Bardeen to C. W. Bardeen, n.d., UWA.

p. 22 **A Little Fifer's War Diary:** C. W. Bardeen (1910).

p. 22 **"conceited, boastful, self-willed":** Ibid., 19.

p. 22 **"My dear John":** C. W. Bardeen (1910), BFC.

p. 22 **"always been an inspiration"**: John Bardeen to Lynn Lester, Christmas 1975. Special thanks to Lynn Lester Maynard for a copy of the letter from her private collection.

p. 22 **continued to be an adventurer:** W. Bardeen private communication to Hoddeson, August 15, 2001.

p. 22 **"I have wanted them"**: A. Bardeen to C. W. Bardeen, n.d., UWA.

p. 23 **"more about them"**: A. Bardeen to C. W. Bardeen, n.d., UWA.

p. 23 **model of social conscience:** C. R. Bardeen to Marie E. Demetre, chair, the Fatherless Children of France, New York City, January 14, 1918, C. R. Bardeen Papers, UWA.

p. 23 **"John opened his bank"**: A. Bardeen to C. W. Bardeen, n.d., UWA.

p. 23 **". . . five and ten cents"**: A. Bardeen to C. W. Bardeen, n.d., UWA.

p. 23 **"kept John brave"**: A. Bardeen to C. W. Bardeen, n.d., UWA.

p. 23 **"doing stunts in fancy diving"**: A. Bardeen to C. W. Bardeen, n.d., UWA.

p. 23 **Boating jaunts:** Ibid.

p. 23 **"development as a father,"**: C. R. Bardeen to C. W. Bardeen, February 8, 1915, UWA.

p. 24 **"I shivered just watching"**: Osterhoudt (1991)

p. 24 **playing every day:** A. Bardeen to C. W. Bardeen, n.d., UWA.

p. 24 **"nearly every afternoon"**: C. R. Bardeen to C. W. Bardeen, August 28, 1910, UWA.

p. 24 **They soon found the car:** William Bardeen interview, August 1, 1964, by Paul Clark, cited in Clark (1967), 44; Henschel (1992).

p. 24 **He began to learn the game:** A. Bardeen to C. W. Bardeen, n.d., UWA.

p. 24 **childhood interest in golf:** See, for example, a letter from C. W. to Charles R. Bardeen, in which he says, "The spirit of outdoors is in the air and I should like to be getting about the links again." May 11, 1916, UWA.

p. 24 **"We discovered a little growth"**: C. R. Bardeen to C. W. Bardeen, March 3, 1918, UWA.

p. 25 **Althea was losing:** C. R. Bardeen's publications on X Rays include "Abnormal Development of Toad Ova Fertilized by Spermatozoa Exposed to the Roentgen Rays," *Journal of Experimental Zoology* 4, no.1 (1907); "Variations in Susceptibility of Amphibian Ova to the X rays at Different Stages of Development," *Anatomical Record* 3, no.163 (1909); "Determination of the Size of the Heart by Means of the X ray," *American Journal of Anatomy* 23 (1918); and "The Value of the Roentgen Ray and the Living Model in Teaching and Research in Human Anatomy," *Anatomical Record* 14, no. 337 (1918).

p. 25 **radiotherapy as a cancer treatment:** Patterson (1987), 64.

p. 25 **Charles pulled as many strings:** C. R. Bardeen to J. C. Bloodgood, January 12, 1920; J. C. Bloodgood, Baltimore to C. R. Bardeen, January 25, 1920, UWA. Dr. Joseph C. Bloodgood was considered one of the nation's premier experts on cancer at the time. Patterson (1987).

p. 25 **radiation itself caused cancer:** Kevles (1997), 33–53, 80.

p. 25 **one in eight American women:** Patterson (1987), 64, 72.

p. 25 **severe influenza epidemics:** Crosby (1977) and (1976). Crosby describes the early 1919 wave of influenza to which the Bardeens were exposed as somewhat less virulent than the two 1918 waves, but nasty all the same.

p. 25 **"seen more of John":** C. R. Bardeen to C. W. Bardeen, March 4, 1919, C. R. Bardeen Papers, UWA.

p. 25 **Althea continued to decline:** C. R. Bardeen to C. W. Bardeen, July 14, 1919, UWA.

p. 26 **"such self control":** Mary Morris to C. R. Bardeen, summer 1919, UWA.

p. 26 **"... massive X ray exposures":** Dr. William A. Thomas to C. R. Bardeen, March 27, 1920; C. R. Bardeen to W. A. Busey, April 14, 1920; C. R. Bardeen to C. W. Bardeen, April 9, 1920, C. R. Bardeen Papers, UWA.

p. 26 **"... that kind of a girl":** Dr. J. L. Yates to C. R. Bardeen, April 12, 1920, UWA.

p. 26 **nerves of steel:** C. R. Bardeen to J. L. Yates, August 28, 1919, C. R. Bardeen Papers, UWA.

p. 26 **"... she suspects it":** C. R. Bardeen to C. W. Bardeen, April 9, 1920, UWA.

p. 26 **"on the way between":** John Bardeen (1977a).

p. 26 **"She said it was delicious":** C. R. Bardeen to C. W. Bardeen, April 9, 1920 and May 12, 1920, C. R. Bardeen Papers, UWA.

p. 26 **"I thought she looked well":** John Bardeen (1977a).

p. 27 **". . . the bravest possible fight":** C. R. Bardeen to C. W. Bardeen, May 12, 1920, UWA.

p. 27 **"her intense feeling":** "Althea Harmer Bardeen," *Madison Democrat*, April 21, 1920.

p. 27 **"put in twelve":** In addition to his university work, he often picked up odd voluntary assignments, such as stocking the library of the new University Club. C. R. Bardeen to C. W. Bardeen, March 4, 1919, UWA.

p. 27 **a temporary fix:** C. R. Bardeen to C. W. Bardeen, May 12, 1920, C. R. Bardeen Papers, UWA.

p. 27 **"So we got along":** John Bardeen (1977a).

3
To Be an Engineer

p. 28 **"I have played some golf":** C. R. Bardeen to C. W. Bardeen, August 25, 1920.

p. 28 **Ruth worked patiently:** Ruth Hames Bardeen McCauley continued to enjoy warm relations with the Bardeen children until her death in September 1979.

p. 28 **Tom, eight years old when Althea died:** Jane Bardeen (1993); W. Bardeen (1993), 266.

p. 28 **John often acted:** Bradley (1992), 2; Henschel (1992).

p. 28 **"stood at the head":** A. Bardeen to C. W. Bardeen, c. 1919; Laura B. Johnson to C. R. Bardeen, May 16, 1921, UWA.

p. 28 **more marks of "good":** Permanent record cards of the Wisconsin High School of the University of Wisconsin Archives; Madison Central High School permanent record card, Admission Records, UWA.

p. 29 **"allowed to go ahead":** C. W. Bardeen to C. R. Bardeen, March 4, 1919, UWA.

p. 29 **"He saw that":** Untitled newspaper article in UIUC-P; also John Bardeen interview by Maynard Brichford, February 8, 1965, UIUC-A.

p. 29 **popular mathematics textbook series:** John Bardeen (1977a). Hart brought *Mathematics in Daily Use* (Hart, et al., 1945) through several editions.

p. 29 **"in the front seat":** Walter W. Hart to John Bardeen, April 7, 1962, UIUC-A.

p. 29 **"express my deep gratitude":** John Bardeen to Walter W. Hart, April 4, 1962, UIUC-A.

p. 30 **"coffin nails":** Henschel (1995).

p. 30 **"some beautiful races":** *Wisconsin Badger* 1927, 312.

p. 30 **playing poker or billiards:** Security Investigation Data For Sensitive Position, FBI security clearance, 1981, UIUC-A; Betsy Bardeen Greytak, Bardeen family interview (1992).

p. 30 **a three-cushion champion:** Burtis (1995).

p. 30 **continued to live at home:** John Bardeen (1977b).

p. 30 **". . . little trouble with Zeta Psi":** Scott H. Goodnight, dean of Men at the University of Wisconsin, to C. S. Marsh, assistant dean at Northwestern University, March 12, 1923.

p. 30 **His college transcript:** Transcript in Bardeen graduate student file, PRIN.

p. 30 **performed better than:** John's undergraduate grade point average (GPA) during his fraternity years appears to have ranged from a low of 0.944 to a high of 2.25, on a scale of A = 3.000, B = 2.000, C = 1.000, D = 0.000, and F = −1.000. Zeta Psi fraternity papers, Wisconsin Library Archives, Madison, WI.

p. 30 **"just floated in and out":** Osterhoudt (1991).

p. 31 **getting a little rowdy:** Osterhoudt (1991).

p. 31 **heebie-jeebies:** Ibid. Bardeen admits to a disorderly conduct arrest in summer 1928 on a personnel security questionnaire of the U.S. Department of Energy filled out in 1982. Bardeen Personnel File, USOPM.

p. 31 **"My father's the dean":** Betsy Bardeen in Bardeen family interview (1992); Henschel (1992).

p. 31 **"end up being a university professor":** John Bardeen (1965b).

p. 31 **academic career sounded stodgy:** Answers to Harvard University Junior Fellowship Questionnaire, October 16, 1967, UIUC-A.

p. 31 **"a lot of mathematics":** John Bardeen (1977a); John Bardeen to Dr. Raymond Damadian, Fonar Corporation, Melville, N.Y., December 5, 1990, UIUC-A.

p. 31 **Weaver, later head of:** John Bardeen (1984b); "Weaver in Wonderland," *Wisconsin Alumnus*, October 1959, 12, 35. In 1954, Weaver became president of the American Association for the Advancement of Science.

p. 32 **"In a detailed theory":** Bardeen's course notes for Weaver's course, "Electrodynamics," p. 1, BFC.

p. 32 **"dignified, formal, and reserved":** Leroy A. Howland to Professor Evans, April 1, 1948, Edward Burr Van Vleck file, UWA.

p. 32 **Van Vleck was a member:** "Memorial Resolution on the Death of Professor E. B. Van Vleck," excerpt from University faculty meeting, October 4, 1943, E. B. Van Vleck file, UWA; Bleaney (1982), 655; Charles Slichter's comments at John Van Vleck's memorial service.

p. 32 **"The University was strong":** John Bardeen (1980c), 77.

p. 33 **The Theory of Electric and Magnetic:** Bleaney (1982); Van Vleck (1932).

p. 33 **"earliest of its kind":** John Bardeen (1980c); also Bardeen's official college transcript, UWA.

p. 33 **Quantum Principles and Line Spectra:** Van Vleck (1926).

p. 34 **"intrigued by physics":** John Bardeen (1977a); (1984b).

p. 34 **"the only opportunities":** unattributed newspaper clipping, no title, UIUC-P.

p. 34 **"a very stimulating":** John Bardeen (1977a, 1984b). Debye's physics background resembled Bardeen's in that he took his undergraduate degree in electrical engineering. Davies (1970).

p. 34 **Principles of Quantum Mechanics:** Dirac (1930).

p. 34 **"much remained mysterious":** John Bardeen (1980c). Bardeen's official transcript, Office of the Registrar, University of Wisconsin, Madison. In 1937, Dirac married the sister of Eugene Wigner, Bardeen's Ph.D. thesis advisor. "Paul Adrien Maurice Dirac," *Biographical Memoirs of Fellows of the Royal Society* (London: The Royal Society, 1986); Kragh (1990).

p. 34 **Other well-known:** John Bardeen (1980c).

p. 34 **"I heard the lectures":** John Bardeen (1977a).

p. 35 **"Being young":** Ibid.

p. 35 **still only twenty":** Ibid.

p. 35 **assistant to Leo J. Peters:** John Bardeen transcript, Office of the Registrar, University of Wisconsin, Madison. Faculty Employment Cards, UWA.

p. 35 **squinted at the world:** Osterhoudt (1991).

p. 35 **field of electrical prospecting:** Nettleton (1940), 2; Peters and Bardeen (1932).

p. 35 **"had a muddy nose":** Thompson (1951), 80.

p. 36 **"from the geological":** Peters and Bardeen (1930), 5.

p. 36 **a research studentship:** John Bardeen (1977a).

p. 36 **"in analytical ability":** L. J. Peters, quoted in E. Bennett to Dean C. S. Slichter, December 11, 1928, UWA.

p. 36 **"a very independent young man":** Weaver, chair of Mathematics Department, to Trinity College, January 11, 1929, UWA.

p. 36 **"Mr. Bardeen is an exceptional":** John Van Vleck to R. H. Fowler, March 18, 1929, Bardeen files, UIUC-P.

p. 36 **"a man of sound judgment":** Ibid.

p. 37 **"The principal aim":** Edward Bennett, "Electric Circuit Equations: Their Derivation and Application," undated course notes, p. 1, BFC.

p. 37 **"modest acceptance":** Edward Bennett, chair of Electrical Engineering Department, to Dean C. S. Slichter, December 11, 1928, UWA.

p. 37 **Bardeen did not find:** John Bardeen (1984b); John Bardeen transcript, Office of the Registrar; Course Catalog, 1929–1930, UWA.

p. 37 **Thornton Fry, an AT&T recruiter:** John Bardeen (1977a).

p. 37 **"These were the days":** Ibid.

p. 38 **"He was tired":** James Affleck, editorial, *Carnegie Mellon Magazine* (Winter 1990). The author of this editorial also wrote, "At the early age of 22 he [Bardeen] had completed most requirements for a Ph.D. at the University of Wisconsin, but, dissatisfied with his dissertation, quit."

p. 38 **Engelbrecht Hall was one:** Description of dormitories provided by University Libraries, Carnegie Mellon University, Pittsburgh, PA.

p. 38 **modest red brick apartment:** Pittsburgh City Directory, 1930, 1931, 1932.

p. 38 **"with such fury":** Markowitz (1976), 4.

p. 38 **"everything from sink drills":** Ibid.

p. 39 **"when the experience":** Thompson (1951), 81; also McAfee (1977), 12; "Introduction," *1900-1976: A Special Issue of the Orange Disc* 22:5, 3–4, 7.

p. 39 **attracting the best and the brightest:** Ibid., 91.

p. 39   **"It was the early days"**: John Bardeen (1977a, 1977b).

p. 39   **"We'd get the results"**: John Bardeen (1977b).

p. 40   **"changes of resistivity"**: Peters and Bardeen (1932), 122.

p. 40   **Peters described the theory:** Thomas A. Elkins, "A Brief History of Gulf's Geophysical Prospecting," Gulf Research and Development Corporation Archives, UIUC-A; Peters (1949).

p. 40   **"Do you think you're any better"**: Osterhoudt (1991).

p. 40   **"Johnny worked in magnetics"**: Ibid.

p. 40   **". . . if I wanted to do geophysics"**: John Bardeen (1977b).

p. 41   **"at most eight or ten"**: John Bardeen (1977b).

p. 41   **Bardeen found the discussions:** John Bardeen (1977a, 1977b, 1984b).

p. 41   **"because there was an outstanding"**: John Bardeen (1977a).

p. 41   **"decided that university life"**: John Bardeen (1965a).

p. 41   **"take a try"**: C. R. Bardeen to Abraham Flexner, Princeton Institute, New York, N.Y., December 22, 1932, PRIN.

p. 41   **an unlikely scenario:** John Bardeen (1984b).

p. 41   **". . . student of outstanding ability"**: John Van Vleck, University of Wisconsin, Madison, to dean of the graduate school, Princeton University, February 8, 1933, PRIN.

p. 41   **a "general rule"**: Warren Weaver, Rockefeller Foundation, New York, N.Y., to Robert K. Root, chair, Graduate School Committee, Princeton University, February 16, 1933, PRIN.

p. 42   **"It was 1933"**: John Bardeen (1977b).

p. 42   **"resist any longer"**: John Bardeen (1984b).

p. 42   **"I quit my job"**: Bardeen (1977a).

p. 42   **his small inheritance:** W. Bardeen (1993), 208.

p. 42   **some additional savings:** Seitz (1998); C. R. Bardeen to C. W. Bardeen, May 5, 1919, C. R. Bardeen Papers, UWA; Employment chronology, July 28, 1941, Bardeen personnel file, USOPM.

p. 42   **"I could take a gamble"**: Harvard Fellows tape.

p. 42   **Gulf Research was then:** Thompson (1951), 91; "This is Gulf Research," public relations documents, Library and Archives Division, Historical Society of Western Pennsylvania, Pittsburgh, PA.

p. 42   **"drop down and see your friend"**: Osterhoudt (1991).

p. 42   **"Bardeen swiveled his chair"**: Ibid.

p. 43   **Jewish physicists and mathematicians:** Fleming and Bailyn (1969). See especially Weiner (1968).

p. 43　**There he met Jane Maxwell:** Jane Bardeen (1991b).

p. 43　**"I think you ought":** Jane Bardeen (1991a).

p. 43　**Elizabeth "Bess" Patterson:** Despite debilitating arthritis, her mother lived to celebrate her fiftieth wedding anniversary and beyond. "Local Couple Well Known," the *Washington (Pennsylvania) Reporter*, June 11, 1956; "Dr. R. J. Maxwell Has Practiced Here 56 Years," the *Washington (Pennsylvania) Reporter*, August 15, 1958.

p. 43　**". . . I'm really interested in science":** Jane Maxwell to Mrs. J. R. Maxwell, May 1, 1927, BFC.

p. 44　**Jane moved to Pittsburgh:** Jane Bardeen (1991a, 1991b).

p. 44　**She was attracted:** Jane Bardeen (1993).

p. 44　**he unobtrusively captured:** Jane Bardeen (1991a).

4
A Graduate Student's Paradise

p. 45　**It was dark:** Seitz (1999).

p. 45　**"the embodiment of scholasticism":** Morse (1977), 56–57.

p. 45　**"Here's a new student":** Seitz (1999).

p. 45　**Fred was fascinated:** Ibid.

p. 46　**Bardeen was more advanced:** Seitz (1998).

p. 46　**about 150 graduate students:** Seitz (1992b).

p. 46　**". . . worried about passing":** Seitz (1999).

p. 46　**"This fellow looked at John":** Ibid.

p. 46　**"They were rags":** Ibid.

p. 46　**Always wear a tie:** Seitz (1981a, 1982b).

p. 46　**find the formalities "insignificant":** Seitz (1999); Morse (1977), 57.

p. 46　**Abraham Flexner had conceived:** To attract the best scholars, Flexner set the Institute's salary more than three times the median of a full professor's salary. The early days of the Institute are described in Sopka (1988), 229–234. See also Sayen (1985), 58–61; Woolf (1980), 32-3; and Flexner (1960), 252.

p. 47　**transferring the Vatican:** Sayen (1980), 81–82; Sopka (1988), 229–234; Woolf (1980).

p. 47　**"One great park":** Sayen (1980), 61–64.

p. 47　**". . . village of puny demigods":** Einstein, quoted in Gleick (1992), 97.

p. 47 **"coffeehouses in the European sense":** Szanton (1992), 133–134.

p. 47 **bowled regularly:** R. Brattain (1993). Bardeen later won a bowling award while working during the Second World War at the Naval Ordnance Laboratory. Years later Charles Jakowatz told Nick Holonyak about bowling with Bardeen and how "tough" he was. Holonyak to L. Hoddeson, January 31, 2001.

p. 47 **referred to as "Slice":** R. Brattain (1993).

p. 47 **"We played until everyone got so sleepy":** R. Brattain (1993.)

p. 48 **when Brattain returned to Bell Labs:** W. Brattain (1964); Hoddeson (1981b).

p. 48 **"herding cattle":** W. Brattain (1964).

p. 48 **Brattain played aggressively:** J. C. Phillips to John Bardeen, April 19, 1988, UIUC-A.

p. 48 **"Walter and I had a common interest":** John Bardeen (1994).

p. 48 **The two also had in common:** Riordan and Hoddeson (1997a); also Bardeen (1977a).

p. 49 **"didn't really talk that much physics":** John Bardeen (1977b).

p. 49 **Bardeen's closest physics friend:** Hoddeson et al. (1992), 186, Herring and Herring (2000).

p. 49 **". . . part of the ring":** Seitz (2000).

p. 49 **The physics and mathematics students:** Seitz (1992a); Wigner (1963).

p. 49 **"everyone who could walk":** Seitz (1981a).

p. 49 **". . . strong synergistic effect":** John Bardeen to S. I. Goldberg, Department of Mathematics and Statistics, Queen's University, Kingston, Canada, April 24, 1981, UIUC-A.

p. 50 **By the time Bardeen:** This "coming-of-age" of American physics was the result of ample institutional and financial support, as well as the attraction of many bright young Americans to the field of physics. Hoch (1983); Sopka (1988), xii–xii, xix; Van Vleck (1964); Weiner (1968).

p. 50 **Princeton had "reluctantly":** Seitz (2001).

p. 50 **Wigner accepted an appointment:** Szanton (1992), 150–153.

p. 51 **". . . the two young Hungarians":** John Bardeen (1984a), 9.

p. 51 **Robertson's course:** John Bardeen (1977b); Bardeen transcript, Graduate Student files, 1936, Box 62, PRIN.

p. 51 **"Robertson's lectures":** Seitz (2001).

p. 51 **"sort of person who":** John Bardeen (1984b).

p. 51 **"prominent in the beer parties":** Herring and Herring (2000).

p. 51 **special lectures and seminars:** John Bardeen (1977b); Seitz (1992a). Also, Rigden (1987).

p. 51 **"most of the time":** John Bardeen (1984b).

p. 52 **"getting a little feeling":** John Bardeen (1972a, 1972b).

p. 52 **"Only a few courses":** John Bardeen (1984b).

p. 52 **"graduate student's paradise":** Seitz (1994), 47.

p. 52 **written prelims were identical:** John Bardeen (1984b).

p. 52 **". . . working with Einstein":** NHK.

p. 52 **immensely difficult projects:** Seitz (1994), 61–62; Seitz (1998).

p. 52 **The Theory of Atomic Spectra:** Condon and Shortley (1935).

p. 52 **"didn't sound too interesting":** John Bardeen (1977a).

p. 53 **"Atomic theory is a cold fish":** Seitz (2001).

p. 53 **"all those infinities":** John Bardeen (1977a, 1984b).

p. 53 **Julian Schwinger, Richard Feynman:** Brown (1983), 311–375; Schweber (1994).

p. 53 **"Maybe they're in your overcoat, Eugene":** Wigner (1981).

p. 53 **"If I drop my keys":** Wigner (1981).

p. 53 **"atoms in crystals":** Wigner (1963, 1981).

p. 54 **Bardeen wondered:** Duke (1993); John Bardeen (1984b).

p. 54 **"too polite for this":** Morse (1977), 98.

p. 54 **"It is my fault":** Szanton (1992), v–vi.

p. 54 **"one of the most remarkable":** Seitz (1992a).

p. 54 **"fish out of water":** Seitz (1998).

p. 54 **". . . he rarely communicated":** Seitz to L. Hoddeson, February 12, 2001.

p. 54 **with his penetrating questions:** John Bardeen (1984b).

p. 54 **". . . simplest possible case":** Ibid.

p. 55 **"very refined mathematical approach":** Seitz (1998). See also John Bardeen (1977a); (1984b); Duke (1993); Seitz (1992a).

p. 55 **In his later years:** John Bardeen (1984b).

p. 55 **"try simplest cases mantra":** P. Anderson (1998).

p. 55   **"very encouraging":** John Bardeen (1984b).

p. 55   **When the work was superceded:** E. Wigner, "Statement Concerning John Bardeen," Graduate School, John Bardeen file, PRIN.

p. 55   **occasionally visit each other:** Sopka (1988), 259. See also Fisk (1976); Slater (1975), 173; W. Shockley to May Shockley, December 12, 1932, Stanford Archives (10/8).

p. 55   **"practically all descendents":** John Bardeen (1984b).

p. 56   **"Conyers knew more solid-state":** Wigner (1984).

p. 56   **"It's John Bardeen":** Jane Bardeen (1991a).

p. 56   **Bruce offered:** Ibid. Also Jane Bardeen to Vicki Daitch, personal communication, 1992. According to Osterhoudt (1991), John occasionally dated other women. A letter in the archives hints of a romance: "I should address you more formally but since I do feel I know you from way back in the 20's I am going to be informal. . . . I worked at the U.W. Clinic when your father was Dean. . . . I'm anxious to know did you marry Elizabeth Kruse?" Helen Brown White to John Bardeen, January 9, 1990, UIUC-A.

p. 56   **Peters and Eckhardt invited:** Jane Bardeen (1991a).

p. 56   **the time John and Jane spent together:** Jane Bardeen (1991b).

p. 56   **Unlike John:** Wisconsin High School of the University of Wisconsin permanent record card, UWA.

p. 56   **developing his own derivation:** Elmer Krack to Sigmund Hammer, February 15, 1981, UIUC-A.

p. 56   **Tom's quick intelligence:** John Bardeen, "Career of Tom Bardeen with Gulf Oil Company," family document. The authors thank Lynn Maynard and Ellen Stiehl for kindly sending us a copy.

p. 57   **"He wouldn't say a word":** Henschel (1995); Jane Bardeen (1991a).

p. 57   **When he was in Pittsburgh:** Jane Bardeen (1991b).

p. 57   **"thought they were safe":** Seitz to L. Hoddeson, phone conversation, November 13, 1999. In this period there were 17 students enrolled for roughly four years each in the math-physics program that Bardeen was in.

p. 57   **"raised the height":** Seitz to L. Hoddeson, February 12, 2001.

p. 57   **". . . using too much intuition":** John Bardeen (1984b).

p. 57   **"noisily and as honored guests":** F. Seitz to C. Herring, cited in Herring and Herrring (2000).

p. 57 **"Without too much encouragement":** Seitz (1994), 64–65.

p. 58 **"No one in my experience":** E. Wigner, "Statement Concerning John Bardeen," c. fall 1934 or early spring 1935, Graduate School, John Bardeen file, PRIN.

p. 58 **blacksmiths, potters:** Hoddeson, et al. (1992), ch. 1.

p. 58 **standing on the edge:** John Bardeen (1977c).

p. 59 **"far more complex things":** Smith (1965).

p. 59 **Max von Laue:** Forman (1969); Hoddeson, et. al. (1992), ch. 1.

p. 60 **"physics of dirt":** Hoddeson, et al. (1992), 88–160; Pauli quote is on p. 181, note 458. The brief survey presented in these pages of the history of the quantum theory of solids is largely based on the treatment found in Hoddeson, et al. (1987).

p. 61 **Wilson's tour de force:** Wilson (1931).

p. 61 **ghostlike notion of the "hole":** Peierls (1929). See also Hoddeson and Riordan (2001).

p. 61 **Semiconductors were enormously controversial:** See reference 218 in Hoddeson, et al. (1992). According to Wilson, the "canard" that silicon is a good metal lingered through the pre–World War II period.

p. 61 **adding impurities ("doping"):** Hoddeson, et al. (1992), 122–123.

p. 62 **Focusing on sodium:** Wigner and Seitz (1933, 1934).

p. 62 **"looked like it would open up":** John Bardeen (1977, 1984b).

p. 62 **"There were so many approximations":** Slater, quoted in Hoddeson, et al. (1992), 188.

p. 62 **a conglomeration of fields:** Weart, "The Solid Community," in Hoddeson, et al. (1992).

p. 62 **a dozen major reviews:** Hoddeson, et al. (1987).

p. 62 **monumental, almost 300-page:** Sommerfeld and Bethe (1933).

p. 62 **"Theory of the Solid State":** Wigner (1981).

p. 63 **In his thesis calculation:** John Bardeen (1937); David Pines, Meeting with Brian Pippard, etc., October 9, 1991.

p. 63 **"Wigner actually did most":** John Bardeen (1936, 1984b); Wigner and Bardeen (1935).

p. 63 ". . . worked practically independently": E. Wigner, c. fall 1934 or early spring 1935, note in John Bardeen's graduate student file, PRIN.

p. 64 **"the man beyond the scholar"**: The Society did not at this point expect to initiate any women into its ranks. Homans and Bailey (1959), 25.

p. 64 **His interviewing committee**: John Bardeen (1977b); Slichter (1992).

p. 64 **"I was placed before"**: Harvard Society of Fellows (1987).

p. 64 **"I'm sure it was Van Vleck"**: Harvard Society of Fellows (1987); Slichter (1995).

p. 64 **"Three years of job security"**: John Bardeen (1984a), 10.

p. 64 **"it was obvious"**: John Bardeen (1984b).

p. 64 **form of pancreatic cancer**: Meek (1935). John's older brother William died of a similar form of cancer in 1986. Henschel (1992, 1995).

p. 64 **buried beside Althea**: Henschel (1995).

p. 64 **"relaxations came through walks"**: Sullivan (1936).

p. 65 **"a quiet giant"**: Harold C. Bradley, "Reminiscences of Doctor Bardeen," in "Dedication of the Bardeen Memorial Laboratories," pamphlet, Charles R. Bardeen biographical file, UWA. Fifty-three years later, when John Bardeen died, Morton Weir, the chancellor of the University of Illinois used words that echoed those of Bradley, "A giant has passed from our midst." *Inside Illinois* 10, no. 10 (February 7, 1991).

p. 65 **"a prodigious worker"**: Clark (1967), 45.

p. 65 **establishment of a preceptorship**: Bradley (1992), 17–8.

p. 65 **Van Vleck's continuing influence**: John Bardeen (1977, 1980b). Also John Bardeen file, PRIN.

5
Many-Body Beginnings

p. 67 **"The idea was"**: John Bardeen (1977b).

p. 67 **". . . their promise of notable contribution"**: Brinton (1959), 67–68.

p. 67 **"a great privilege and inspiration"**: Bardeen to James B. Fisk, Bell Laboratories, October 29, 1973, UIUC-A.

p. 67 **". . . ordinary small talk of academic life"**: Homans and Bailey (1959), 29, 31–32.

p. 67 **"general broadening influence":** Harvard Society of Fellows (1987).

p. 67 **Ivan Getting:** John Bardeen (1977b).

p. 67 **theorist James Fisk:** Harvard Society of Fellows (1987).

p. 68 **Bardeen also befriended:** Harvard Society of Fellows (1987).

p. 68 **Fisk would be instrumental:** John Bardeen (1977b).

p. 68 **"for the nominal sum":** Garrett Birkhoff to Charles Slichter, October 19, 1992. The authors thank Slichter for sharing this letter with us. Garrett's father, George Birkhoff, a Harvard mathematics professor, was a senior fellow.

p. 68 **Eddington's authoritative text:** John Bardeen (1977b); Eddington (1923).

p. 68 **"To find out any physical quantity":** Eddington (1923).

p. 69 **immersed himself in the physics literature:** John Bardeen (1980b), 80.

p. 69 **"It was still possible":** John Bardeen (1977b).

p. 69 **100 or so German-Jewish physicists:** Weiner (1968).

p. 69 **Bethe's no-frills practical style of physics:** Bethe (1972).

p. 69 **the two theorists conferred:** John Bardeen (1977b).

p. 69 **"close correlation between":** Mott and Jones (1936); Wilson (1936). Also see John Bardeen (1980b).

p. 70 **help to establish solid-state:** Seitz (1940); Mott and Jones (1936); and Wilson (1936).

p. 70 **"tight-binding method":** Bardeen and Van Vleck (1939).

p. 70 **"I had more interaction with Van Vleck":** John Bardeen (1977b).

p. 70 **"... Slater remembered me":** John Bardeen (1984a).

p. 70 **"... prosaic, matter-of-fact type":** Slater (1975), 42.

p. 70 **Bridgman had managed:** Hoddeson, et al. (1992), 41.

p. 70 **Bridgman encouraged Slater:** Hoddeson and Baym (1980); Hoddeson, et al. (1992), 4.

p. 71 **"I was convinced":** Slater (1975), 5–6.

p. 71 **"Well here's something that needs":** Herring and Herring (2000).

p. 71 **"grinding out" results:** Herring (1992a).

p. 71 **preferred "to just keep butting away":** Herring and Herring (2000).

p. 71 **"strongest intellectual contacts":** Herring's "Recollections," in Mott (1980), 65–76.

p. 71 **"that the things he [Bardeen] was working on"**: Herring and Herring (2000).

p. 72 **"... influenced by the Hollywood culture"**: Seitz (1994), 67.

p. 72 **"two desperadoes were loose"**: Ibid., 67.

p. 72 **"We went to joint seminars"**: John Bardeen (1977b).

p. 72 **"dangling bonds"**: Shockley eventually wrote a paper in 1939 predicting surface states whenever two electronic bands within a finite periodic structure intersect. Hoddeson, et al. (1992), 468; Tamm (1932); Shockley (1939).

p. 72 **surface states also arise**: John Bardeen (1977b). In chemical language, the surface states are "the extra-orbitals leftovers. The other valence bonds are filled with electrons," Bardeen explained.

p. 72 **Shockley's thesis calculation**: John Bardeen (1977b); (1984b).

p. 72 **"empty lattice test"**: Shockley (1974).

p. 73 **Shockley received an enviable**: Riordan and Hoddeson (1997a), 81. Also see Hoddeson (1981b).

p .73 **"residue of stories"**: Herring and Herring (2000).

p. 73 **observation, not theory**: Walter (1990), 31.

p. 73 **"fierce in his inner disdain"**: Edwin C. Kemble, "Apostle of Ruthless Logic," lecture given at Percy Williams Bridgman Memorial Meeting, October 24, 1961, Harvard University, Cambridge, Massachusetts, AIP.

p. 73 **"indulged in no elaboration"**: Edward M. Purcell, "The Teacher and Experimenter," lecture given at Percy Williams Bridgman Memorial Meeting, October 24, 1961, Harvard University, Cambridge, Massachusetts, AIP.

p. 73 **"the new field of high pressure"**: Walter (1990), 49.

p. 74 **"... height of his productivity"**: John Bardeen (1980b), 81; John Bardeen to Maila Walter, July 28, 1981, UIUC-A.

p. 74 **phase change in cesium**: John Bardeen (1980b), 82. Also John Bardeen to Maila Walter, July 28, 1981, UIUC-A.

p. 74 **"This was one of the first cases"**: John Bardeen (1977b).

p. 74 **"Bridgman was very impressed"**: Harvard Society of Fellows (1987).

p. 74 **"The alkali metals were"**: John Bardeen (1938); (1980b), 81; (1977b).

p. 75 **"John never forgave"**: Seitz to L. Hoddeson, February 12, 2001.

p. 75 **Bardeen ever mentioning:** Bridgman (1927, 1936); Benjamin (1955), esp. 1–20; also Bridgman (1980).

p. 75 **"Operationalism":** Walter (1990), 2–4.

p. 75 **"The concept is synonymous":** Bridgman (1927), 5.

p. 75 **"penumbra" surrounding the operations:** See for example, Margenau (1956), 39.

p. 75 **emphasis on usefulness:** Schweber (1986).

p. 76 **"selectively recruit":** Zuckerman (1983).

p. 76 **"John just hangs on":** A. H. Bardeen to C. W. Bardeen, n.d., UWA.

p. 77 **"It was clear to me":** Bethe (1981), quoted in Hoddeson, et al. (1992), 210; John Bardeen (1980b).

p. 77 **X-ray emission spectrum:** Hoddeson, et al. (1992), 196, 199; Jones et al. (1934).

p. 77 **answered in 1951 by Viktor Weisskopf:** Weisskopf (1951).

p. 77 **Not until 1957:** Luttinger and Kohn showed that the occupation number has a sharp discontinuity at the Fermi surface, like a cliff that slopes gently down, then falls drastically, then slopes more gently again near the bottom. Landau (1957); Luttinger (1960).

p. 77 **". . . too elaborate a calculation":** John Bardeen (1977c).

p. 78 **in 1936 Bardeen could not:** John Bardeen (1937).

p. 78 **"random phase approximation":** John Bardeen (1980b).

p. 78 **"the most influential years":** Harvard Society of Fellows (1987).

p. 78 **Holy Grail:** Niels Bohr, Wolfgang Pauli, Werner Heisenberg, Felix Bloch, Lev Landau, Leon Brillouin, W. Elsasser, Yakov Frenkel, Ralph Kronig, and Einstein were among the many who tried and failed to develop a detailed first-principles theory of superconductivity. For a summary of research on superconductivity, from its discovery up to the Second World War, see Dahl (1992). For other references see Hoddeson, et al. (1992), 588, note 21, 141–153; also Hoddeson, et al. (1987), Section 4: "Superconductivity, 1929–1933."

p. 78 **"Only a number of hypotheses":** Sommerfeld and Bethe (1933), 555, 558.

p. 79 **"Every theory of superconductivity":** Hoddeson, et al. (1992), 144.

p. 79 **Léon Brillouin associated:** Hoddeson, et al. (1992), ch 2.

p. 79 **"have not materially":** Shoenberg (1938), ix.

p. 79 **a breakthrough in superconductivity:** London and London (1933).

p. 80 **The Londons had fled:** For insights into the London brothers, especially Fritz, see the biography of Fritz London by Gavroglu (1995); also Dahl (1992); and Hoddeson, et al. (1992).

p. 80 **existing current will flow forever:** London and London (1933), 348.

p. 80 **suggested that there is "rigidity":** London (1935).

p. 81 **"Every time I saw":** Jane Bardeen (1991b).

p. 81 **". . . you should support your wife":** John Bardeen (1977a).

p. 81 **the reproductive patterns of wasps:** Jane Bardeen to Mrs. J. R. Maxwell, July 2, 1936, BFC; Jane Bardeen (1991b).

p. 81 **"Dinner at the Wayside Inn":** Jane Maxwell to Mrs. J. R. Maxwell, June 28, 1936, BFC.

p. 81 **they drove to Pennsylvania:** Jane Bardeen to Mrs. J. R. Maxwell, July 2, 1936, BFC.

p. 81 **". . . I decided to try presence":** Jane Bardeen (1991a).

p. 81 **"resigning from Tech":** Jane Bardeen to Maxwell family, August 15, 1937, BFC.

p. 82 **had any "plans":** Jane Bardeen (1991b).

p. 82 **" 'that nice young man' ":** Jane Maxwell to Mrs. J. R. Maxwell, November 29, 1937, BFC.

p. 82 **"more philosophical about":** Jane Maxwell to Mrs. J. R. Maxwell, November 29, 1937, BFC.

6
Academic Life

p. 83 **a serious love interest:** Jane Maxwell to Mrs. J. R. Maxwell, May 26, 1938, BFC; Nier (1993).

p. 83 **"Isn't that sporting":** Jane Maxwell to Mrs. J. R. Maxwell, May 26, 1938, BFC.

p. 83 **"all of her Dana Hall":** Jane Bardeen to Mrs. J. R. Maxwell, December 6, 1938.

p. 84 **"on the map":** Gray (1951), 3, 415–418. Tate, who came to Minnesota as an instructor, was appointed dean in 1937.

p. 84 **"so highly of his work":** John Van Vleck to Johns Hopkins, n.d., UMP.

p. 84 **Minnesota offered Bardeen:** J. W. Buchta to John Bardeen, G31 Lowell House, March 29, 1938, UMP. According to an employment application in Bardeen's personnel file of the USOPM, he was earning $3,000 per year at Gulf.

p. 84 **when Harvard offered Nier:** Nier (1993).

p. 84 **lecture on the physics of metals:** Bardeen spoke there on the theory of "Electrical conduction and other transport phenomena." Francis Bitter and Foster Nix spoke on experiment, while Seitz and Mott spoke on theoretical subjects. Later, Bardeen expanded the material from his talk into a review article. John Bardeen (1940, 1980b).

p. 84 **"The date will probably be":** Jane Maxwell to Mrs. J. R. Maxwell, May 26, 1938; February 28, 1940, BFC.

p. 84 **"Johnny called tonight":** Jane Maxwell to Sue Maxwell, June 2, 1938, BFC.

p. 85 **"My mother would be pleased":** Jane Bardeen (1991b).

p. 85 **By getting married:** Jane Bardeen (1993b).

p. 85 **he moved his practice:** "Dr. R. J. [sic] Maxwell Has Practiced Here 56 Years," the *Washington (Pennsylvania) Reporter*, August 15, 1958, 36.

p. 85 **Reverend Dr. Lippencott:** Jane Bardeen (1993b).

p. 85 **Jane's immediate family attended:** Jim was following the model of his father and training to become a doctor. "Dr. R. J. [sic] Maxwell." the *Washington (Pennsylvania) Reporter*, August 15, 1958.

p. 85 **"traditional white and veil":** Jane Maxwell to Sue Maxwell, June 2, 1938, BFC.

p. 85 **fringes shimmied:** Henschel (1995).

p. 85 **the only snapshots:** Jane Bardeen (1993).

p. 85 **"That didn't please John":** Jane Bardeen (1993).

p. 85 **"We drove as far as Canton":** Jane Bardeen to Maxwell family, July 26, 1938, BFC.

p. 86 **"lousy luck with a tire":** Jane Bardeen to Maxwell family, July 26, 1938, BFC.

p. 86 **at the tools:** W. Bardeen private communication to L. Hoddeson, August 16, 2001.

p. 86 **"John was pretty angry":** Jane Bardeen to Maxwell family, July 26, 1938, BFC.

p. 86 **"stowed the stuff":** Jane Bardeen to Maxwell family, July 26, 1938, BFC.

p. 86 **"They killed the fatted calf"**: Jane Bardeen to Mrs. J. R. Maxwell, July 26, 1938, BFC.

p. 87 **"I think we are going"**: Ibid.

p. 87 **"found more happiness"**: Jane Bardeen to Maxwell family, July 26, 1938, BFC.

p. 87 **Heading west**: Jane Bardeen to Mrs. J. R. Maxwell, August 9, 1938, BFC.

p. 87 **"very good coffee"**: Jane Bardeen to Mrs. J. R. Maxwell, August 3, 1938, BFC.

p. 87 **thank-you notes**: Jane Bardeen to Mrs. J. R. Maxwell, August 1, 1938, BFC.

p. 87 **"Can't you see them"**: Jane Bardeen (1993).

p. 87 **contracted tuberculosis**: Jane Bardeen to Mrs. J. R. Maxwell, July 26, 1938, BFC.

p. 87 **made herself "scarce"**: Jane Bardeen (1993).

p. 87 **"It is a desperate fight"**: Jane Bardeen to Mrs. J. R. Maxwell, August 16, 1938, BFC.

p. 88 **"about four or five miles"**: Jane Bardeen to Mrs. J. R. Maxwell, August 22, 1938, BFC.

p. 88 **"We go so hard"**: Ibid.

p. 88 **"rooted in the cellar"**: Jane Bardeen to Betty Maxwell, September 3, 1938, BFC.

p. 88 **the few household goods**: Jane Bardeen to Betty Maxwell, September 12, 1938, BFC.

p. 88 **"a swell job of sanding"**: Ibid.

p. 88 **"go into John's capital"**: Jane Bardeen to the Maxwell family, October 1938, BFC.

p. 89 **". . . brings me tea with sugar"**: Ibid.

p. 89 **"Don't let this baby"**: Jane Bardeen (1993).

p. 89 **". . . going to produce a litter"**: Jane Bardeen to Betty Maxwell, January 19, 1939, BFC.

p. 89 **"The most amazing present"**: Jane Bardeen to Mrs. J. R. Maxwell, February 28, 1939, BFC. The present was from Jo Gerken and Florence Rumbaugh.

p. 89 **a letter of recommendation**: Helen Temple Cooke to Jane Maxwell Bardeen, June 26, 1939, BFC.

p. 89 **"a genetics project"**: Jane Bardeen (1991a).

p. 89 **"I'm really enjoying cooking"**: Jane Bardeen to Mrs. J. R. Maxwell, February 28, 1939; November 11, 1939, BFC.

p. 90   *". . . John really opened up"*: Jane Bardeen to Mrs. J. R. Maxwell, November 11, 1940, BFC.

p. 90   *"John hates to take medicine"*: Jane Bardeen to Mrs. J. R. Maxwell, February 28, 1939, BFC.

p. 90   **faculty billiards championship:** Young (1972).

p. 90   **the faculty always won:** Nier (1993).

p. 90   *"25 to 3 or something"*: Ibid.

p. 90   *"We beat them"*: Ibid.

p. 91   **the department's star athletes:** Ibid.

p. 91   **Bardeen could be quietly grumbling:** Ibid.

p. 91   *"Tell them what you know"*: Jane Bardeen (1993).

p. 91   *". . . a little extra money"*: John Bardeen (1977c).

p. 91   *"John had a hard week"*: Jane Bardeen to Mrs. J. R. Maxwell, August 7, 1939, BFC.

p. 91   *"a batch of blackberry jam"*: Jane Bardeen to Mrs. J. R. Maxwell, August 23, 1939, BFC.

p. 91   *"vacations are for concentrating"*: Jane Bardeen to Mrs. J. R. Maxwell, March 28, 1939, BFC.

p. 91   *"about the only time"*: Jane Bardeen to Susan Maxwell, January 10, 1940, BFC.

p. 91   **expensive dental work:** Jane Bardeen to Mrs. J. R. Maxwell, October 19, 1938, BFC.

p. 92   *"These difficulties have been almost"*: John Bardeen (1940), 88, 90.

p. 92   *"a perfect periodic lattice"*: Ibid., 90–94, 96.

p. 92   *"some phenomena for which"*: Ibid., 96, 111.

p. 93   *"Any labor pains"*: Jane Bardeen to Mrs. J. R. Maxwell, May 4, 1939, BFC.

p. 93   *". . . a big, fat girl"*: Jane Bardeen to Mrs. J. R. Maxwell, March 28, 1939, BFC.

p. 93   **made backup plans:** Ibid.

p. 93   *"a fine job on the baby's basket"*: Jane Bardeen to Mrs. J. R. Maxwell, May 4, 1939, BFC.

p. 93   *"He was so much fun"*: Jane Bardeen to Susan Maxwell, November 15, 1939, BFC.

p. 93   **little boy's babble:** Nier (1993).

p. 93   **frequently ended up playing:** Helen Buchta Gustafson to Jane Bardeen, April 15, 1991, BFC.

p. 93   *"Johnny and I went gallivanting"*: Jane Bardeen to Mrs. J. R. Maxwell, April 6, 1940, BFC.

p. 93 "... red cinnamon candy hearts": Jane Bardeen to Mrs. J. R. Maxwell, February 16, 1940, BFC.

p. 93 "The kind of love": Jane Bardeen to Mrs. J. R. Maxwell, December 15, 1939, BFC.

p. 94 John's "first love": Jane Maxwell to Mrs. J. R. Maxwell, November 29, 1937, BFC.

p. 94 "The closets are wonderful": Jane Bardeen to Mrs. J. R. Maxwell, August 18, 1940, BFC.

p. 94 "If Germany wins": Jane Bardeen to Dr. J. R. Maxwell, May 19, 1940, BFC.

p. 94 "Both of us feel": Jane Bardeen to Mrs. J. R. Maxwell, December 15, 1939, BFC.

p. 94 "try to be economical": Jane Bardeen to Mrs. J. R. Maxwell, November 11, 1940, BFC.

p. 94 "as excited as a small boy": Jane Bardeen to Betty Maxwell, March 22, 1940, BFC.

p. 94 he drove to Pittsburgh: Jane Bardeen to Dr. J. R. Maxwell, May 19, 1940, BFC.

p. 94 a colleague would sometimes present: Jane Bardeen to Susan Maxwell, January 10, 1940, BFC.

p. 95 an atomic bomb: Hoddeson, et. al (1993).

p. 95 separating the isotopes: Bardeen and Nier (1941); Nier (1993). Also John Bardeen (1977b).

p. 95 "On Friday afternoon": Booth et al., (1969), 28; Rhodes (1986), 332.

p. 95 "... a Hollywood version": Jane Bardeen to Mrs. J. R. Maxwell, May 10, 1940, BFC; *Minneapolis Star-Journal*, May 5, 1940.

p. 96 "John is teaching": Jane Bardeen to Mrs. J. R. Maxwell, November 11, 1940, BFC.

p. 96 "knew a lot about magnetism": Harvard Society of Fellows (1987).

p. 96 Lynn phoned from Washington: Jane Bardeen to Mrs. J. R. Maxwell, February 6, 1941, BFC.

p. 96 "John had been slated": Jane Bardeen to Maxwell family, November 11, 1940, BFC.

p. 96 "It is a completely new experience": Jane Bardeen to Betty Maxwell, March 22, 1940, BFC.

p. 96 "... less courageous than Florence": Jane Bardeen to Mrs. J. R. Maxwell, February 6, 1941, BFC.

p. 96 **Helen caused additional worry:** Jane Bardeen to Mrs. J. R. Maxwell, December 6, 1938; Jane Bardeen to Betty Maxwell, November 29, 1940, BFC.

p. 96 **"All of us are heartsick":** Jane Bardeen to Mrs. J. R. Maxwell, November 29, 1940, BFC.

p. 96 **"will ever be well again":** Ibid.

p. 97 **"Glenis would arrive":** Jane Bardeen to Dr. J. R. Maxwell, January 14, 1941, BFC.

p. 97 **"Being an only child":** Ibid.

p. 97 **She remembered laughter:** Morrison (1995); Jane Bardeen to Mrs. J. R. Maxwell, December 6, 1938; January 19, 1939; February 28, 1939, BFC.

p. 97 **"I shall not see the doctor":** Jane Bardeen to Mrs. J. R. Maxwell, February 6, 1941, BFC.

p. 97 **Heisenberg's uncertainty principle:** John Bardeen (1980b).

p. 98 **The disparity between these energies:** John Bardeen (1941, 1980b).

p. 98 **"shook his head":** Seitz (2001).

p. 98 **". . . getting a small energy gap":** John Bardeen (1980b).

7
Engineering for National Defense

p. 99 **"John and I both feel":** Jane Bardeen to Dr. and Mrs. J. R. Maxwell, April 10, 1940, BFC.

p. 99 **"Now one bond unites":** Keegan (2000), 55; Gilbert (1989), 64.

p. 100 **British radar:** Buderi (1996).

p. 100 **Battle of Britain:** Weinberg (1994).

p. 100 **National Defense Research Committee:** Zachery (1997); Hart (1998).

p. 101 **the government network:** For more detail, see Kevles (1978).

p. 101 **to expand the NOL:** Smaldane (1977), 167–169; also Rowland and Boyd (1953), 18.

p. 101 **Rumbaugh had been brought:** John Bardeen to George Hamlin, Naval Surface Warfare Center, Dahlgren, Virginia, July 31, 1985, UIUC-A; Anspacher (1992).

p. 101 **By appropriately selecting the current:** Rowland and Boyd (1953), 71; Bennett (1987).

p. 101 **the NOL aggressively recruited:** Rowland and Boyd (1953).

p. 101 **persuaded Bardeen to join:** Anspacher (1992); Lt. Cmdr. Ralph D. Bennett, U.S.N.R., to John Bardeen, March 19, 1941; John Bardeen personnel file, USOPM. Bennett mentioned that "Wetzel, Johnson and others" had brought up Bardeen's name as a likely candidate to help them with the kinds of problems they faced.

p. 102 **persuaded Buchta to spare Bardeen:** Correspondence between John Bardeen and Lt. Cmdr. Ralph D. Bennett, U.S.N.R., Naval Ordnance Laboratory, Washington, D.C., March 19, 1941; March 28, 1941; April 9, 1941; April 17, 1941; April 26, 1941; May 13, 1941; May 16, 1941; May 19, 1941; and June 4, 1941, USOPM.

p. 102 **hiring civilian scientists:** "Putting Science to Sea in World War II: The Development of the Modern Naval Ordnance Laboratory," *On the Surface* 10, no. 14 (Special R&D Technology Edition, July 10, 1987) 7–9.

p. 102 **the navy increased its offer to $17:** Jay Buchta to Lt. Cmdr. R. D. Bennett, May 16, 1941, USOPM.

p. 102 **"You can't have this baby":** Jane Bardeen (1993).

p. 102 **overburdened the city's hospitals:** Stevens (1941), 51–52.

p. 103 **boarded the train to Pittsburgh:** Jane Bardeen (1991a).

p. 103 **"I always liked having a brother":** Jane Bardeen (1993).

p. 103 **the train ride felt completely different:** Jane Bardeen to Mrs. J. R. Maxwell, February 6, 1941, BFC.

p. 103 **Automobiles crawled along:** See Brinkley (1988) and Stevens (1941).

p. 103 **part of Fairfax Village:** Anspacher (1992).

p. 103 **Jane's sister Betty:** Jane Bardeen (1993).

p. 103 **"hottest place in Washington":** John Bardeen to George Hamlin, Naval Surface Warfare Center, Dahlgren, Virginia, July 31, 1985; Mitchell (1987).

p. 103 **avoid the fumes from a paint shop:** John Bardeen to George Hamlin, Naval Surface Warfare Center, Dahlgren, Virginia, July 31, 1985. The authors thank George Hamlin for sharing these documents from his personal correspondence files.

p. 104 **"objectionable conditions":** R. H. Park to Secretary of the Navy Frank Knox, November 20, 1942, NARA, RG 298, Box 2.

p. 104 **NOL also suffered less interference:** Rear Adm. W. R. Purnell to Rear Adm. J. A. Furer, coordinator of R & D, September 6, 1942, NARA, RG 298, Box 96.

p. 104 **physicist Charles Kittel:** Nick Holonyak to L. Hoddeson, February 2, 2001; Charles Kittel to Jane Bardeen, February 11, 1991, BFC.

p. 104 **". . . swear he was sound asleep":** Anspacher (1992).

p. 104 **"everything John's group worked on":** Charles Kittel to Jane Bardeen, February 11, 1991, BFC.

p. 104 **They tested the scale models:** Gilbarg (1992).

p. 104 **". . . no intellectual snobbery":** Anspacher (1992).

p. 105 **LaCoste-Romberg gravity meter:** Bardeen and Kendall (1941).

p. 105 **A more promising design:** John Bardeen (1942); Bardeen and Keithley (1942).

p. 105 **target areas of submarines:** Bardeen et al. (1942).

p. 105 **classify the magnetic signatures:** Bardeen and Shortly (1942).

p. 105 **Bardeen often consulted with:** John Bardeen travel orders dated March 4, 1942; March 14, 1942; April 6, 1942; April 21, 1942; June 23, 1942; September 21, 1942; November 25, 1942; March 31, 1943; and June 15, 1943, John Bardeen personnel file, USOPM.

p. 105 **visits to naval stations:** John Bardeen travel orders dated November 29, 1941; February 4, 1943; March 10, 1943; May 3, 1943; May 24, 1943; July 5, 1943; August 3, 1943; and August 16, 1943, John Bardeen personnel file, USOPM; Seattle reference is from John Bardeen (1977c).

p. 105 **his work on torpedoes:** John Bardeen (1977c); John Bardeen travel orders dated May 15, 1943 and July 27, 1943, John Bardeen personnel file, USOPM.

p. 105 **". . . spent a very enjoyable afternoon":** John Bardeen to Stephen Brunauer, Clarkson College, Potsdam, New York, July 30, 1980. There are travel orders dated June 30, 1943, for Bardeen to visit the Institute for Advanced Study on or around July 2, 1943, USOPM.

p. 106 **the Field Analyst section:** John Bardeen (1977c); Kaplan (1946), 35.

p. 106 **Bardeen headed one of five:** Kaplan (1946), 38.

p. 106 **"If he wanted to send work":** Jane Bardeen (1991a).

p. 107 **Bardeen became head:** Kaplan (1946), 40–45.

p. 107 **He planned and directed:** Job Classification Sheet for John Bardeen, September 20, 1943, USOPM.

p. 107 **Not only did compartmentalization:** Kaplan (1946), 45.

p. 107 **"project orientation":** John Bardeen to A. H. Hausrath, August 16, 1967, UIUC-A. Bardeen's own work status changed from contract to civil service in June 1943 when the navy changed its policy regarding contract work. His new title was Principal Physicist, but his duties, almost entirely administrative, hardly changed. Memos from Personnel Officer R. F. Cautley to Bardeen, June 10, 1943, from F. L. Reichmuth, commandant, NOL Navy Yard, Washington, D.C., to the assistant secretary of the navy; from Bardeen to purchase officer via officer-in-charge, NOL, September 30, 1943, USOPM.

p. 107 **to keep technical details from anyone:** Jane Bardeen (1991a).

p. 107 **fifty-three hours:** Affidavit—Occupational Classification (Industrial) for John Bardeen, 8 December 30, 1943, USOPM.

p. 107 **gave up smoking:** William Bardeen, private communication to L. Hoddeson, November 19, 1999.

p. 108 **"Eugene got worried":** Seitz (2001).

p. 108 **". . . said 'start learning'":** William F. Whitmore to John Bardeen, January 11, 1988, UIUC-A.

p. 108 **bureaucratic straightjacket:** John Bardeen to chief of the Bureau of Ordnance via commandant, U.S. Navy Yard, October 31, 1942, USOPM.

p. 108 **his superiors noticed:** Reports of Efficiency Rating on John Bardeen, August 29, 1944; March 31, 1945; and September 11, 1945, USOPM.

p. 108 **". . . NOL's low-score golfers":** "Dr. John Bardeen is World Famous As Versatile Theoretical Physicist," *Report*, July 1945, 8. Charles V. Jakowatz who bowled both with and against Bardeen during those years later told Nick Holonyak that Bardeen was "tough." N. Holonyak to L. Hoddeson, January 31, 2001.

p. 108 **He won a medal:** John Bardeen to George Hamlin, Naval Surface Warfare Center, Dahlgren, Virginia, July 31, 1985, UIUC-A.

p. 109 **"he'd get scolded":** Jane Bardeen (1993).

p. 109 **Bill and Charlotte sent Glenis home:** G. W. Beach to John Bardeen, February 18, 1943; Williams and Rae, Barristers and

Solicitors, to John Bardeen, March 3, 1943; Donald Beach to John Bardeen, December 8, 1942; Williams and Rae to John Bardeen, March 5, 1943, BFC.

p. 109 **developing a drinking problem:** John Bardeen, "Career of Tom Bardeen with Gulf Oil Company," n.d., family document. The authors thank Lynn Maynard and Ellen Stiehl for kindly sending us a copy.

p. 109 **Security regulations prevented:** John Bardeen to Professor Sigmund Hammer, Department of Geology and Geophysics, University of Wisconsin-Madison, February 25, 1981, UIUC-A.

p. 109 **"I am thinking of you constantly":** James Bardeen to Jane Bardeen, November 30, 1943, BFC.

p. 109 **the best possible medical care:** Jane Bardeen (1993).

p. 109 **"Is she pretty?":** Jane Bardeen (1993).

p. 109 **"help your mommy":** John Bardeen to Bill Bardeen, April 1943, BFC.

p. 109 **". . . help bring the family back":** John Bardeen to Jay Buchta, June 10, 1944, UMP.

p. 110 **"invited to have another":** Jane Bardeen (1993).

p. 110 **"With my present family responsibilities":** John Bardeen to Jay Buchta, May 6, 1945, UMP.

p. 110 **bitterly disappointed:** Telegram from Jay Buchta to John Bardeen, May 24, 1945, Bardeen

p. 110 **the university did not yet recognize:** Alfred O. C. Nier, Acting Chair, Department of Physics, to Dean J. W. Buchta, April 18, 1951, Bardeen personnel file, UMP; Al Nier (1993); Jane Bardeen (1991a).

p. 110 **The field grew so rapidly:** Hoddeson, et al. (1992), ch. 9. Slater pushed for the application to solid-state research of the compact helium liquefier that Collins developed during the war for portable oxygen generators to be used on submarines and airplanes. The subsequent wide use of the Collins liquefier enabled low-temperature physics studies to be conducted at many institutions.

p. 110 **"The plans are":** John Bardeen to Jay Buchta, May 6, 1945, Bardeen personnel file, UMP.

p. 111 **created the new solid-state department:** Kelly (1943).

p. 111 **cutting-edge research on semiconductors:** Torrey and Whitmer (1948), viii. The classic work on this radar program is Guerlac (1987). See also Hoddeson (1994); Henry Torrey to L. Hoddeson, June 6, 1993. The Telecommunications Research Establishment played a similar role in England, coordinated with the General Electric Company; British Thompson Houston, Ltd.; and Oxford University.

p. 111 **"All of this art":** Kelly (1943).

p. 111 **modeled his new solid-state department:** Kelly (1950).

p. 111 **"He thought that":** Harvard Society of Fellows (1987).

p. 111 **Bardeen learned more:** Technical Employment Manager R. A. Deller, Bell Telephone Laboratories, to John Bardeen, May 26, 1945, UIUC-A; John Bardeen to Officer-in-Charge W. G. Schindler, via L. H. Rumbaugh and personnel office, NOL, Washington, D.C., June 26, 1945, USOPM.

p. 112 **"$550 per month":** According to an application for employment as a federal consultant, Bardeen was making $11,400 at Bell Labs by the time he was ready to leave in March 1951, only six years later. USOPM.

p. 112 **not intended to interfere:** Technical Employment Manager R. A. Deller, Bell Telephone Laboratories, to John Bardeen, May 26, 1945, UIUC-A.

p. 112 **"It was a difficult choice":** John Bardeen to Jay Buchta, June 11, 1945, UMP.

p. 112 **"BTL appears to offer":** Ibid.

p. 112 **"It might be a good idea for Minnesota":** Ibid.

p. 112 **"which have little chance":** John Bardeen to Officer-in-Charge Capt. W. G. Schindler, via L. H. Rumbaugh and personnel office, NOL, Washington, D.C., June 26, 1945, USOPM.

p. 113 **"no more will be trained":** Ibid.

p. 113 **"the armed services have control of scientific talent":** Ibid.

p. 113 **"Dr. Bardeen's decision":** L. H. Rumbaugh to officer-in-charge, NOL, July 3, 1945, USOPM.

p. 113 **"really worked on":** Marlowe (1992).

p. 113 **"I consider Dr. Bardeen's request":** Officer-in-Charge Capt. W. G. Schindler, NOL, to John Bardeen, via personnel officer and Dr. L. H. Rumbaugh, July 4, 1945, USOPM.

p. 113 **"outstanding contributions":** Memo to the Board on Awards for Civilian Service from Officer-in-Charge Capt. W. G. Schindler, USN, NOL, November 29, 1945, USOPM.

p. 114 **"accomplishments are not considered":** Secretary of the Navy James V. Forrestal to Rear Adm. F. L. Reichmuth, commandant, Navy Yard, Washington, D.C., June 11, 1946; G. B. Davis, commandant, Potomac River Naval Command, to John Bardeen, September 4, 1946, USOPM.

8
The Transistor

p. 115 **They had shipped:** Jane Bardeen to Mrs. J. R. Maxwell, October 14, 1945, BFC.

p. 115 **Betsy was now:** Ibid.

p. 115 **"The nightmare really started":** Ibid.

p. 116 **"found 35 miles":** Ibid.

p. 116 **"a stinker":** Ibid.

p. 116 **"scoured the bathroom":** Ibid.

p. 116 **"Baths are infrequent":** Jane Bardeen to Mrs. J. R. Maxwell, October 25, 1945, BFC.

p. 117 **"look brighter next time I write":** Jane Bardeen to Mrs. J. R. Maxwell, October 14, 1945, BFC.

p. 117 **"shoveling out the dirt":** Jane Bardeen to Mrs. J. R. Maxwell, October 25, 1945, BFC.

p. 117 **Watchung Mountain Reservation:** Bardeen family interview (1992); W. Bardeen (1995), BFC.

p. 117 **"comfortable presence":** Betsy Bardeen Greytak, Bardeen family interview (1992), BFC.

p. 117 **". . . color films of the children":** Jane Bardeen to Mrs. J. R. Maxwell, April 25, 1948, BFC.

p. 117 **John's half-sister Ann:** Ann Bardeen Henschel, Bardeen family interview (1992), BFC.

p. 117 **"He thought women":** John Bardeen (1977a).

p. 118 **she accepted the internship:** Ann Bardeen Henschel, Bardeen family interview (1992); Henschel (1992).

p. 118 **found time for squash:** Bardeen family interview (1992).

p. 118 **he bowled often enough:** Jane Bardeen to Mrs. J. R. Maxwell, January 17, 1949, BFC.

p. 118 **an excuse to quit the club:** Jane Bardeen (1993).

p. 118 **"a good match":** John Bardeen (1994), 73.

p. 118 **"her cooking would have been":** Herring and Herring (2000).

p. 118 **Herring's sport was tennis:** Ibid.

p. 119   **a warm and gracious woman:**   Herrring and Herring (2000); Seitz (2001).

p. 119   **"in a gentle way to hold the reigns":**   Seitz (2000).

p. 119   **Jane grew less tolerant:**   Bardeen family interview (1992).

p. 119   **". . . a very significant regression":**   Pearson (1992), 240.

p. 119   **"Aren't we having a nice winter season?":**   Herring and Herring (2000).

p. 119   **to risk revealing:**   Jane Bardeen (1991a).

p. 119   **"he wanted some part of himself":**   Jane Bardeen (1993b); Bardeen family interview (1992).

p. 120   **"be doing some physics":**   Jane Bardeen to Mrs. J. R. Maxwell, October 25, 1945, BFC.

p. 120   **"first love":**   Jane Bardeen (1991a); Jane Maxwell to Mrs. J. R. Maxwell, November 29, 1937, BFC.

p. 120   **struggle with hearing loss:**   Herring and Herring (2000).

p. 120   **In the first decades:**   The Bell Telephone Laboratories was officially incorporated on January 1, 1925, but the research groups out of which the corporation emerged had been created over a decade earlier. Hoddeson (1981c), 512–544.

p. 120   **Those whose labs overlooked:**   Ibid.; see also Fagen (1975). For the information about doughnuts and coffee, the authors are grateful to Terry Eisinger, personal communication to L. Hoddeson, September 1995.

p. 120   **a "country lab experiment":**   Mervin Kelly memorandum to Oliver Buckley, January 1945, Box 53, pp. 18, 25, AT&T.

p. 121   **"a gigantic technological warren":**   Bernstein (1984), viii.

p. 121   **showed up with the head cold:**   Jane Bardeen to Mrs. J. R. Maxwell, October 25, 1945, BFC.

p. 121   **regular initiation ritual:**   N. Holonyak to L. Hoddeson, February 3, 2001.

p. 121   **". . . this is only fair":**   undated newspaper clipping, UIUC-P.

p. 121   **small, fourth-floor office:**   See Bell Labs phone directories for 1945 and 1946, AT&T.

p. 122   **"You'll find that":**   Herring and Herring (2000).

p. 122   **"layoff days":**   W. Brattain (1964).

p. 123 **the Purdue group had built:** Guerlac (1987); Hoddeson (1994); Seitz and Einspruch (1998), esp. ch. 11; Torrey and Whitmer (1948).

p. 123 **"Now I know":** F. Seitz to L. Hoddeson, February 12, 2001.

p. 123 **connecting the East and West coasts:** Hoddeson (1981c)

p. 124 **evolved into Bell Telephone Laboratories:** Hoddeson (1981c).

p. 125 **used as a "click reducer":** N. Holonyak to L. Hoddeson, February 3, 2001.

p. 125 **"out in the elements":** Wooldridge (1976).

p. 125 **"orders of magnitude away":** Ibid.

p. 125 **copper oxide amplifier:** Shockley described the design, which would now be called a "Schottky-gate-field-effect" transistor, in his laboratory notebook on December 29, 1939. See also Riordan and Hoddeson (1997a).

p. 125 **"I laughed at him":** W. Brattain (1964).

p. 126 **Ohl noticed that a particular:** Riordan and Hoddeson (1997a), 88–89; also (1997b).

p. 126 **The work that led to the transistor:** Bardeen BNB 20780 October 23, 1945, p. 3, AT&T.

p. 127 **"I cannot overemphasize":** W. Brattain (1964).

p. 128 **"much bomb damage":** John Bardeen (1987), 70–71.

p. 128 **"I told [Muriel]":** John Bardeen to Jane Bardeen, July 9, 1947, BFC.

p. 128 **"John and Bill turned up":** Muriel Kittel to Jane Bardeen, February 12, 1991, BFC.

p. 129 **". . . heads at opposite ends":** John Bardeen (1987), 70–71. Also W. Shockley to Ralph Bown, July 21, 1947, STAN.

p. 129 **"Sorry we can't be together":** John Bardeen to Jane Bardeen, July 15, 1947, BFC.

p. 129 **to go rock climbing:** John Bardeen to Jane Bardeen, July 19, 1947, BFC.

p. 129 **"One of my shoes":** John Bardeen to Jane Bardeen, July 20, 1947, BFC.

p. 130 **"to get to Paris next Sunday":** John Bardeen to Jane Bardeen, July 19, 1947, BFC.

p. 130 **"the hottest day":** John Bardeen to Jane Bardeen, August 1, 1947, BFC.

p. 130   "spoke to him in our high school French": John Bardeen (1977d).

p. 130   "As you must have gathered": John Bardeen to Jane Bardeen, August 1, 1947, BFC.

p. 131   "very hard work": Ibid.

p. 131   "I'm a lazy physicist": W. Brattain (1963).

p. 131   "to anybody in the group": W. Brattain (1964).

p. 131   "Wait a minute": Ibid.

p. 131   "We could vary the photo emf": Ibid.

p. 131   his own work on dislocations: John Bardeen (1978).

p. 131   He told no one else: Shockley (1976), 609. See also Riordan, et al. (1999), S336–S345.

p. 132   "Come on John": W. Brattain (1963).

p. 132   "The geometry was essentially": W. Brattain (1964).

p. 134   "a simple way in which": John Bardeen (1978).

p. 134   "I told my driving group": W. Brattain (1964).

p. 134   "high-back-voltage germanium": W. Brattain BNB 18194, December 8, 1947, p. 171. Also see Shockley (1976), 610.

p. 134   "John Bardeen was great at": W. Brattain (1964).

p. 135   "This is the opposite": W. Brattain BNB 18194, December 8, 1947.

p. 135   "Bardeen suggests that the surface field": Ibid., 176–177.

p. 136   negative ions in the electrolyte: W. Brattain (1964). See also Riordan and Hoddeson (1997a), 134.

p. 135   "I can remember the green color": W. Brattain (1964), 30.

p. 136   "This voltage amplification was independent": W. Brattain BNB 18194, December 15, 1947, p. 192; W. Brattain (1964).

p. 137   "We knew that something different was happening": Harvard Society of Fellows (1987).

p. 137   "holes were flowing": John Bardeen (1957).

p. 137   "What we didn't know then": W. Brattain (1963).

p. 137   "really could not have occurred": N. Holonyak to L. Hoddeson, February 2, 2001.

p. 137   closed his eyes to: Bondyopadhyay (1998).

p. 137   abstruse mathematical paper: Davydov (1938).

p. 138   put forth by Holonyak: Holonyak (1992).

p. 138   "There was a period in which": W. Brattain (1964).

p. 138   "our first thought was": John Bardeen (1963).

p. 138   **"that holes could flow through the bulk"**: John Bardeen (1963). That holes flow into the bulk was later mentioned in Bardeen and Brattain's joint paper presented at the November 1948 National Academy of Sciences meeting, and in their article on the point-contact transistor submitted to the *Physical Review* in December 1948. They wrote, "[H]oles may flow either through the surface layer or through the body of the semiconductor." Bardeen and Brattain (1949), 1211; John Bardeen (1978). See also W. Brattain (1964); Riordan and Hoddeson (1997a), 153–155; (1997b).

p. 138   **"The observed effect"**: John Bardeen (1963).

p. 139   **"I slit carefully"**: W. Brattain (1964).

p. 140   **"I had an amplifier"**: W. Brattain (1963).

p. 140   **"It was one of those days"**: Jane Bardeen (1991c).

p. 140   **"at each level of supervision"**: W. Brattain (1964).

p. 140   **"The circuit was actually spoken over"**: W. Brattain BNB 21780, December 24, 1947, pp. 7–8.

p. 141   **"When A [the gold electrode] is positive"**: John Bardeen BNB 207880, December 24, 1947, p. 72.

9
The Break from Bell

p. 142   **"swept in from the Atlantic"**: *New York Times,* December 27, 1947, 1–2.

p. 142   **"great sport (?) stumbling"**: Jane Bardeen to Mrs. J. R. Maxwell, January 1, 1948; January 7, 1948, BFC. The quizzical parenthetical question mark is Jane's.

p. 143   **"thought he would have"**: John Bardeen (1990a).

p. 143   **he took the train to Manhattan:** Riordan and Hoddeson (1997a), 142–144.

p. 144   **"until the patent arrangements"**: Jane Bardeen to Mrs. J. R. Maxwell, January 1, 1948.

p. 144   **"Your personalities are as different"**: Harry C. Hart to John Bardeen and Walter Brattain, April 13, 1964, BFC.

p. 144   **"that he could write a patent"**: W. Brattain (1974), 25.

p. 144   **"turned up in a number"**: John Bardeen (1990a).

p. 145   **John Shive demonstrated:** Riordan and Hoddeson (1997a), 153–155.

p. 145   **an n-p-n rather than a p-n-p:** Ibid., 147–151.

p. 145   **"went off by himself"**: W. Brattain (1964).

p. 145 **"jumped in with both feet"**: John Bardeen (1978).

p. 145 **with Kelly's secrecy order in place:** Harry C. Hart to John Bardeen and Walter Brattain, April 13, 1964, UIUC-A.

p. 146 **"I think if somebody"**: W. Brattain (1974), 28.

p. 146 **"it is probable that"**: Handel (1998), 8–9.

p. 146 **Bardeen and Brattain sent a letter:** Bardeen and Brattain (1948).

p. 146 **"very strongly that most restrictions"**: Walter Brattain quoted in *Time* magazine, June 29, 1959, 58.

p. 146 **"John, you're just the man"**: W. Brattain, "How the Transistor was Named," n.d., WHIT.

p. 147 **"Pierce, that is it!"**: Ibid.

p. 147 **"We have called it the Transistor"**: Bown, cited in Riordan and Hoddeson (1997a), 164.

p. 147 **"Everything went well"**: John Bardeen (1990a).

p. 148 **"chauffeured home through the blazing heat"**: Jane Bardeen to Maxwell family, July 1, 1948, BFC.

p. 148 **Bown's presentation and Shockley's articulate handling:** Riordan and Hoddeson (1997a), 164.

p. 148 **"we read the *New York Times*"**: John Bardeen (1990a).

p. 148 **" a device called"**: *New York Times*, June 30, 1948, 46.

p. 148 **"engineers believe it will"**: *New York Herald Tribune*, July 1, 1948.

p. 148 **"would just buy a germanium dial"**: W. Brattain (1964).

p. 148 **"A radio set without"**: Jane Bardeen to Mrs. J. R. Maxwell, April 25, 1948, BFC.

p. 148 **a carefully posed image:** Riordan and Hoddeson (1997a), 167.

p. 148 **"Walter sure hates"**: Holonyak (1993b).

p. 149 **"get a haircut"**: Jane Bardeen to Mrs. J. R. Maxwell, January 17, 1949, BFC.

p. 149 **he "has been under considerable pressure"**: Jane Bardeen to Mrs. J. R. Maxwell, October 26, 1948, BFC.

p. 149 **"Dad was very proud"**: Bardeen family interview (1992).

p. 149 **"How Dry I Am"**: Holonyak (1992), 41–42.

p. 149 **transistorized hearing aids:** Betsy Bardeen Greytak, Bardeen family interview (1992).

p. 149 **In recognition of Alexander:** Riordan and Hoddeson (1997a), 205.

p. 150 **"wanted to have everything"**: Herring and Herring (2000).

p. 150 **"a great measure of confidence"**: Herrring and Herring (2000).

p. 150 **offices on a different floor:** Bell Labs phone books, 1948–1949, AT&T.

p. 150 **"probably knew was a blind alley"**: Michael Riordan, quoted in Del Guercio (1998).

p. 150 **"but these solutions"**: John Bardeen to M. Kelly, May 24, 1951, BFC.

p. 151 **"in a highly emotional state"**: Ibid.

p. 151 **"I am still interested in spending"**: John Bardeen to Alvin M. Weinberg, September 23, 1948, AT&T. See also John Bardeen to James Fisk, April 22, 1949, STAN.

p. 151 **"You make a very strong case"**: John Bardeen to A. M. Weinberg, April 22, 1949, AT&T. The authors would like to thank Michael Riordan for sharing this letter with us.

p. 151 **"has been in contact"**: John Bardeen to J. B. Fisk, April 22, 1949, Stanford Archives. The authors are grateful to Michael Riordan for sharing this letter with us.

p. 151 **"The rising anti-intellectual sentiment"**: Jane Bardeen to Mrs. J. R. Maxwell, May 25, 1949, BFC.

p. 152 **"conditions if anything"**: John Bardeen to M. Kelly, May 24, 1951, BFC.

p. 152 **"he wanted to study diffusion"**: Herring and Herring (2000); Herring (1974).

p. 152 **"stopped in at a luggage store"**: Herring and Herring (2000).

p. 153 **bought Jane a corsage:** Jane Bardeen to Mrs. J. R. Maxwell, April 10, 1948, BFC.

p. 153 **"Dad and I went skating"**: James Bardeen to Mrs. J. R. Maxwell, January 4, 1949, BFC.

p. 153 **"Even had candles"**: Jane Bardeen to Mrs. J. R. Maxwell, May 25, 1949, BFC.

p. 153 **"could be very determined"**: Jane Bardeen to Mrs. J. R. Maxwell, April 18, 1948, BFC.

p. 153 **never lost his love of sweets:** Jane Bardeen to Mrs. J. R. Maxwell, January 17, 1949, BFC.

p. 153 **"One day V^2"**: Walter Kohn to Jane Bardeen, April 4, 1991, BFC.

p. 153 **"It showed me a great scientist":** Ibid.

p. 154 **the "Bardeen number":** Ravin Bhatt to Jane Bardeen, February 15, 1991, BFC.

p. 154 **". . . infinity on that scale":** Bhatt (1992).

p. 154 **"Bardeen was fed up":** For Brattain reference see Riordan and Hoddeson (1997a). See also P. Anderson (1992a).

p. 154 **With Fisk's approval:** John Bardeen to M. Kelly, May 24, 1951, BFC.

p. 155 **Bardeen received a phone call:** John Bardeen, handwritten note, May 16, 1950, UIUC-A. See also John Bardeen (1973).

p. 155 **Serin, an experimental physicist:** Torrey (1983); Maxwell (1983).

p. 155 **Emanuel Maxwell:** Hoddeson, et al. (1992), 548–549.

p. 155 **"Serin coming to me":** Torrey (1983).

p. 155 **"These results indicate that electron–lattice interaction":** John Bardeen (1950a).

p. 156 **To secure priority:** John Bardeen (1950a); (1950b), 167–168; Hoddeson, et al. (1992), 549–550.

p. 156 **simultaneous discovery:** The classic discussion of simultaneous discovery in science is Kuhn (1989).

p. 156 **several researchers, including Bardeen:** Fröhlich (1950).

p. 156 **When Fröhlich learned:** Ibid.

p. 156 **"It was very common":** Herring and Herring (2000).

p. 157 **"Although there were mathematical difficulties":** John Bardeen (1973), 15–57.

p. 157 **"There are very few":** John Bardeen to M. Kelly, May 24, 1951, BFC.

p. 157 **the discovery of the Lilienfeld patents:** The fact that Welker had applied for a patent for the junction transistor was apparently not known in the United States. Handel (1998).

p. 157 **"We got the patent!":** Betsy Bardeen Greytak, Bardeen family interview (1992).

p. 158 **"I am convinced I want to go back":** Seitz (1992b).

p. 158 **"decided to change the physics curriculum":** Seitz (1994), 194–196.

p. 158 **Unfortunately, there was no money:** Seitz (1997).

p. 158 **Everitt assured Seitz:** Frederick Seitz to John Bardeen, February 19, 1951; Seitz (1992b).

p. 158 **"keep cool":** Seitz (1993).

p. 158　**Everitt pieced together:** Seitz (1994), 219; del Guercio (1998).

p. 159　**"was not quite enough":** Seitz (1993).

p. 159　**"I have found more":** F. Seitz to John Bardeen, February 19, 1951, UIUC-A.

p. 159　**Bardeen leaned toward accepting:** G. M. Almy to John Bardeen, March 31, 1951; W. L. Everitt to John Bardeen, April 16, 1951, UIUC-P.

p. 159　**"One Friday," recalled Brattain:** Riordan and Hoddeson (1997a), 191.

p. 159　**"There was a reorganization here":** Ibid., 191; John Bardeen to G. M. Almy, April 6, 1951, UIUC-P.

p. 159　**"Oh, don't you bother":** Fisk, quoted in Seitz (1992b). See also Riordan and Hoddeson (1997a), 191.

p. 159　**"that in no case would you be asked":** William L. Everitt to John Bardeen, April 16, 1951, UIUC-P.

p. 160　**"everyone I have spoken to":** F. Seitz to John Bardeen, April 27, 1951, UIUC-A.

p. 160　**Seitz enlisted his wife Betty:** F. Seitz to John Bardeen, April 5, 1951, UIUC-A.

p. 160　**Bardeen accepted the Illinois offer:** Record of Training and Professional Experience, Application for position at University of Illinois, Bardeen file, UIUC-P. By 1955, his salary would grow to $13,500.

p. 160　**warm letter of welcome:** F. Seitz to John Bardeen, April 28, 1951, BFC.

p. 160　**"at least for the next year or two":** John Bardeen to Mervin Kelly, May 24, 1951, BFC.

p. 160　**"My difficulties stem":** Ibid.

p. 161　**"was well aware of the situation":** John Bardeen to M. Kelly, May 24, 1951, BFC.

p. 161　**"I seriously considered":** Ibid.

p. 161　**"Before making the decision":** Ibid.

p. 162　**"To summarize":** Ibid.

p. 162　**"And when Bardeen makes up his mind":** W. Brattain (1974), 33.

p. 162　**recognized the hazards:** F. Seitz to L. Hoddeson, February 12, 2001.

p. 162　**"glass ceiling":** Riordan and Hoddeson (1997a), 225.

p. 162 **high "mental temperature":** More detail on this episode can be found in Riordan and Hoddeson (1997a), 225–253.

p. 162 **". . . could stop here for a few days":** Seitz (2000).

p. 163 **"She did it for me":** Seitz (2001).

p. 163 **"not in any serious way":** Seitz (2000).

p. 163 **" . . . sleight of hand tricks":** Seitz (2000); F. Seitz to Professor and Mrs. John Bardeen, October 18, 1989.

p. 163 **"the traitorous eight":** For a more complete story, see Riordan and Hoddeson (1997a), 225–253.

p. 163 **and achievied notoriety:** For more information on Shockley see the manuscript by Joel Shurkin, *Broken Genius.*

p. 164 **the house sold for much more:** Seitz said that he had heard this story from several sources. Seitz (1998).

p. 164 **Driving first to Madison:** Bill Bardeen recalls they had the DeSoto from 1951 until 1958, when Jim took it along to college. W. Bardeen, private communication to L. Hoddeson, May 15, 2000.

p. 164 **"To John Bardeen":** Holonyak (1992), 39; Shockley (1950).

10
Homecoming

p. 165 **Homecoming:** The authors are grateful to Fernando Irving Elichirigoity for contributions to this chapter on the history of the University of Illinois Department of Physics.

p. 165 **"gasped because":** Jane Bardeen, quoted in the *Champaign-Urbana News Gazette,* October 10, 1991.

p. 166 **subdivisions replacing soybeans:** Hansen (1974), 159–161; Baker and Miller (1984), 103.

p. 166 **"I don't like big-city living":** John Bardeen (1965b).

p. 166 **a room with "Hollywood" beds:** Jane Bardeen to Betty Maxwell, April 1, 1962, BFC.

p. 166 **Jane couldn't wait:** Jane Bardeen (1991a); (1993).

p. 167 **"What has he done now?":** Bill Bardeen, talk given at John Bardeen's memorial service, February 8, 1991; W. Bardeen (1995).

p. 167 **were about science:** Bardeen family interview (1992).

p. 167 **"his instrument was":** Betsy Bardeen Greytak, talk given at John Bardeen's eightieth birthday celebration, May 7, 1988, UIUC-A.

p. 167 **"He would wrestle":** Betsy Bardeen Greytak, talk given at John Bardeen's eightieth birthday celebration, May 7, 1988, UIUC-A.

p. 167 **had nearly 500 members:** Alice Townsend Barlow, "From Caddies to Carts: Champaign Country Club Historical Reminiscences, 1904–1996," pamphlet.

p. 167 **Bardeens often entertained:** Holonyak (1998b).

p. 168 **"my golf":** Jane Bardeen (1993).

p. 168 **"a man of almost fewer words":** Jane Bardeen (1991a).

p. 168 **"interpreted my lack of interest":** Jane Bardeen to Maxwell family, August 4, 1956, BFC.

p. 168 **Her contrite husband:** John Bardeen to Jane Bardeen, August 7, 14, 17, 20, 1956; Jane Bardeen to Maxwell family, August 10, 1956, BFC.

p. 168 **"Thanks a million":** Jane Bardeen to John Bardeen, August 7, 1956, BFC.

p. 168 **the Score Club:** Jane Bardeen (1993).

p. 168 **"helpful to newcomers":** Ibid.

p. 168 **Jane participated in its affairs:** Robert L. Kabel to Vicki Daitch, October 25, 1993; *Champaign-Urbana Courier*, May 10, 1962 , UIUC-A.

p. 169 **"took the kids down to the First Presbyterian Church":** Jane Bardeen (1993).

p. 169 **"might have considered the Unitarians":** Ibid.

p. 169 **"horse around":** Ibid.; also William Bardeen, private communication to Lillian Hoddeson, January 2002.

p. 169 **John was resolutely secular:** Jane Bardeen (1993).

p. 169 **"I am not a religious person":** John Bardeen to Sergei Kapitsa, July 6, 1988. UIUC-P.

p. 169 **the department had been built up:** Lazarus (1987), 2; Warne (1991), 10.

p. 170 **Loomis also was not thrilled:** Seitz (1991), 119; Lazarus (1987), 3.

p. 170 **He was, however, challenged:** I. Rabi, "Dedication of the Loomis Laboratory of Physics," February 14, 1980, 12.

p. 170 **last on the list:** Lazarus (1987), 4.

p. 170 **"I love subways":** Ibid., 7.

p. 170 **"young, competent but relatively unproven"**: Bradley (1992).

p. 170 **he hired Gerald Kruger:** Warne (1991), 17; Lazarus (1987), 7–8. On the history of the betatron at the University of Illinois, see Kingery, et al. (1967), 66–77.

p. 170 **hired twelve young physicists:** Warne (1991), 19.

p. 170 **median age of the department:** Lazarus (1987), 11.

p. 171 **World War II scattered:** Five years later, during the Korean War, he returned for two more years to set up MIT's Lincoln Laboratory. Frederick Seitz, Dedication of the Loomis Laboratory of Physics, 18. Lazarus (1987), 3; Warne (1991), 26–27; Lazarus (1993).

p. 171 **"old-boy network":** Lazarus (1993).

p. 171 **discrimination against Jews:** Ibid.

p. 171 **"Loomis made the decision":** Goldwasser (1992).

p. 172 **"the department swung":** Russel (1995).

p. 172 **"a lot of money":** Seitz (1994), 194–196.

p. 172 **"many of the qualities of the sea":** Ibid., 195–201.

p. 172 **"There were no rivalries":** Ibid., 194–196.

p. 172 **Lazarus, who was appointed:** Seitz, "Biographical Notes," p. 98, in Mott (1980); Lazarus (1993).

p. 172 **one of the top two:** Lazarus (1993). Eventually several other universities established major solid-state research groups. They included Charles Kittell's group at the University of California-Berkeley and Karl Lark-Horowitz's group at Purdue.

p. 172 **"one of the busiest machines":** Warne (1991), 25.

p. 173 **"Loomis list":** Alpert (1992); Lazarus (1993).

p. 173 **His "wonderful parties":** Alpert (1992).

p. 173 **At lunch they would meet:** DeBrunner (1992); Drickamer (1992).

p. 173 **Abundant monetary resources:** A reporter once asked Bardeen whether accepting funds from a government agency directed his research toward problems of military interest. He said he did not think so. He noted that nearly all government funding of research at the university was initiated by the researchers, not by the government. Polleti (1972).

p. 173 **friends of Emmanuel Piore:** Sapolsky (1990), 39, 49, 64.

p. 173 **"Everyone was Manny's friend":** Lazarus (1993).

p. 173 **Most modest requests:** Lazarus (1993).

p. 173 **hired a military plane:** John Bardeen to Jane Bardeen, September 9, 1953, BFC.

p. 173 **"given very special attention":** Seitz (1994), 221.

p. 173 **"I've never seen so many flashbulbs":** John Bardeen to Jane Bardeen, September 24, 1953, BFC.

p. 173 **"like 'carrying coal to Newcastle'":** John Bardeen, draft of informal talk (1953), UIUC-A.

p. 174 **"So I went to the station":** Nakajima (1993).

p. 174 **"one of the great events":** John Bardeen to George Hatoyama, November 11, 1953, UIUC-A.

p. 174 **Kazuo Iwama:** George Hatoyama to John Bardeen, December 16, 1955, UIUC-A.

p. 174 **Kikuchi took over as director:** Makoto Kikuchi to Nick Holonyak, Jr., August 2, 1974. The authors would like to thank Nick Holonyak for sharing this document with us.

p. 174 **frequent and honored guest at Sony:** See, for example, John Bardeen to George Hatoyama, November 1, 1955, and December 23, 1955; George Hatoyama to John Bardeen, December 16, 1955, UIUC-A.

p. 174 **Bardeen often sent:** John Bardeen to Harold Clark, April 22, 1957; John Bardeen to George Hatoyama, April 23, 1957; George Hatoyama to John Bardeen, May 1, 1957, UIUC-A.

p. 175 **"learned very effectively":** Holonyak (1991b).

p. 175 **". . . get the experiments organized":** John Bardeen to J. B. Fisk, August 9, 1951, UIUC-A.

p. 175 **"about eight or nine or ten inches":** Holonyak (1991a).

p. 175 **He again used Bill Shockley's text:** Shockley (1950).

p. 175 **"I remember him pointing":** Holonyak (1992).

p. 176 **"opportunity to get into something fresh":** Holonyak (1993b).

p. 176 **"as animated as Bardeen is dry":** Lex Peterson, "John Bardeen: Quiet Brilliance," *Champaign-Urbana News Gazette,* January 9, 1981.

p. 176 **"voluble and extroverted":** Bardeen family interview (1992).

p. 176 **"he missed his partner":** Holonyak (1993a). After his move from Bell Labs to Walla Walla, Washington, Brattain tried to interest Bardeen to come visit him. "I can guarantee a [golf] game almost any month except January." Brattain to John Bardeen, November 12, 1965, UIUC-A.

p. 177   **an informal paternal fashion:** Holonyak (1993a); Handler (1993).

p. 177   **The behemoth computer:** Warne (1991), 25.

p. 177   **"didn't need anything lavish":** Holonyak (1991b).

p. 178   **"I thought that even theorists":** John Bardeen to Nick Holonyak, December 13, 1987, UIUC-A.

p. 178   **"The semiconductor topic":** The story is related in a letter from Hisashi Shichijo to Nick Holonyak, October 30, 1987 and Nick Holonyak to Hisashi Shichijo, November 2, 1987. Shichijo cites an article in *Nikkei Electronics*.

p. 178   **by the time Schrieffer:** Nick Holonyak to Hisashi Shichijo, November 2, 1987. Schrieffer arrived in the fall of 1953 with a B.S.; Holonyak left in September 1954 with a Ph.D.

p. 178   **"For whatever reason":** Bray (1993b).

p. 179   **"picked up a pair of pliers":** Holonyak (1991b).

p. 179   **". . . Russia is training scientists":** John Bardeen to Local Board No. 17, Newark, New Jersey, September 10, 1954, UIUC-A.

p. 179   **"knew what were important problems":** Handler (1993).

p. 179   **"he'd come by about twice":** Ibid.

p. 180   **"If you can possibly do it":** John Bardeen to Nick Holonyak, Jr., September 9, 1955, UIUC-A.

p. 180   **"I know the indirect pressure":** John Bardeen to Nick Holonyak, October 16, 1955, UIUC-A.

p. 180   **"is a very nice fellow":** John Bardeen to Nick Holonyak, September 26, 1956, UIUC-A.

p. 180   **"he was always talking of his wife":** George Hatoyama to John Bardeen, May 1, 1957, UIUC-A.

p. 180   **"I think there are some real advantages":** John Bardeen to Nick Holonyak, Jr., November 2, 1959, personal collection of Nick Holonyak.

p. 180   **his muffled speech:** Bray (1993a).

p. 180   **he pressed the chalk:** Bhatt (1992).

p. 180   **"and when our previous class":** A. Anderson (1992).

p. 181   **"found a dark blue suit":** Handler (1993); Bob and Anne Schrieffer, "Remembrances of Jane Bardeen," Jane Maxwell Bardeen Memorial, April 5, 1997.

p. 181   **"He would simply repeat":** Bray (1993b).

p. 181   **ended up dropping:** A. Anderson (1992).

p. 181 **"He was fascinating":** Holonyak (1991b).

p. 181 **Bardeen viewed student participation:** John Bardeen to Alan J. Pifer, Carnegie Corporation of New York, New York City, January 19, 1971, UIUC-A.

p. 181 **"You can figure it out":** Hess (1991); Pao (1994).

p. 181 **"mutual influences":** Zuckerman (1983), 249.

p. 181 **"not only cognitive substance":** Ibid., 240–250.

p. 181 **"take on the spirit":** Bradley (1992), 14–15.

p. 182 **"would not impart motivation":** Bray (1993b).

p. 182 **"it would sound like":** Schrieffer (1992a); see also Handler (1993); Bhatt (1992).

p. 182 **"would sort of outline":** Bray (1993b).

p. 182 **"Often he would sit":** Schrieffer (1992a).

p. 183 **"one phrase or one sentence":** Sah (1993).

p. 183 **"he always *appeared* too busy":** Ibid.

p. 183 **"The best strategy":** Dan Mattis to Vicki Daitch, February 4, 1993.

p. 183 **"Bardeen was extremely accessible":** Allender (1993).

p. 183 **"always would go to Bardeen":** Pao (1994). Pao met Bardeen as an undergraduate at the University of Illinois in the late 1950s. After finishing a master's degree in electrical engineering in 1960, he spent some years in industry and then returned to the university hoping to finish a Ph.D. under Bardeen. As Bardeen was then spending a great deal of time in Washington, D.C., Pao turned to Chih-Tang Sah, who had studied with Bardeen when he was an undergraduate at the University of Illinois. Pao (1994).

p. 183 **"He was the one I":** Lazarus (1992). Also Alpert (1992).

p. 183 **"You mean to tell me":** Coleman (1992).

p. 183 **"one of the reasons":** Alpert (1992).

p. 184 **"depends on who":** Ibid.

p. 184 **"after it was over":** Russell (1995).

p. 184 **"I then escaped feeling":** A. Anderson (1992); Ansel Anderson to Vicki Daitch, January 30, 2002.

p. 187 **"very bright original":** John Bardeen, talk given at memorial service for Bill McMillan, October 17, 1984, UIUC-A.

p. 188 **"work on deriving phonon spectra":** John Bardeen to W. L. McMillan, July 14, 1978, UIUC-A.

p. 188 **"It was very informal":** Bardeen family interview (1992).

p. 188 **"I always immediately ran to Jane":** Hess (1991).

p. 188 **John and Jane stopped by:** Ludwig Tewordt to Vicki Daitch, February 10, 1994.

p. 188 **in the kitchen serving drinks:** Hess (1991).

p. 189 **play "loon" with his grandson:** Elizabeth Bardeen Greytak, talk given at John Bardeen memorial service, February 8, 1991, BFC.

p. 189 **"Uncle John's" undivided attention:** Tom Bardeen, personal communication with Vicki Daitch, May 1998.

p. 189 **an "absolutely superb" experience:** Goldwasser (1992).

p. 189 **"opening doors, but not trying":** James Bardeen, talk given at John Bardeen memorial service, February 8, 1991, UIUC-A.

11
Cracking the Riddle of Superconductivity

p. 190 **"I believe that":** John Bardeen to Rudolf Peierls, July 17, 1951, Peierls Papers, Oxford.

p. 190 **painstaking route:** Pines (1993).

p. 191 **"(1) Derivation of":** John Bardeen handwritten notes, October 23, 1951, UIUC-A.

p. 191 **new field theory tools:** Schweber (1994); Brown and Hoddeson (1983), 311–375.

p. 191 **". . . becoming clear that field theory might be useful":** John Bardeen, "Talk on Superconductivity," 1951 unpublished notes, UIUC-A.

p. 192 **Bohm's original theory:** Hoddeson, et al. (1992), 534–535.

p. 192 **treating the electron–electron interactions:** Hoddeson, et al. (1992), ch. 8.

p. 192 **"intermediate coupling method":** Pines (1981).

p. 193 **formulated the polaron problem:** Lee, Low, and Pines (1953).

p. 193 **Bardeen and Pines found that:** Bardeen and Pines (1955).

p. 193 **"ordered phase in which":** John Bardeen (1956).

p. 194 **"A framework for an adequate theory":** Ibid.

p. 194 **"a model in my own mind":** Schrieffer (1974).

p. 194 **"doing numerical calculations":** Ibid. (1974).

p. 195 **"I passed a gentleman":** Ibid. (1974).

p. 195   **"Why don't you think":** Ibid. (1974).

p. 195   **"How old are you?":** Ibid. (1974).

p. 196   **"OK, fine":** Ibid. (1974).

p. 196   **"versed in field theory":** John Bardeen (1973a).

p. 196   **"one of the active":** Harvard Society of Fellows (1987).

p. 196   **"the latest and most fashionable theoretical techniques":** Cooper (1987), 376.

p. 196   **"that didn't matter, that he'd teach me everything":** Cooper (2000).

p. 197   **"my problem I was going to solve":** Cooper (2000); Cooper (1987), 376.

p. 197   **". . . didn't like the geography":** Cooper (2000).

p. 197   **"by and large perturbative":** Schrieffer (1974).

p. 197   **"we didn't want to use":** Ibid.

p. 197   **two recent texts:** London (1950); Shoenberg (1952).

p. 197   **"quantum structure on a macroscopic":** London (1950), esp. 142–155, quotes on p. 150.

p. 198   **"if we got something":** Cooper (2000).

p. 198   **"You sort of knew what that was like":** Ibid.

p. 198   **". . . one of the ways people delude themselves":** Ibid.

p. 198   **"what happens if you have":** Ibid.

p. 199   **"it seemed clear that":** Cooper (1987), 377.

p. 199   **Cooper "was very excited":** Schrieffer (1974).

p. 199   **"I was reasonably excited":** Cooper (1987), 377.

p. 199   **"We went through a period":** Cooper (2000).

p. 199   **"Bardeen couldn't figure out":** Ibid.

p. 199   **"each contributed parts":** Ibid.

p. 199   **"all the way back":** Miller (1993).

p. 200   **"the enormous fun it was":** Schrieffer (1992a).

p. 200   **"one of the most affable":** Schrieffer (1974).

p. 200   **"Institute for Retarded Study":** Ibid.

p. 200   **"wheel around their chairs":** Ibid.

p. 200   **"He was very stubborn":** Cooper (2000).

p. 201   **"the smallest weapon available":** Schrieffer (1974).

p. 201   **"to divide up the problem":** Ibid.

p. 201   **"those suggestions were exactly":** Ibid.

p. 201   **"in an isolated simply connected superconductor":** Bardeen, et al. (1957b).

p. 202   **"had a feeling" the condensation involved:** Schrieffer (1974).

p. 202 **"to think of the normal"**: Ibid.

p. 202 **Fermi liquid theory**: Ibid.

p. 203 **"year in the wilderness"**: Cooper (2000).

p. 203 **"We tried many techniques"**: Schrieffer (1974).

p. 203 **many couples are doing the Frug**: Ibid.

p. 203 **"We were feeling a little bit downtrodden"**: Ibid.

p. 203 **"Oh, I just wanted to mention"**: Schrieffer (1974).

p. 204 **"I guess I better go shave"**: Elizabeth Bardeen Greytak quoted in, Julie Wurth, "Nobel-winner awards go to UI," *The Champaign-Urbana News Gazette*, October 10, 1991, BFC; Bardeen family interview (1992); Handler (1993).

p. 204 **"The children were jubilant"**: Jane Bardeen, notes on Nobel, 1956, BFC.

p. 204 **"a normal Thursday evening"**: Ibid.

p. 204 **"It was really thrilling"**: Jane Bardeen in Bardeen family interview (1992).

p. 204 **"Everybody was elated"**: Handler (1993).

p. 204 **"Husband of Former Local Girl"**: "Husband of Former Local Girl Nobel Prize Winner," *Washington Observer*, November 2, 1957.

p. 204 **"I can't tell you how proud I am"**: Wigner to Bardeen, handwritten note on a memo from Wigner to Members of the Council of the APS, November 8, 1956, BFC.

p. 204 **"I like to recall"**: John Van Vleck to John Bardeen, November 2, 1956, BFC.

p. 204 **"Gretchen and I got a thrill"**: Walter Osterhoudt to John Bardeen, March 5, 1957, BFC.

p. 205 **"he felt he didn't deserve a Nobel"**: Lazarus (1992).

p. 205 **"I suspect that"**: John Bardeen to E. J. W. Vewey, 29 November 1956, UIUC-A.

p. 205 **especially from Feynman**: Feynman (1957). The conference was held from September 17–21, 1956.

p. 205 **"When one works on it"**: Feyman quoted in Cooper (1987), 378.

p. 205 **"really worked, day and night"**: Jane Bardeen, notes on Nobel, 1956, BFC.

p. 205 **"Yes, only I don't have as much time for it"**: Tommy von Foerster to Vicki Daitch, October 21, 1993, UIUC-A.

p. 205 **"mixed feelings"**: Schrieffer (1974).

p. 205   **"three-way phone call":** Jane Bardeen, notes on Nobel, 1956, BFC.

p. 206   **something "significant":** Jane Bardeen (1993).

p. 206   **"The girls all piled in":** Seitz (2000).

p. 206   **royal blue silk faille gown:** Jane Bardeen, notes on Nobel, 1956, BFC. See also *Illini Week*, February 19, 1987.

p. 206   **heard him swear ferociously:** P. Anderson (1992a).

p. 206   **"We didn't expect that":** Jane Bardeen (1993); Jane Bardeen, notes on Nobel, 1956, BFC.

p. 206   **"expedition to buy a vest":** W. Brattain (undated).

p. 206   **"He not only did":** Ibid.

p. 207   **"a shopper's paradise":** Jane Bardeen to the Maxwell family, December 5, 1956, BFC.

p. 207   **"low hilly country":** W. H. Brattain (undated).

p. 207   **"stickpin with a single crystal":** Ibid.

p. 207   **When they arrived in Stockholm:** Emmy Shockley, personal communication to Michael Riordan, May 16, 1996, cited in Riordan and Hoddeson (1997a), 328, note 244.

p. 207   **"hit on the idea of a bottle of quinine":** W. H. Brattain (undated).

p. 208   **Brattain fretted:** Ibid.

p. 208   **"the King stands to receive":** Jane Bardeen to the Maxwell family, December 12, 1956, BFC.

p. 208   **"went first and made all his bows properly":** W. H. Brattain (undated).

p. 208   **It had been rumored:** Jane Bardeen to the Maxwell family, December 5, 1956, BFC.

p. 208   **The dinner was not cancelled:** W. H. Brattain (undated). Also interview with Jane Bardeen for Champaign-Urbana's *The Courier*, December 31, 1975.

p. 208   **"It was a grand time":** W. H. Brattain (undated).

p. 208   **"conversation was not difficult":** Jane Bardeen to the Maxwells, December 12, 1956, BFC.

p. 208   **"fabulous life":** Ibid.

p. 209   **"like living in a different world":** John Bardeen to Nick Holonyak, December 13, 1956, UIUC-A.

p. 209   **"would go out just before":** Nick Holonyak in remarks made at John Bardeen Memorial service, February 8, 1991; Holonyak (1991b).

p. 209   **"sharing in the nicest":** Slichter (1992).

p. 209   **finally bought a television:** W. Bardeen (1995).

p. 209   **always rooting with enthusiasm:** Bardeen family interview (1992).

p. 209   **It happened on the university golf course:** Bray (1993b).

p. 209   **"almost as good as the Nobel":** "Supercold Superprize," *The National Observer*, December 9, 1972, UIUC-P.

p. 209   **"worth more than one hole-in-one":** Edward Jordan to Jane Bardeen, n.d., BFC; Edward Jordan, "Introductory Comments," IEEE Student Section talk, undated.

p. 210   **"was desperately important to him":** James Bardeen in Bardeen family interview (1992).

p. 210   **off in another world:** Betsy Bardeen Greytak, Bardeen family interview (1992).

p. 210   **"I realized that the algebra":** Schrieffer (1974).

p. 210   **"I solved the gap equation":** Ibid.

p. 211   **"I knew immediately we could":** Cooper (2000).

p. 211   **"Let's go and talk to John":** Schrieffer (1974).

p. 211   **"he thought that there was":** Ibid.

p. 211   **"How would you like to write a paper together":** Cooper (2000).

p. 211   **"We had the experimental number":** Schrieffer (1974).

p. 211   **"felt that this was the right direction":** Ibid.

p. 211   **"almost all the pieces":** Ibid.

p. 212   **"done all these calculations for normal metals":** Cooper (2000).

p. 212   **"using all sorts of complicated field theory":** Schrieffer (1974).

p. 212   **Bardeen's colleagues knew something:** Lazarus (1992).

p. 212   **"the most concentrated, intense and incredibly fruitful work":** Cooper (1987), 378.

p. 212   **"I know that you object":** John Bardeen to S. A. Goudsmit, February 15, 1957, UIUC-A.

p. 213   **"such that if one of the pair":** Bardeen, Cooper, and Schrieffer, (1957a).

p. 213   **It leads to an energy-gap:** Ibid.

p. 213   **"John was somehow off":** Schrieffer (1974).

p. 213   **"He was really excited":** Ibid.

p. 213   **"Well, I think we've figured out superconductivity":** Charles Slichter, remarks at John Bardeen's memorial service, February 8, 1991.

p. 214 **Slichter became "so heavily involved":** Cooper (1987), 378.

p. 214 **Experiments by Glover and Tinkham:** Hoddeson, et al. (1992), 604, 606, refs 320 and 378.

p. 214 **"Come on now. The wave function looks right":** Schrieffer (1974).

p. 214 **"He wanted to make sure that the young people":** Ibid.

p. 215 **"in the space of a few months":** Herring and Herring (2000).

p. 215 **young Billy became fascinated:** Bill Brattain to Jane Bardeen, February 13, 1991, BFC. "Dr. Brattain, Chatham, Dies," *Newark Evening News*, April 12, 1957.

p. 215 **"I could work any way I wanted":** Cooper (1987), 379.

p. 216 **this masterpiece of modern physics:** Bardeen, Cooper, and Schrieffer (1957b).

p. 216 **"In case what follows":** A. B. Pippard to John Bardeen, September 11, 1957, UIUC-A.

p. 216 **"mostly those who have tried":** John Bardeen to Jane Bardeen, September 1, 1957, BFC.

p. 216 **"There was considerable interest":** John Bardeen to Leon Cooper, September 19, 1957, UIUC-A.

p. 216 **"defending the fort":** Schrieffer (1974).

p. 217 **"In formulating our theory":** John Bardeen to A. B. Pippard, September 27, 1957, UIUC-A.

p. 217 **lack of gauge invariance:** Anderson (1958a, 1958b); Pines and Schrieffer (1958).

p. 217 **"certainly gives the answer":** John Bardeen to Philip Anderson, October 11, 1957, UIUC-A.

p. 217 **"broken symmetry":** Nambu and Jona-Lasinio (1961); L. Brown, R. Brout, T. Y. Cao, Peter Higgs, and Y. Nambu, "Panel Session: Spontaneous Breaking of Symmetry," in Hoddeson, et al. (1997),. 478–522.

p. 217 **"Particle physicists are so desperate":** Brown et al, panel on broken symmetry, in Hoddeson, et al. (1997).

p. 217 **"came through Copenhagen":** Schrieffer (1974).

p. 217 **"Unfortunately, it's wrong":** Robert Schrieffer remarks at John Bardeen's memorial service, February 8, 1991.

p. 218 **"big ideas" of physics:** For the nuclear problem, see A. Bohr, B. R. Mottelson, and D. Pines (1958); A. B. Migdal (1959); and S.T. Belyaev, (1959). For neutron stars, see Migdal (1959); Ginzburg and Kirzhnits (1964); and G. Baym, C. J. Pethick, and D. Pines, (1969). For the superfluidity of ^{3}He, see Oscheroff, et al., (1972) and Leggett (1975).

p. 218 **"a tremendous universal"**: Lazarus (1992).

12
Two Nobels Are Better Than One Hole in One

p. 219 **". . . an office with Buddha"**: Bernard Serin, cited in personal communication from Enid Sichel to L. Hoddeson, May 8, 1998. The authors would like to thank Professor Sichel for sharing her recollection. According to the 1958–1959 University of Illinois Staff Directory, Serin's campus address was 416 Physics Laboratory, while Bardeen's was 307.

p. 219 **"I think that this"**: "Journal of Glenn T. Seaborg," Lawrence Berkeley Laboratory Technical Information Department, Berkeley, California.

p. 219 **"We felt greatly honored"**: Jane Bardeen to Dr. J. R. Maxwell and Beth Maxwell, May 13, 1962, BFC.

p. 220 **"every nook from basement to attic"**: University of Illinois, Department of Physics (1980), 59.

p. 220 **he could see the afternoon sun:** Bray (1993b).

p. 220 **officemate could receive:** Frederick Lamb to L. Hoddeson, private communication, October 3, 2000.

p. 221 **coauthored a research monograph:** Baym and Kadanoff (1962). Here gauge invariant means having the freedom to make certain formal changes in the electromagnetic potentials without changing the electric or magnetic fields.

p. 221 **three most important centers:** The Landau-Ginzburg theory, the derivation of Landau-Ginzburg from BCS, and the Landau Fermi liquid theory were among the great projects of the Landau Institute during this period. See Hoddeson, et al. (1992), ch. 8, esp. 538–541, 558–572.

p. 221 **Bardeen served as the leader:** Baym (1992).

p. 222 **the heart of a problem:** John Wheatley to John Bardeen, October 25, 1976. Hess (1991).

p. 222 **McDonald has studied:** McDonald (2000 draft) and (2001). The authors are indebted to McDonald for sharing his research with us prior to its publication. See also McDonald and Kautz (1999). Among those interviewed by McDonald were Morrel

Cohen, Pierre-Gilles de Gennes, William Little, Robert L. Powell, Geoffrey Sewell, and Harold Weinstock.

p. 223 **"cutting in"**: Walter A. Harrison to Vicki Daitch, November 14, 1993.

p. 223 **"are not paired"**: John Bardeen (1961).

p. 223 **Phillips joined his colleagues:** Cohen, et al. (1962), 316–318.

p. 223 **"pairing does not extend"**: John Bardeen (1962), 148–149.

p. 223 **"a disconcerting experience"**: P. Anderson (1970), 23.

p. 224 **"was fascinated by the idea of broken symmetry"**: Josephson (1974), 251.

p. 224 **"perfectly possible for"**: McDonald (2001), 48.

p. 224 **argued that the probability:** Josephson (1974), 251.

p. 224 **"It was some days"**: Ibid., 252.

p. 224 **"We were all"**: P. Anderson (1970), 23–24.

p. 224 **"The embarrassing feature"**: Josephson (1974), 252.

p. 225 **"new effects are predicted"**: P. Anderson (1970), 23–24. Josephson's classic paper is Josephson (1962).

p. 225 **submitted his paper:** John Bardeen (1962), 148–149.

p. 225 **"wanted macroscopic quantum phenomena"**: Derek Martin, private communication to Donald G. McDonald, cited in McDonald (2000).

p. 225 **"I introduced Josephson to Bardeen"**: Ivar Giaever, private communication to Donald G. McDonald, cited in McDonald (2000).

p. 225 **"was crowded late in the afternoon"**: McDonald (2001), 49–50.

p. 226 **"Bardeen was outspokenly skeptical"**: Wolfgang Klose to Vicki Daitch, July 11, 1995, UIUC-A.

p. 226 **"Bardeen's basic error"**: Brian Josephson, cited in McDonald (2001), 50.

p. 226 **The issues would be fully spelled out:** Anderson and Rowell (1963); Rowell (1963).

p. 226 **"Your evidence, particularly the effect"**: John Bardeen to Philip Anderson, February 26, 1963, UIUC-A.

p. 226 **"several startling accompaniments"**: Shapiro (1963), 82.

p. 226 **second-order phase transition:** Josephson (1966).

p. 227 **"sent off for little photographs"**: Robert Schrieffer, remarks made at John Bardeen's memorial service, February 8, 1991.

p. 227 **"Makes one feel humble":** John Bardeen to A. Abrikosov, L. Gor'kov, and I. Khalatnikov, June 4, 1968, UIUC-A.

p. 227 **transistor radio inside:** Makoto Kikuchi to John Bardeen, May 18, 1976; Bardeen to Kikuchi, May 25, 1976; George Hatoyama to John Bardeen, June 4, 1976; Bardeen to Hatoyama, June 7, 1968, UIUC-A.

p. 227 **play tricks with it:** Bray (1993b).

p. 227 **"I will have to take it apart":** John Bardeen to George Hatoyama, December 18, 1985, UIUC-A.

p. 228 **a controversial problem:** Bray (1993a).

p. 228 **"not just to throw you into the wolves":** Bray (1993a, 1993b).

p. 228 **"I was sitting in coffee hour":** Salamon (2000).

p. 228 **"we jumped in":** Bray (1993b).

p. 228 **"excess conductivity":** Ibid. Bray's thesis was titled "Fluctuation Conductivity from Charge Density Waves in Pseudo-one-dimensional Systems." David William Allender's thesis was titled "Model for an Exciton Mechanism of Superconductivity, in Planar Geometry."

p. 229 **"So we went to John":** Baym (1992).

p. 229 **"Well, it's just the same as":** Ibid.

p. 229 **"So we went back to John":** Ibid. The reference was to Bardeen and Schrieffer (1961). B and H are discussed on page 188.

p. 229 **"I was going to see Bob Schrieffer":** Baym (1992).

p. 229 **"Well it's just the difference":** Frederick Lamb, private communication to L. Hoddeson, October 2, 2000.

p. 230 **"The Comstock Prize is more significant":** John Bardeen to C. Kittel, November 28, 1967.

p. 230 **"contributions to the theory":** C. Kittel to F. Seitz, December 19, 1967.

p. 230 **"came to realize":** Slichter (1995).

p. 230 **"an area that has had":** John Bardeen to Nobel Committee for Physics of the Royal Academy of Sciences, January 27, 1967, UIUC-A.

p. 231 **He renewed the nominations:** Ibid.; John Bardeen to the Nobel Committee for Physics of the Royal Academy of Sciences, January 4, 1971, and January 25, 1973; A. B. Pippard to John Bardeen, January 3, 1973; John Bardeen to A. B. Pippard, February 15, 1973, UIUC-A.

p. 231   **"John was a very politically savvy person":** Slichter (1995).

p. 231   **"I didn't quite believe him":** "Bardeen Physicists Galore," *Chicago Tribune*, October 21, 1972, clipping in John Bardeen file, UIUC-P.

p. 231   **"I've got my wife working":** John Bardeen (1973b); John Bardeen to Robert L. Weber, January 12, 1979.

p. 231   **". . . had that door for years":** *Champaign-Urbana News Gazette*, October 20, 1972, clipping in UIUC-P.

p. 231   **the champagne:** Pines and Slichter (1972).

p. 231   **"At a few minutes before":** Simon (1973).

p. 232   **"spoke softly, slowly":** Paula Peters (1972).

p. 232   **"does not miss home games":** Simon (1973).

p. 232   **"another Nobel Prize with two other guys":** Nick Holonyak, private communication to L. Hoddeson, February 4, 2001.

p. 232   **"Sorry you could not get your garage door open":** Walter Brattain to John Bardeen, October 30, 1972, UIUC-A.

p. 232   **"Jane and I have been going over":** John Bardeen to W. Brattain, November 7, 1972, UIUC-A.

p. 232   **". . . feeling of all club members":** George A. Russell, president, Champaign Country Club to John Bardeen, November 17, 1972, UIUC-A.

p. 232   **". . . keep these things in perspective":** Letter to the editor by John Smetana, Urbana, Illinois, no name of newspaper listed, n.d., clipping in UIUC-P.

p. 233   **"Being eighth-graders":** Shannon, Illinois, eighth grade class, to John Bardeen, October 27, 1972, UIUC-A.

p. 233   **"I would be so pleased":** Sister Rita Michael Aguillard to John Bardeen, November 21, 1972, UIUC-A.

p. 233   **Bardeen wrote back that:** John Bardeen to Sister Rita Michael Aquillard, December 28, 1972, UIUC-A.

p. 233   **"very pretty beading":** Jane Bardeen to Betty Maxwell, November 24, 1972, BFC.

p. 233   **"so we arrived at O'Hare":** Jane Bardeen, notes on "Nobel Journey 1972," BFC.

p. 233   **"to catch our breath":** John Bardeen to Aage Bohr, November 7, 1972; Jane Bardeen, notes on "Nobel Journey 1972," BFC.

p. 233 **"almost no time to shop":** Jane Bardeen, notes on "Nobel Journey 1972," BFC.

p. 234 **"We were quite a crew":** "Bardeen Compares Nobel Trips," *The Courier, Champaign-Urbana,* December 21, 1972.

p. 234 **thank-you and season's greetings card:** Typed mailing list and card in BFC.

p. 234 **". . . profits show a sharp drop":** John Bardeen to Carl G. Vernersson, December 28, 1972. See also Carl Vernersson to John Bardeen, November 20, 1972, UIUC-P. Three years later Bardeen and other Nobel Prize winners would return to Stockholm for a grand celebration of the seventy-fifth anniversary of the Nobel Prize.

p. 234 **stopped off in Switzerland:** John Bardeen to Olli v. Lounsasmaa, November 8, 1972, UIUC-A.

p. 234 **"John Bardeen's second Nobel Prize":** Jim Fisk to John Bardeen, November 22, 1972.

p. 234 **"Naturally, all of us at Wisconsin Telephone":** H. B. Groh, president, Wisconsin Telephone Company to John Bardeen, December 12, 1972, UIUC-A.

p. 234 **". . . warm feeling for my home state":** John Bardeen to H. B. Groh, January 3, 1973, UIUC-A.

p. 234 **the quality of the exchanges:** Bray (1993a); Goldwasser (1992).

p. 234 **They were unable to agree:** Jim Phillips to John Bardeen, September 11, 1972; John Bardeen to Jim Phillips, September 26, 1972, UIUC-A.

p. 235 **an editorial blunder:** Bray (1993a); Phillips (1972); Allendar, et al. (1973).

p. 235 **"inexcusable set of blunders":** Bray (1993a).

p. 235 **"An author certainly has a right":** John Bardeen to Samuel Goudsmit, January 5, 1973, UIUC-A.

p. 235 **"the author must have been in contact":** Samuel Goudsmit to John Bardeen, January 25, 1973, UIUC-A.

p. 235 **"Poor McMillan never knew":** Hess (1991).

p. 235 **"semiconductors with different":** Allender, et al. (1973); P. Anderson (1992); Inkson and Anderson (1973).

p. 235 **"agreed to disagree":** P. Anderson (1992).

p. 235 **"I got better while":** Bray (1993a).

p. 236 **"It was something my father would remember":** Allender (1993).

p. 236 **He rarely used a cart:** Werstler and Werstler (1992).

p. 236 **"mental toughness":** Graham (1990), xviii.

p. 236 **A player's game suffers:** Graham (1990), 16.

p. 237 **"He'd get out there":** Bray (1993b).

p. 237 **"never rushed things":** Coleman (1992).

p. 237 **respectably in the low 80s:** Ibid.

p. 237 **numerous golfing friendships:** Werstler (1992).

p. 237 **One foursome consisted of:** Robert Kabel to Vicki Daitch, December 20, 1993, UIUC-A.

p. 237 **"never gave me the feeling":** Werstler (1992).

p. 237 **"Call him again!":** The authors would like to thank Robert Kabel for sharing the card with us. Robert Kabel to Vicki Daitch, October 25, 1993.

p. 237 **His scientific renown:** Robert Kabel used the word "ordinary" to describe Bardeen, as did another member of the golfing community: attorney Art Lerner. Art Lerner private communication to L. Hoddeson, c. March, 1992.

p. 237 **"Say John, you know I've been meaning to ask":** Cited in Simon (1973).

p. 237 **the large and garrulous Louis Burtis:** Coleman (1992).

p. 238 **"Boy, things aren't very good":** Drickamer (1992).

p. 238 **"If it couldn't be John":** Bhatt (1992). Bhatt's thesis was titled "Electronic Instabilities and the Martensitic Transition in A-15 Compounds" (1976).

p. 238 **Physics 463:** Bhatt (1992).

p. 239 **"great distaste for any formal":** R. O. Simmons to P. Handler, N. Holonyak, and D. Pines, January 27, 1975, UIUC-P.

p. 239 **"There's been BCS and APS":** "Symposium on Frontiers in Condensed Matter Physics to Honor Professor John Bardeen," October 15, 1976, dinner menu, UIUC-A.

p. 239 **"really encyclopedic knowledge of solid-state":** J. Miller (1993); Pines (1993).

p. 240 **"You feel more self-confidence":** Young (1972).

p. 240 **...almost garrulous:** A. Anderson (1992); Baym (1992). Frederick Lamb, private communication to L. Hoddeson, October 3, 2000. Baym traces Bardeen's new talkativeness to the receipt of his second Nobel Prize.

13
A Hand in Industry

p. 241   **A Hand in Industry:** This chapter draws on material prepared in 1992 by Fernando Irving Elichirigoity and on a University of Illinois senior thesis by Andrew Dribin, "John Bardeen and Xerox: Bridging the Gap between Science and the Empirical Art." Dribin (1997). The authors are grateful to both Elichirigoity and Dribin for their contributions.

p. 241   **"whether there is a technological basis":** Hess (1991).

p. 241   **"I choose another problem":** "John Bardeen to retire," *The Daily Illini*, February 14, 1975, UIUC-P.

p. 242   **"face always lit up":** Leonard (1992).

p. 242   **"never once mentioned":** Gallo to the editor, *Physics Today*, April 1992, 15.

p. 242   **Haloid Company had invested:** Harold Clark, "The Early Association of John Bardeen with the Haloid Company," unpublished manuscript. The authors would like to thank Erik Pell for generously sharing this document with us.

p. 242   **Haloid's president, Joseph Wilson:** Dessauer (1971), 9. Wilson was the grandson of one of Haloid's four founders. Except where indicated, our story of the early years of Xerox comes from Dessauer's book.

p. 242   **Paul Selenyi:** R. Radnai, and R. Kunflavi, *Physics in Budapest* (Amsterdam: North Holland, 1988), 64, cited in Dribin (1997), 78.

p. 243   **application for a patent:** Dessauer (1971).

p. 243   **Haloid bought the patents:** Ibid. Carlson remained involved and later raised the money to buy back some of his own rights in electrophotography.

p. 243   **Dessauer considered it unlikely:** Harold Clark, "The Early Association of John Bardeen with the Haloid Company," unpublished manuscript.

p. 243   **copies that Princeton had required:** Bardeen's file in PRIN contains a form letter dated December 11, 1935, explaining that "it will be necessary for the candidate to file two copies of his dissertation in the office of the Graduate School."

p. 244   **"a key to our development":** Joseph R. Wilson, quoted in the *Rochester Democrat and Chronicle*, September 25, 1960, BFC.

p. 244 **"challenge its short-run position"**: Peter Bart, "Advertising: TV Series on U.N. Stirs Debate," *The New York Times*, April 10, 1964.

p. 244 **"in a few laboratory rooms"**: John's Bardeen's lecture to the Ministry of Science and Higher Education in Tehran, Iran, cited in Dribin (1997), 17.

p. 244 **"the most successful product"**: *Fortune* magazine, quote cited in *International Dictionary of Company Histories* (Chicago: St. James Press, 1991).

p. 244 **innovation, for which he filed a patent:** Patent 3041166, "Xerographic Plate and Method," February 12, 1958, UIUC-A; Pell (1995).

p. 244 **Bardeen's presence as a consultant:** Dessauer and Pell (1992).

p. 244 **"would listen intensely"**: J. D. Wright to Morton Weir, February 1, 1991, BFC.

p. 245 **"the business of research"**: Charles B. Duke to A. C. Anderson, March 21, 1991. Copy provided to the authors by C. B. Duke.

p. 245 **"levels of effort"**: John Bardeen to G. E. Pake, Report of visit to Webster, New York, November 29–December 1, 1978 with TAP, UIUC-A.

p. 245 **"invention does not occur"**: John Bardeen (c. 1968).

p. 245 **"Those doing basic research"**: John Bardeen to Dr. N. B. Hannay, vice-president, Bell Laboratories, March 8, 1973, BFC. The term "bean counters" appears to have been widely used among Xerox's technical staff to deride executives whose efforts to save money undermined the company's long-term investment in research and development. Pell (1995); Smith and Alexander (1988), 162.

p. 246 **". . . to control the flow of electrons"**: John Bardeen (c. 1968); John Bardeen (1965a), 65.

p. 246 **"finding something in science"**: John Bardeen (c. 1968).

p. 246 **"Basic research is defined"**: John Bardeen (1965a), 56–57.

p. 247 **"really no sharp dividing line"**: John Bardeen (c. 1968).

p. 247 **"areas where significant advances"**: John Bardeen (1965a). See also John Bardeen to B. D. Thomas, September 18, 1964; John Bardeen to Jerome Weisner, September 18, 1964; John Bardeen to John Dessauer, February 1958, UIUC-A.

p. 247   **"scientific dust bowl"**: The industrial laboratories (the "cat") would never consume all the interesting scientific problems of solid-state physics (the "cream"), as there would be a constant flow of new ones. John Bardeen (1965a), 68; Pippard (1961), 38.

p. 247   **"the most difficult to find"**: John Bardeen (c. 1968).

p. 247   **"to publish their results"**: John Bardeen (c. 1968).

p. 247   **Bardeen was elected**: Joe Wilson to John Bardeen, February 23, 1961, UIUC-A; Lewis (1991).

p. 247   **"the outstanding problem in computers"**: John Bardeen to John Dessauer, October 6, 1961, UIUC-A.

p. 248   **"build one of the finest industrial laboratories"**: Charles Duke to Ansel Anderson, March 21, 1991.

p. 248   **he became the driving force**: Goldman (2000).

p. 248   **PARC had barely opened**: Hiltzik (1999), 56.

p. 248   **threat from the "bean counters"**: Goldman (2000); Hiltzik (1999), 56–57.

p. 249   **the Alto computer**: Smith and Alexander (1988), 14.

p. 249   **"Dallas turned out to grow"**: Jack Goldman, quoted in Smith and Alexander (1988), 163.

p. 249   **"It seems to me"**: John Bardeen to C. Peter McColough, May 14, 1973, UIUC-A.

p. 250   **"be difficult to attract"**: Ibid.

p. 250   **It was too late**: Archie McCardell to John Bardeen, May 22, 1973, UIUC-A.

p. 250   **Bardeen was deeply troubled**: Smith and Alexander (1988), ch. 14.

p. 250   **"The problem is not so much"**: John Bardeen to George Pake, December 18, 1980, UIUC-A.

p. 250   **"It is your name"**: John Bardeen to John Dessauer, January 2, 1976, UIUC-A.

p. 250   **Holonyak noticed his mentor's name**: Holonyak (2001).

p. 250   **Bardeen resigned from his GE**: John Bardeen to A. H. Markham, June 30, 1961. See also John Fisher to John Bardeen, July 29, 1959; John Bardeen to John Fisher, August 28, 1959, UIUC-A.

p. 251   **mentoring a former student**: Pao (1994). Information on Supertex is from *Corporate Technology Directory*, 1989 edition.

p. 251   **Silicon Valley workers accepted**: Pao (1994).

p. 251   **John's shaky hands**: Ibid.

p. 251   **"John Bardeen is a great figure in science"**: "Sony gift honors Nobel laureate," *Champaign-Urbana News-Gazette*, October 6, 1989.

p. 252 **". . . he wanted to have Nick on the Bardeen Chair"**: Mikoto Kikuchi to Jane Bardeen, February 26, 1993, BFC.

p. 252 **"last day in Tokyo"**: Nakajima (1993).

p. 252 **the Midwest Electronics Research Center**: John Bardeen, "Plans for an Industrial Research Visitor's Program," lecture given at Industry-University Research Forum, April 18, 1963, UIUC-A.

p. 252 **the center sponsored symposia**: John Bardeen to John A. Kennedy, Illinois Governor's Office, July 24, 1964, UIUC-A.

p. 252 **"When I took John along"**: Coleman (1992).

p. 252 **"low keyed, a little on the shy side"**: Ibid.

p. 253 **"The fact that you think"**: Charles Gallo to John Bardeen, January 30, 1986, UIUC-A.

p. 253 **university, in turn, gained**: Coleman (1992). John Bardeen, "Comments on Industry-University Interactions," enclosed in a letter to Floyd Ingersoll, Illinois Foundation Seeds, Inc., September 30, 1982.

14
Citizen of Science

p. 254 **"the fact that there are all too few people"**: Bardeen (1960). Illinois State Normal University, May 1960, UIUC-A.

p. 254 **"is doing a very effective job"**: Ibid.; Hoxie (1990).

p. 254 **appointed James Killian**: Bardeen (1960); see also Rigden (1987), 247–250. Rabi was an important participant in the creation of PSAC.

p. 255 **"They looked for generalists"**: Slichter (1995).

p. 255 **"all sorts of questions"**: Ibid.; Press release and typed copy of talk by John Bardeen at Illinois State Normal University, May 1960, UIUC-A.

p. 255 **"it might be best"**: John Bardeen to Earl G. Droessler, March 10, 1961, UIUC-A.

p. 256 **"the problems of ICAS"**: John Bardeen to Jerome Wiesner, March 10, 1961, UIUC-A.

p. 256 **"What is needed"**: John Bardeen to George B. Kistiakowsky, May 31, 1960, UIUC-A.

p. 256 **"There was a time"**: Bardeen (1960).

p. 256 **"I feel strongly"**: John Bardeen to Robert E. Green, August 5, 1963, UIUC-A.

p. 256 **"We thought it was pretty important"**: Seaborg (1995). See also White House press release, February 28, 1959, in Bardeen personnel file, USOPM.

p. 256 **"Everyone who was on it"**: Slichter (1995).

p. 257 **"... express my deep gratitude"**: John Bardeen to President Dwight D. Eisenhower, January 27, 1961, UIUC-A.

p. 257 **The work of PSAC enhanced:** The authors would like to thank Glenn T. Seaborg for sending us his "National Service with Ten Presidents of the United States" Seaborg (n.d.).

p. 257 **PSAC's accomplishments included:** Rigden (1987), 250–251.

p. 257 **"the committee as a whole":** John Bardeen to President John F. Kennedy, January 21, 1963, UIUC-A.

p. 257 **series of handwritten notes:** Harvey Brooks, President's Science Advisory Committee, memo, "Issues on Research and Development in the Federal Government," n.d., UIUC-A.

p. 258 **"is a topic in which":** John Bardeen to G. B. Kistiakowsky, October 7, 1959, UIUC-A. Bardeen also told Kistiakowsky that he was on the Executive Committee of the Graduate College at Illinois and that he had "participated in study groups on engineering education."

p. 258 **"In the support of basic research":** Harvey Brooks, "Issues on Research and Development in the Federal Government," UIUC-A.

p. 258 **"in the long run make it impossible":** John Bardeen to the Hon. Emilio Q. Daddario, House of Representatives, Washington, D.C., March 20, 1970.

p. 258 **"Seaborg report":** Glenn T. Seaborg, "Journal of Glenn T. Seaborg," Lawrence Berkeley Laboratory Technical Information Department, n.d. The authors thank Glenn Seaborg for generously sharing these documents with us.

p. 259 **"factors determining the desirable rate":** John Bardeen, notes on PSAC memo, "Jobs to be Done on the Basis of the Kistiakowsky Paper," UIUC-A.

p. 259 **"My own feeling":** John Bardeen to I. I. Rabi, November 18, 1960, UIUC-A.

p. 259 **"not a Communist":** USOPM, Official Personnel Folder on John Bardeen.

p. 259 **Bardeen's personal distaste:** Charles Piozet, Special Assistant to the Chief of Industrial Relations, Department of the Navy to John Bardeen, October 31, 1952, Bardeen personnel file, USOPM.

p. 260 **"endanger the common defense":** Notifications of Personnel Action dated June 26, 1952, and June 26, 1953, John Bardeen personnel file, USOPM; Alan T. Waterman, Director NSF to John Bardeen, June 2, 1952; John Bardeen to Alan T. Waterman, June 24, 1952, John Bardeen personnel file, USOPM.

p. 260 **Bardeen subsequently served:** John Bardeen personnel files, USOPM. Others on the Special Commission on Weather Modification included Adrian R. Chamberlain, Qilliam G. Colman, John C. Dreier, Leoni Hurwicz, Thomas F. Malone, Arthur W. Murphy, Sumner T. Pike, William S. Von Arx, Gilbert F. White, and Karl M. Wilbur, according to the list in Bardeen's government personnel file.

p. 260 **President's Commission on the Patent System:** President Lyndon B. Johnson to John Bardeen, July 23, 1965; John Bardeen to Alfred C. Marmor, President's Commission on the Patent System, July 29, 1965; Lawrence Fleming to John Bardeen, August 29, 1966; John Bardeen to Alfred C. Marmor, comments on draft of committee report, August 29, 1966, UIUC-A.

p. 260 **"... rapid and free exchange of technical information":** John Bardeen to Alfred C. Marmor, August 29, 1966, UIUC-A.

p. 260 **"the grace period is valuable":** John Bardeen to Alfred C. Marmor, President's Commission on the Patent System, September 29, 1966 and October 17, 1966, UIUC-A.

p. 260 **"Medallic History of Science":** Vincent J. Higgins to John Bardeen, January 9, 1979, UIUC-A.

p. 261 **charming town of St. Andrews:** Robertson (1974); Jack Allen to David Pines, April 28, 1992, UIUC-A.

p. 261 **John rushed to her side:** Jane Bardeen (1994).

p. 261 **"There was hardly any pain":** Jane Bardeen to Betty Maxwell, July 12, 1967, BFC.

p. 261 **"lost no time in locating a projector":** John Bardeen to John Dessauer, July 25, 1967, UIUC-A.

p. 261 **"John was a bit chagrined":** Jack Allen to Vicki Daitch, October 6, 1993; John Bardeen to John Dessauer, July 25, 1967, UIUC-A.

p. 261 **the "Bardeen clubs":** Jack Allen to Vicki Daitch, October 6, 1993; Enid Sichel to L. Hoddeson, May 9, 1998, UIUC-A.

p. 262 **"had the best holiday":** Jack Allen to Vicki Daitch, October 6, 1993, UIUC-A.

p. 262 **"longer range technological":** John Bardeen to T. Suguwara, February 24, 1971, UIUC-A.

p. 262 **The contracts were cancelled:** John Bardeen to T. Suguwara, February 24, 1971, UIUC-A; "Washington Pulls Plug on 4 Helium Contracts After a Year of Talks," *Wall Street Journal,* February 1, 1971, UIUC-A.

p. 262 **"expressing sorrow and anger":** Jack Allen to Vicki Daitch, October 6, 1993, UIUC-A.

p. 262 **"The original basis":** John Bardeen to Hugh Odishaw, National Academy of Sciences, Washington, D.C., December 8, 1970, UIUC-A.

p. 262 **U.S. helium conservation program:** Jack Allen to Vicki Daitch, October 6, 1993, UIUC-A.

p. 263 **"it seems impossible":** John Bardeen to Charles Laverick, August 30, 1974, UIUC-A.

p. 263 **"the energy crisis has been known":** John Bardeen (1973b).

p. 263 **"My greatest concern is the environment":** Hackett (1989).

p. 263 **Population Crisis Committee:** James W. Riddleberger, Population Crisis Committee, Washington, D.C., to John Bardeen, June 12, 1970; John Bardeen to James W. Riddleberger, July 16, 1970, UIUC-A.

p. 263 **Foreign aid and birth control:** "John Bardeen: Quiet Brilliance," *Champaign-Urbana News Gazette,* January 9, 1981.

p. 263 **he supported Planned Parenthood:** Betsy Bardeen, private communication to Vicki Daitch, October 1994.

p. 264 **"involves using the APS organization as a weapon":** Conyers Herring, "On the Wisdom of Cancelling a Meeting in a 'Disapproved' City," draft of statement to the American Physical Society, UIUC-A.

p. 264 **Even if the APS Council:** John Bardeen to Elliott Lieb, MIT, November 5, 1968; Jay Orear to John Bardeen, President, American Physical Society, 29 October 1968; John Bardeen to Richard J. Daley, Mayor of Chicago, October 17, 1968; W. W. Havens to John Bardeen, November 14, 1968, UIUC-A; Drickamer (1992).

p. 264 **"If we were scheduling":** John Bardeen to William C. H. Joiner, November 20, 1968, UIUC-A.

p. 264   **Bardeen was greatly relieved:** Drickamer (1992).

p. 264   **"my physics has been neglected":** John Bardeen to Reiner Kummel, January 17, 1969, UIUC-A.

p. 264   **found it unacceptable:** Jack Allen to Vicki Daitch, November 15, 1993, UIUC-A.

p. 265   **"I regret very much":** John Bardeen to Dr. T. Sugawara, General Secretary, Twelfth International Conference on Low Temperature Physics, Tokyo, Japan, March 26, 1970, UIUC-A.

p. 265   **"... remotest connection with weapons research":** The military laboratories represented included the U.S. Naval Research Lab and the Royal Military College of Canada. John Bardeen to Dr. T. Sugawara, June 1, 1970; John Bardeen to Jack Allen, St. Andrews, Scotland, July 7, 1970, UIUC-A.

p. 265   **"so few cases are involved":** John Bardeen to Dr. T. Sugawara, June 1, 1970 and June 11, 1970, UIUC-A.

p. 265   **contacted Jack Allen:** John Bardeen to Jack Allen, July 7, 1970, UIUC-A.

p. 265   **whether he would say a few words:** John Bardeen to Jack Allen, July 27, 1970, UIUC-A.

p. 265   **ran without a hitch:** Jack Allen to Vicki Daitch, November 15, 1993, UIUC-A.

p. 266   **"grave reservations about attending":** John Bardeen to Dr. J. So'lyom, Central Research Institute for Physics, Budapest, Hungary, August 13, 1976, UIUC-A.

p. 266   **"there have to date been few restrictions":** John Bardeen to Senator Thomas C. Hennings, Jr., April 22, 1959, UIUC-A.

p. 266   **"I don't remember the reason":** John Bardeen to Thomas H. Johnson, March 24, 1982, UIUC-A.

p. 266   **"but he pushes himself too hard":** Jane Bardeen to Betty Maxwell, March 27, 1974, BFC.

p. 266   **"they were unable to find a specific cause":** John Bardeen to John D. Hoffman, March 8, 1973, UIUC-A.

p. 266   **rising inflation, the war in Vietnam:** Ibid.

p. 267   **"a liberating influence":** B. L. R. Smith (1990), 3–4.

p. 267   **"... I am useless in a group":** Keyworth (2000).

p. 267   **"but his heart wasn't in it":** Slichter (1992).

p. 268   **"John was very, very useful":** Keyworth (2000).

p. 268   **White House was less enthusiastic:** Marshall (1985).

p. 268   **"while this country is the leader":** John Bardeen to

James G. Ling, November 24, 1982. The WHSC Federal Laboratory Review Panel consisted of Bardeen, David Packard (chair), James G. Ling (executive secretary), Mrs. Minh-Triet Lethi (policy analyst), D. Allan Bromley, Donald S. Fredrickson, Arthur K. Kerman, Edward Teller, and Albert D. Wheelon. "Report of the White House Science Council," Appendix A, May 1983, UIUC-A.

p. 269 **"Such a far-reaching proposal"**: John Bardeen (1986a); Marshall (1985).

p. 269 **"there was no point"**: Slichter (1992).

p. 269 **"he was being used for his name"**: Goldwasser (1992).

p. 269 **"I do not feel"**: John Bardeen to George Keyworth, April 11, 1983, UIUC-A.

p. 269 **"thought that I might be helpful"**: Ibid.

p. 269 **"when I resigned I was more concerned"**: John Bardeen to Eugene P. Goldberg, Professor, Materials Science and Pharmacology, Director, Biomedical Engineering Center, University of Florida, Gainsville, February 12, 1986, UIUC-A. Draft 4 of letter on Marshall's (1985) article states Bardeen's point of view is more bluntly. John Bardeen, letter in regard to Marshall's article on Keyworth's resignation, Draft 4, December 23, 1985, UIUC-A.

p. 269 **"impotent and obsolete"**: John Bardeen (1986b).

p. 269 **all but unanimously opposed:** Ibid.; John Bardeen to George J. Keyworth, Office of Science and Technology Policy, White House, Washington, D.C., April 11, 1983, UIUC-A.

p. 270 **exacerbate the arms race:** A. Anderson (1992).

p. 270 **"pushing this petition"**: Lazarus (1992).

p. 270 **"As noted in my letter"**: John Bardeen to Hans A. Bethe, February 3, 1986, UIUC-A.

p. 270 **". . . a $25 billion price tag"**: Ibid.

p. 270 **Bardeen and Bethe's coauthored editorial:** Hans Bethe to John Bardeen, April 9, 1986, UIUC-A; John Bardeen to Hans Bethe, April 18, 1986, UIUC-A; Hans A. Bethe and John Bardeen, "Back to Science Advisors," *New York Times*, Op-Ed page, May 17, 1986. Bethe and Gottfried were members of the Union of Concerned Scientists. Bardeen published a number of editorials in addition to the one with Bethe. See, for example, John Bardeen (1986a), 203, 231.

p. 271 **"even if composed largely of scientists"**: Bethe and Bardeen (1986); also Herken (1992).

p. 271 **"the most important issue"**: John Bardeen to Professor Joseph A. Burton, the Arms Control Association, Washington, D.C., April 21, 1986, 7–22, UIUC-A.

p. 271 **"militarization of space"**: Bardeen (1986b).

p. 271 **"was very careful about what he endorsed"**: Goldwasser (1992).

p. 271 **Bardeen was so cautious:** Betsy Bardeen Greytak, private communication to Vicki Daitch, October 1994.

p. 271 **"In agreement with general principles"**: Bardeen to Peace Research Institute, October 27, 1962, BFC. Also, Peace Research Institute to John Bardeen, October 26, 1962.

p. 272 **"the death penalty has a place"**: Paul Cornil, European Committee on Crime Problems, to John Bardeen, January 20, 1965, UIUC-A.

p. 272 **"I do not believe"**: John Bardeen to Paul Cornil, March 5, 1965, UIUC-A.

p. 272 **"I do not feel that I should lend my name"**: John Bardeen to Pierre-Frantz Chapou, February 26, 1981, UIUC-A.

p. 272 **"didn't say a word"**: Bray (1993a).

p. 272 **"being of the mind"**: Jane Bardeen (1993).

p. 272 **"I did it for the Jews"**: Ibid.

p. 272 **The "Janes" also designed:** Ibid. "The Urban League of Champaign County, Inc.: Building Toward Equality," brochure, n.d.

p. 273 **"once you break the door"**: Ibid.

p. 273 **he occasionally made comments:** Hess (1991).

p. 273 **"You can't measure intelligence"**: Bardeen's comments written by hand on a draft of Karen Fitzgerald's obituary of William Shockley for the *IEEE Spectrum*, UIUC-A.

p. 273 **local steering committee:** Flyer produced by Scientists and Engineers for Johnson/Humphrey, UIUC-A.

p. 273 **"eradicate poverty"**: Illinois Scientists and Engineers for Johnson/Humphrey, "Statement of Principles," October 6, 1964, UIUC-A.

p. 273 **"In this and all other areas"**: Ibid.

p. 273 **the Johnson-Humphrey ticket:** Cameron B. Satterthwaite to members, Scientists and Engineers for Johnson/Humphrey, Champaign-Urbana Chapter, December 22, 1964, UIUC-A.

p. 274 **"were really heartbroken"**: C. J. Chang to Vicki Daitch, November 21, 2001.

p. 274 **"played an important role":** John Bardeen to E. L. Goldwasser, October 21, 1981, UIUC-A; Goldwasser (1992); C. J. Chang to Vicki Daitch, November 21, 2001.

p. 274 **"You can't cut back continually":** "Prof. Bardeen—Research Top Priority," *Technograph*, October 1972, UIUC-P.

p. 274 **"people who do well in course work":** John Bardeen to Gerald Almy, April 25, 1966, UIUC-A.

p. 275 **"better science education for a larger":** Goldwasser (1992).

p. 275 **active in national attempts:** "Marjorie Bardeen Named 'Outstanding Woman Leader,'" *FermiNews*, April 21, 1989, BFC.

p. 275 **"Natural scientists are not all geniuses":** "Panel Shifts Approach on Natural Science," *The Christian Science Monitor*, May 28, 1958.

p. 276 **need special clearances:** John Bardeen to G. B. Kistiakowsky, May 31, 1960; David Z. Beckler to John Bardeen, June 7, 1960, the Dwight D. Eisenhower Library, PSAC Records, Series III, Box 6, Correspondence B folder.

p. 276 **clearances suddenly came through:** John Bardeen to James Bardeen, September 6, 1960; John Bardeen to Jane Bardeen, September 7, 1960, BFC.

p. 276 **moved "slowly here":** John Bardeen to Jane Bardeen, September 10, 1960, BFC.

p. 276 **Peter Kapitza, Isaac Khalatnikov:** John Bardeen, "Report on Trip to Prague and the Soviet Union, August 25–September 25," UIUC-A.

p. 276 **came to consider Bardeen a second mentor:** Abrikosov (1993); John Bardeen to Lawrence C. Mitchell, Staff Director, Section on USSR and Eastern Europe, National Academy of Sciences, Washington, D.C., September 26, 1969, UIUC-A.

p. 276 **"we are familiar with the very outstanding":** John Bardeen, handwritten notes, n.d., UIUC-A.

p. 276 **Everyone was "very friendly":** John Bardeen to Jane Bardeen, December 1, 1963, and December 5, 1963, BFC.

p. 276 **"a distinguished theorist":** John Bardeen and David Pines, "Report of Trip to Moscow to Attend a Conference on Solid State Theory, December 1–12, 1963," UIUC-A.

p. 277 **"We know the result now":** Lev Gor'kov to David Pines, February 6, 1991, BFC.

p. 277 **"one of the greatest distinctions":** John Bardeen to G. I. Marchyuk, July 6, 1988, UIUC-A.

p. 277 **rare foreign membership:** John Bardeen to A. P. Alexandrov, President, and C. R. Skriabin, Secretary, USSR Academy of Sciences, Moscow, October 1, 1982, UIUC-A.

p. 277 **"going into some real depth":** John Bardeen to Jane Bardeen, September 12, 1975, BFC.

p. 277 **"one of the few of our group":** John Bardeen to Jane Bardeen, September 25, 1975, BFC.

p. 277 **"I'm certain much of the warmth":** Samuel C. Chu to Vicki Daitch, December 21, 1991.

p. 277 **"enjoy some of the beauties of China.":** John Bardeen to Jane Bardeen, September 25, 1975, BFC.

p. 277 **". . . This house needs you":** Jane Bardeen to John Bardeen, September 25, 1975, UIUC-A.

p. 278 **"given VIP treatment":** John Bardeen's Summary Report of Visit to Peoples Republic of China as Exchange Lecturer, April 30–May 30, 1980, UIUC-A.

p. 278 **"introduce Chinese physicists":** Ibid.

p. 278 **"Science should set an example":** "Bardeen Calls for a New Style of Life," *Delhi Statesman*, January 11, 1977, BFC.

p. 278 **Project on the History:** John Bardeen to Robert N. Noyce, April 10, 1980, UIUC-A.

p. 278 **improve the resulting book:** Hoddeson, et al. (1992).

p. 278 **a science advisor for Elizabeth Ante'bi's:** John Bardeen to Elizabeth Ante'bi', *Editions Hologramme*, Neuilly-sur-Seine, France, June 2, 1982, UIUC-A, C-13, C-81.

p. 279 **"This fund is being established":** Bardeen to J. David Ross, December 29, 1972. Jane, too, was generous, donating funds to provide for a $10,000 endowment over time to the Citizens Library in Washington, Pennsylvania, in memory of her mother, Elizabeth Patterson Maxwell.

p. 279 **"More than anyone else":** "Nobel Prize-Winner Establishes Duke Physics Research Fund," *Chapel Hill Newspaper*, March 29, 1973, 8, BFC.

p. 279 **"Fritz was so happy":** Edith London to John Bardeen, 13 April 1973. UIUC-A.

p. 279 **"for the basic ideas":** John Bardeen to Edith London, May 21, 1973, UIUC-A.

p. 279 **Superconducting Super Collider:** See Riordan, et al. (forthcoming).

p. 280 **"It wasn't a political gesture":** Goldwasser (1992).

p. 280 **"The Super Collider should proceed":** John Bardeen and Robert Schrieffer, "Response to *New York Times* article of April 28, 1987," (Editorial, "Super Hasty on the Super Collider," *New York Times*, April 28, 1987), draft, May 6, 1987, published May 14, 1987, UIUC-A.

p. 280 **"Walter made a good pick":** John Bardeen to Jane Bardeen, June 7, 1958, BFC.

p. 280 **"please send the date of Bill's":** John Bardeen to Jane Bardeen, May 30, 1958, BFC.

p. 281 **"it has been nice":** John Bardeen to Jane Bardeen, October 5, 1978, BFC.

p. 281 **doing "good work":** John Bardeen to Jane Bardeen, October 11, 1978, BFC.

p. 281 **"a little research" on family history:** John Bardeen to Jane Bardeen, June 15, 1958, BFC.

p. 281 **"It was like going back":** Douglas Scalapino to Jane Bardeen, February 28, 1991; Woodward (1979).

p. 281 **he had paintings at home:** Bill Bardeen to John Bardeen, November 22, 1970, BFC.

p. 281 **John liked traveling best:** Betsy Bardeen Greytak, private communication to Vicki Daitch.

p. 281 **"It is going to be a long time away from home":** John Bardeen to Jane Bardeen, September 2, 1975, BFC.

p. 281 **left his raincoat:** John Bardeen to Jane Bardeen, August 27, 1960, BFC.

p. 281 **"I was able to get $500":** John Bardeen to Jane Bardeen, September 2, 1975, BFC.

p. 282 **"the only thing that got really cold":** "C-U Awaits Relief from Blizzard," *Champaign-Urbana Courier*, January 27, 1978.

p. 282 **"I was pretty sure they'd come along":** "Blizzard Traps Bardeens on I-57," *Illinois Alumni News*, March 1978, UIUC-P.

p. 282 **National Inventors Hall of Fame:** Alfred L. Haffner, Jr., to John Bardeen, January 28, 1974, UIUC-A.

p. 282 **His fellow honorees included:** "John Bardeen Among 21 to Get Medal of Freedom," *Champaign-Urbana News-Gazette*, January 2, 1977, BFC.

p. 283   **"Dr. Bardeen is contributing to the city"**: "Prize-Winning Professor Awarded Key to City from Champaign Mayor," *Daily Illini*, October 4, 1989.

15
Pins and Needles and Waves

p. 284   **Pins And Needles:** The authors are grateful to Joseph Tillman for allowing this discussion of Bardeen's work on charge density waves to draw freely on his 1995 senior thesis at the University of Illinois.

p. 284   **"And this gray spirit"**: Tennyson (1987), 11–12.

p. 285   **typical of the exceptionally creative:** Csikszentmihalyi (1996), 211–212.

p. 285   **"band begins to march"**: Brown and Grüner (1994), 51–56.

p. 285   **"Peierls instability"**: Peierls (1930); (1964), 108–112.

p. 285   **related to superconductivity:** The ground states of both are described as a coherent superposition of pairs—pairs of electrons in superconductivity and electron-hole pairs in the CDW case. Fröhlich (1954), 296. Campell (1998).

p. 286   **Ong and Pierre Monceau:** P. Monceau et al. (1976).

p. 286   **London's picture:** Bardeen (1990a).

p. 286   **potential importance to industry:** Thorne (1996), 42.

p. 286   **analogous to Josephson tunneling:** John Bardeen (1980a); John Bardeen to Joel Moses, March 3, 1987.

p. 286   **Bardeen envisioned a collective tunneling:** Tucker, et al. (1982), 25.

p. 286   **"If you have a brick wall"**: Grüner (2000).

p. 287   **Among the prominent physicists:** Most of the references to the classical model are from Grüner (1983), 183–197.

p. 287   **The Aerospace Corporation:** The visit appears to have occurred on November 11, 1974. John Bardeen to Ivan Getting, October 29, 1974.

p. 288   **"Hello, this is John"**: Tucker (1995).

p. 288   **"It was only the second example"**: Tucker (1995).

p. 288   **"reaching for something that was exotic"**: Ibid.

p. 288   **Tucker's first student:** J. Miller (1993).

p. 289   **"We used to hold group meetings"**: Tucker (1995).

p. 289   **"Throwing theoretical spaghetti"**: Tucker (1995).

p. 289 **"...the system is still noisy"**: Bardeen to P. Monceau, February 3, 1982.

p. 289 **"Maybe he was just carried away"**: Grüner (2000).

p. 289 **"I'm really working hard"**: Salamon (2000).

p. 289 **"The first set of experiments"**: Grüner (2000).

p. 290 **Karlheinz Seeger:** Seeger had been one of the early workers to study TTF-TCNQ. John Bardeen (1987); Karlheinz Seeger to Vicki Daitch and Lillian Hoddeson, September 21, 1995. The authors are grateful to Seeger for contributing much information about his work on CDWs with Bardeen.

p. 290 **"didn't have one minute"**: McMillan (1995).

p. 290 **"I feel a deep sense"**: John Bardeen (1985).

p. 291 **"excellent agreement is found"**: John Bardeen to the editors of *Physical Review Letters*, May 21, 1985.

p. 291 **"greatly upset Bardeen"**: Tucker (1995). See also Bray (1993a).

p. 291 **"the opportunity to work"**: Lyding (2000).

p. 291 **"... nice working with Bardeen"**: Ibid.

p. 292 **It had never been easy:** Zawadowsky (1998).

p. 292 **friends avoided discussing:** Leonard (1992); Tucker (1995); Zawadowsky (1998).

p. 292 **"Tucker always seemed on edge"**: Frederick Lamb, priv. comm to Hoddeson, October 2, 2000.

p. 292 **"... you *are* the establishment"**: Douglas Scalapino to Vicki Daitch, December 10, 1991.

p. 292 **his own serious reservations:** Tucker (1995).

p. 292 **"... ticket to Stockholm"**: Lyding (2000).

p. 293 **"lemming effect"**: A. B. Pippard to L. Hoddeson, summer 1987.

p. 293 **"... didn't hang together"**: Salamon (2000).

p. 293 **"John is trying to do"**: Salamon (2000); Salamon and Bardeen (1987).

p. 294 **"wasn't at all wedded"**: Salamon (2000).

p. 294 **"Boy, I'm glad this high-Tc came out"**: Ibid.

p. 294 **"It is inconceivable to me"**: John Bardeen to H. Takayama and H. Matsukawa, May 29, 1987.

p. 294 **his hot temper:** Holonyak (1993a).

p. 294 **Tucker tried to avoid Bardeen:** John Tucker to John Bardeen, dated November 8, 1987, sent May 5, 1988.

p. 295   his own "strong pinning" theory: John Tucker to John Bardeen, October 23, 1987.

p. 295   "in strong pinning, every impurity stops": John Tucker to L. Hoddeson, March 17, 2001.

p. 295   "I think he felt betrayed": Salamon (2000).

p. 295   everything necessary to support the physics: Tucker (1995).

p. 295   Tucker called for help: Tucker (1995); J. Miller (1993).

p. 295   "Don't let people": Holonyak (1998a).

p. 296   "an ugly ceremony": Karlheinz Seeger to Vicki Daitch and Lillian Hoddeson, September 21, 1995.

p. 296   "general feeling at the conference": Robert Thorne to John Bardeen, October 8, 1987.

p. 296   "I don't think that's true": Lyding (2000).

p. 296   "would lead to some kind of calm": Lyding (2000).

p. 296   "talk fast and think": Ibid.

p. 297   "is confirmed by many experiments": John Bardeen to John Tucker, October 24, 1987.

p. 297   "messenger boy": Lyding (2000).

p. 297   "diatribe": Ibid.; and Lyding's personal notes from the period.

p. 297   "seem to be narrowing": John Bardeen to John Tucker, November 6, 1987, UIUC-A.

p. 297   ". . . correct as it stood": John Tucker to John Bardeen, dated November 8, 1987, sent May 5, 1988.

p. 297   "I hope that we can make progress": John Bardeen to John Tucker, November 13, 1987.

p. 297   ignoring available experimental facts: John Tucker to John Bardeen, November 13, 1987.

p. 297   "I am glad that we agree": John Bardeen to John Tucker, November 23, 1987.

p. 298   "Your statement that we now agree": John Tucker to John Bardeen, November 25, 1987.

p. 298   "Because of the change": John Bardeen to NSF, January 4, 1988.

p. 298   "with no strings attached": Lyding (2000).

p. 298   "You have by no means": John Bardeen to John Tucker, January 5, 1988.

p. 298   "No plausible classical model": John Bardeen (1989a).

p. 299   "I know you're recruiting": Bhatt (1992).

p. 299 **"A considerable body of evidence"**: John Bardeen (1989a).

p. 299 **"few other theorists have worked"**: John Bardeen to Gene L. Wells, April 18, 1989.

p. 299 **"Everyone is so busy writing"**: Nick Holonyak to L. Hoddeson, February 3, 2001.

p. 299 **last of Bardeen's single-authored papers:** John Bardeen (1990).

p. 299 **"mattered a lot to him"**: Betsy Greytak (2000a); John Bardeen (1990b).

p. 300 **It is not impossible:** See e.g., J. Miller (1993).

p. 300 **"the referee was totally out of line"**: Lazarus (1992).

p. 300 **Psychologists claim the behavior:** Csikszentmihalyi argues that such people are not workaholics. "For most of them work is not a way to avoid a full life, but rather is what makes a life full." Csikszentmihalyi (1996), 224.

p. 300 **"... a lot of criticism"**: Tucker (1995).

16
Last Journey

p. 301 **He had not been feeling well:** Nick Holonyak private communication to L. Hoddeson, February 4, 2001.

p. 301 **"I come in most every day"**: John Bardeen (1990a).

p. 302 **"the junction transistor is a bipolar transistor"**: Holonyak, quoted in John Bardeen (1990a).

p. 302 **"Brattain and I have the basic patent"**: John Bardeen (1990a). Bardeen is referring to his patent 2,524,033 (October 2, 1950) filed on February 26, 1948, prior to the Bardeen and Brattain transistor patent, 2,524,035, which was filed on June 17, 1948.

p. 303 **"of course he was very excited"**: John Bardeen (1990a).

p. 303 **"a tragedy that he got involved"**: Peterson (1981). For a detailed account of Shockley's arguments, including documents prepared by Shockley, see Pearson (1992).

p. 303 **"could have accomplished much more"**: John Bardeen to Eric Weiss, September 6, 1989, UIUC-A.

p. 303 **"the existence of tragic genetic deficiencies"**: William Shockley, "Sperm Banks and Dark-Ages Dogmatism," Position paper presented at the Rotary Club of Chico, California, April 16,

1980, 227; "Interview with William Shockley," *Playboy*, August 1980, both from Pearson (1992).

p. 304   **"if you've been jacking off"**: Holonyak (1998a).

p. 304   **"enjoyed teaching undergraduates"**: John Bardeen (1994), 81.

p. 304   **"A part of my life is gone"**: *Illini Week*, undated and untitled clipping, UIUC-P.

p. 304   **". . . I'd drop anything and everything"**: Holonyak (1998b).

p. 304   **"While history will remember Walter"**: John Bardeen (1988).

p. 304   **"in memory of a long-time friend"**: John Bardeen to Whitman College, Walla Walla, Washington, Walter H. Brattain Lectureship, November 11, 1987, UIUC-A.

p. 305   **"his scientific ideas and where"**: Greytak (2000a).

p. 305   **"wanted to emphasize what he'd done"**: Pines (1993).

p. 305   **He could no longer read:** Holonyak (1998b); Greytak (2000b).

p. 305   **"coming to Japan this fall"**: George Hatoyama to John Bardeen, October 1, 1990, BFC.

p. 305   **"stop, take his shoe off"**: Werstler (1992).

p. 305   **dangerous plaque had begun to build:** Greytak (2000b).

p. 305   **"and now *this*!"**: Holonyak (1998b).

p. 305   **"Probably a few names"**: "Professor Among 'Most Influential' Americans," *The Daily Illini*, September 6, 1990.

p. 306   **his *Physics Today*:** John Bardeen (1990b).

p. 306   **"He wanted me to be sure and get a copy"**: J. McMillan (1995).

p. 306   **When the local doctors:** Holonyak (1998b).

p. 306   **trust the local doctors:** Greytak (2000b).

p. 306   **"He learned to swim"**: Andrew Greytak, January 12, 1990, BFC.

p. 307   **might have talked a little physics:** Holonyak (1998a).

p. 307   **"I was struck right away with his analytical"**: Sugarbaker (2001).

p. 307   **The surgery went smoothly:** Ibid.

p. 308   **"How's your day"**: Sugarbaker (2001).

p. 308   **"found him unresponsive"**: Ibid.

p. 308   **"very thought-provoking incident"**: Ibid.

p. 308   **"a man with a lot of presence"**: Ibid.

p. 309   **several hundred friends:** *Inside Illinois* 10, (February 7, 1991), 10.

p. 309   **"children have all ended up":** Statements by William Bardeen in "John Bardeen Memorial," *Illini Union,* February 8, 1991.

p. 309   **"I suppose you would not consider":** Statements by William Bardeen in "John Bardeen Memorial," *Illini Union,* February 8, 1991.

p. 309   **"he rough-housed":** Statements by James Bardeen in "John Bardeen Memorial," *Illini Union,* February 8, 1991.

p. 309   **"Most of all, my father enjoyed his grandchildren":** Statements by Elizabeth Bardeen Greytak in "John Bardeen Memorial," *Illini Union,* February 8, 1991.

p. 310   **"quiet, genuinely modest":** Statements by David Pines in "John Bardeen Memorial," *Illini Union,* 8 February 1991.

p. 310   **"John was a wonderful colleague":** Ibid.

p. 310   **"Holonyak! Who in the hell":** Statements by Nick Holonyak in "John Bardeen Memorial," *Illini Union,* 8 February 1991.

p. 310   **"I was walking down the hallway":** Statements by Charles Slichter in "John Bardeen Memorial," *Illini Union,* 8 February 1991.

p. 311   **"And there was John in a set of tails":** Statements by Robert Schrieffer in "John Bardeen Memorial," *Illini Union,* 8 February 1991.

p. 311   **planted a tree on the fifteenth tee:** Burtis (1995); Werstler (1992).

p. 311   **"As the luncheon ended":** Fred Seitz to Jane Bardeen, March 6, 1991, BFC.

p. 311   **"... a lot to Louis Ridenour.":** Seitz (2001).

p. 311   **"John was so quiet and modest":** Walter Osterhoudt to Jane Bardeen, February 4, 1991, BFC.

p. 312   **"I will always remember the good times":** John Tucker to Jane Bardeen, February 1991, BFC.

p. 312   **"learning to be a widow":** Gretchen Osterhoudt to Jane Bardeen, May 11, 1994, BFC.

p. 313   **"I'm going to die":** Norma Marder, "Mama's Stroke, a memoir," forthcoming.

p. 313   **"Every time we attend a funeral":** Jane Bardeen to Betty Maxwell, March 27, 1974, BFC.

p. 313 **low stone monument:** William Bardeen, private communication to Lillian Hoddeson, December 2000.

17
Epilogue: True Genius and How to Cultivate It

p. 314 **True Genius:** Much of the material in this epilogue is based on Hoddeson (2001).

p. 314 **For centuries:** For a brief historical overview of the field, see R. S. Albert and Mark A. Runco (1999).

p. 314 **the vast literature:** A large number of the references can be found in Amabile (1996); Boden (1996); Csikszentmihalyi (1996); Gardner (1993); Holton (1973, 1978); Howe (1999); Miller (1984, 1996, 2001); Simonton (1988, 1990, 1999); Sternberg (1999); Sulloway (1996); Weisberg (1993); and Zuckerman (1977). A recent section on creativity in the *American Psychologist*, Sternberg and Dess (2001), featuring pieces on Charles Darwin, Linus Pauling, Thomas Young, and Claude Monet, suggests that the field is moving in the direction of using historical cases studies.

p. 314 **comparisons across many cases:** See e.g., the studies of Michael Faraday (Nersessian, 1984; Tweney, 1985), Charles Darwin (Gruber, 1981; Simonton, 1999), and Sigmund Freud (Sulloway, 1983).

p. 315 **Terman built on Galton's work:** Sternberg (1999).

p. 315 **association with a domain:** See, e.g., Csikszentmihalyi (1996), 22–50.

p. 315 **"paradigms" of their domain:** Kuhn (1962), 1–23.

p. 316 **"high intellectual traits":** Cox cited in Albert and Runco, p. 27.

p. 316 **the genius profile:** See e.g., Feist (1999), or Howe (1999).

p. 316 **importance of talent and intelligence:** Howe (1999), 188–205.

p. 317 **trained to achieve master-level performance:** Ericsson, et al. (1993); Weisberg (1999).

p. 317 **right kind of intelligence:** Gardner (1993), p. 20.

p. 317 **"Only one who bursts":** Waley (1989) Book VII(8), 124.

p. 317 **"intrinsic motivation":** Amabile (1996).

p. 318 **"flow into his being":** Schrieffer (1974, 1992a).

p. 318 **childish ways of focusing:** Gardner (1993); Gopnik, et al. (1999), Einstein (1949).

p. 318 **childhood traumas:** Alice Miller (1990).

p. 319 **engage in constructive competition:** Pines (1993).

p. 319 **"in competition basically with himself":** Schrieffer (1974).

p. 319 **work actively into their senior years:** Csikszent-mihalyi (1996), 210–233.

p. 319 **"He's the only man":** Lazarus (1992).

p. 319 **search through a "problem space":** Simon (1978); A. Newell, and H. Simon (1972); Langley, et al. (1987).

p. 320 **more complex ("chunked"):** Chase and Simon (1973); Reimann and Chi (1989); Chi, et al. (1981).

p. 320 **learned from Wigner:** Bardeen (1977, 1984).

p. 320 **his full attention:** Schrieffer (1974, 1992a).

p. 320 **Althea Bardeen probably preconditioned:** Harmer (1936), 335.

p. 321 **"You reduce a problem":** Bardeen (1984b).

p. 321 **"bully through":** Frederick Seitz, private communication to Lillian Hoddeson, April 2001.

p. 321 **try a range of approaches:** Pines (1993).

p. 321 **scientific study of divergent thinking:** Guilford (1950).

p. 321 **multiple intelligences:** Gardner (1985).

p. 322 **the brainstorming (or "blockbusting"):** de Bono (1992).

p. 322 **Countless examples of this approach:** Hoddeson, et al. (1993); Hoddeson (1987); Bradshaw (1992).

p. 322 **"experiment came first":** Pines (1993).

p. 322 **"to create laws":** Abrikosov (1993).

p. 322 **". . . formalism could lead one astray":** Schrieffer (1992b), 47.

p. 322 **"I don't generally work":** Holonyak (1993b).

p. 323 **such moments "epiphanies":** Lederman and Teresi (1993), 7.

p. 323 **a clever dance analogy:** Schrieffer (1974).

p. 323 **analogy as a mapping:** Gentner (1989).

p. 323 **he and Brattain mapped well-studied features:** Bardeen Bell Labs Notebook 20780, December 20, 1947.

p. 325 **how metaphors can themselves act as models:** Arthur Miller (2001), 217–262; see also Lakoff and Johnson (1980).

p. 325 **"that electron-lattice interactions are":** Bardeen's handwritten notes, 1950, UIUC-P.

p. 325 **"knows a lot about":** Brian Ross, personal communication, February 2001. See also Medin and Ross (1996), "Expertise," 469–91.

p. 325 **"had one wonderful quality":** Wigner (1981).

p. 326 **community of scholars conducting studies:** See, e.g., Nersessian (1984); Tweney (1985); Gruber (1981); Simonton (1999); Sulloway (1983); Langley, et al. (1987).

p. 326 **third dimension of the genius profile:** See e.g., Csikszentmihalyi (1996); Gardner (1993); Sternberg (1988, 1999); Weisberg (1993); Zuckerman (1977).

p. 326 **"hang on" to problems:** Althea Harmer Bardeen to Charles William Bardeen, n.d., BFC.

p. 326 **"bulldog tenacity":** Seitz in Del Guercio, et al. (1998).

p. 326 **"The greatest opportunity":** C. W.'s inscription in the book he wrote about his boyhood adventures in the Civil War, presented to John on his tenth birthday.

p. 327 **skipped three grades had generally benefited him:** Bardeen (1977a).

p. 327 **"allowed to go ahead":** Charles William Bardeen to Charles Russell Bardeen, March 4, 1919.

p. 328 **"Nobel–bound" mentors:** Zuckerman (1977), 62, 96–143, 200–202.

p. 328 **"accomplishments are a good bit of luck":** Bardeen's hand-edited copy of Karen Fitzgerald's profile of him for the IEEE *Spectrum*, Bardeen papers, U. of I.

p. 328 **"on the ground floor":** Harvard Society of Fellows (1987).

p. 329 **In experiments the parameters:** e.g., Ward (1999); Feist and Gorman (1998); Langley, et al. (1987); Dunbar (1997); Csikszentmihalyi (1996).

p. 329 **"In hindsight, the path taken may look straight":** de Waal quoted in *History Newletter* (Spring 2001), 23, Center for History of Physics of the American Institute of Physics, excerpted from de Waal, "The Ape and the Sushi Master," as quoted in *Chronicle of Higher Education*, 30 March 2001, B6.

p. 329 **legendary stories:** Weisberg (1986), 27–33.

Index

A

*J*OHN BARDEEN was an unassuming man, a humble, soft-spoken Midwesterner whose life was filled with simple pastimes like a Sunday picnic with the family or a good game of golf. He was also a giant of modern physics, an extraordinary hero of twentieth century science. His seminal work earned him the distinction of being the only person ever to win two Nobel Prizes in physics—both awarded for discoveries that were breathtaking in scope and responsible for advancing the course of human history.

Bardeen ranks among the most imaginative and inspired scientists of our time. But his genius was quiet and unobtrusive, hidden behind the façade of an average man, which is perhaps why we know so little about him. Equally as influential on our culture as Albert Einstein or Richard Feynman, Bardeen, unlike those popular icons of physics, had no desire to mug for the cameras or to prove himself eccentric, irreverent, and offbeat. As eccentricity and outsize personalities had come to symbolize the true nature of genius and creativity, Bardeen remained cloaked in obscurity.

Without Bardeen's first Nobel Prize-winning discovery— the transistor—the electronics revolution, which brought us desktop computers, supercomputers, and microelectronics, would still be the stuff of science fiction. His second great breakthrough—the theory of superconductivity, which for years had stumped Einstein, Feynman, and many others— promises to revolutionize twenty-first century technology with high speed "mag-lev trains," supercolliding atom smashers, and other fantastic technological wonders.

Yet despite these achievements, this astonishing though decidedly modest Midwesterner was often overlooked by the media as well as the public, simply because he differed radically from the popular stereotype of genius. Through an exploration of his science as well as his life, a fresh and thoroughly engaging portrait of genius and the nature of creativity emerges. This fascinating biography provides a whole new perspective on what it truly means to be a genius.

2030